Genomics: Applications in Human Biology

Genomics: Applications in Human Biology

Edited by Joel White

SYRAWOOD
PUBLISHING HOUSE
New York

Published by Syrawood Publishing House,
750 Third Avenue, 9th Floor,
New York, NY 10017, USA
www.syrawoodpublishinghouse.com

Genomics: Applications in Human Biology
Edited by Joel White

International Standard Book Number: 978-1-64740-392-8 (Hardback)

Cataloging-in-publication Data

Genomics : applications in human biology / edited by Joel White.
p. cm.
Includes bibliographical references and index.
ISBN 978-1-64740-392-8
1. Human genome. 2. Genomics. 3. Human genetics. I. White, Joel.
QH431 .H86 2023
611.018 166--dc23

TABLE OF CONTENTS

PREFACE

This book was inspired by the evolution of our times; to answer the curiosity of inquisitive minds. Many developments have occurred across the globe in the recent past which has transformed the progress in the field.

The complete set of DNA instructions found in a cell is known as the genome. The specialty within biology which deals with the function, mapping, structure, evolution and editing of genomes is known as genomics. Human genome refers to the base pairs of deoxyribonucleic acid (DNA) that comprise the entire set of chromosomes in the human body. It is a collection of long polymers of DNA, which contain both the coding and non-coding DNA regions. The structure, organization, chemical modifications and sequencing of these polymers provide the necessary machinery for communicating the information stored in the genome. All human biological research heavily relies on the sequenced human genome, because it is an essential reference for comparing all human genomes. Understanding the human genome provides insight into the disease risks or health tendencies of humans. It is also helpful in understanding the origin of the human species and the links that exist among human subpopulations. This book contains some path-breaking studies on genomics. It elucidates the concepts and innovative models around prospective developments with respect to its applications in human biology. This book will serve as a reference to a broad spectrum of readers.

This book was developed from a mere concept to drafts to chapters and finally compiled together as a complete text to benefit the readers across all nations. To ensure the quality of the content we instilled two significant steps in our procedure. The first was to appoint an editorial team that would verify the data and statistics provided in the book and also select the most appropriate and valuable contributions from the plentiful contributions we received from authors worldwide. The next step was to appoint an expert of the topic as the Editor-in-Chief, who would head the project and finally make the necessary amendments and modifications to make the text reader-friendly. I was then commissioned to examine all the material to present the topics in the most comprehensible and productive format.

I would like to take this opportunity to thank all the contributing authors who were supportive enough to contribute their time and knowledge to this project. I also wish to convey my regards to my family who have been extremely supportive during the entire project.

Editor

1

Genome-Wide Analysis of Alpharetroviral Integration in Human Hematopoietic Stem/ Progenitor Cells

Arianna Moiani [1], Julia Debora Suerth [2], Francesco Gandolfi [1], Ermanno Rizzi [3], Marco Severgnini [3], Gianluca De Bellis [3], Axel Schambach [2] and Fulvio Mavilio [1,*]

[1] Genethon, 1bis Rue de l'Internationale, 91020 Evry, France; E-Mails: amoiani@genethon.fr (A.M.); fgandolfi@genethon.fr (F.G.)

[2] Institute of Experimental Hematology, Hannover Medical School, Carl-Neuberg-Str.1, D-30625 Hannover, Germany; E-Mails: suerth.julia@mh-hannover.de (J.D.S.); schambach.axel@mh-hannover.de (A.S.)

[3] Institute for Biomedical Technologies, Consiglio Nazionale delle Ricerche, Milan 20132, Italy; E-Mails: ermanno.rizzi@itb.cnr.it (E.R.); marco.severgnini@itb.cnr.it (M.S.); gianluca.debellis@itb.cnr.it (G.B.)

* Author to whom correspondence should be addressed; E-Mail: fmavilio@genethon.fr;

Abstract: Gene transfer vectors derived from gamma-retroviruses or lentiviruses are currently used for the gene therapy of genetic or acquired diseases. Retroviral vectors display a non-random integration pattern in the human genome, targeting either regulatory regions (gamma-retroviruses) or the transcribed portion of expressed genes (lentiviruses), and have the potential to deregulate gene expression at the transcriptional or post-transcriptional level. A recently developed alternative vector system derives from the avian sarcoma-leukosis alpha-retrovirus (ASLV) and shows favorable safety features compared to both gamma-retroviral and lentiviral vectors in preclinical models. We performed a high-throughput analysis of the integration pattern of self-inactivating (SIN) alpha-retroviral vectors in human $CD34^+$ hematopoietic stem/progenitor cells (HSPCs) and compared it to previously reported gamma-retroviral and lentiviral vectors integration profiles obtained in the same experimental setting. Compared to gamma-retroviral and lentiviral vectors, the SIN-ASLV vector maintains a preference for open chromatin regions, but shows no bias for transcriptional regulatory elements or transcription units, as defined by genomic annotations and epigenetic markers (H3K4me1 and H3K4me3 histone modifications). Importantly, SIN-ASLV integrations do not cluster in hot spots and target potentially dangerous genomic loci, such as the EVI2A/B, RUNX1 and LMO2 proto-oncogenes at a virtually random frequency. These characteristics predict a safer profile for ASLV-derived vectors for clinical applications.

Keywords: retroviral integration; alpha-retroviral vectors; gene therapy; human hematopoietic cells

---

1. Introduction

Transplantation of hematopoietic stem cells genetically modified by retroviral vectors has proven its clinical efficacy in a number of seminal clinical trials for the treatment of severe monogenic disorders [1–8]. However, some of these studies also showed the genotoxic risks associated with the insertion of foreign DNA in the human genome, which limit the clinical application of integrating vectors (reviewed in [9]). Several efforts have been made to improve the safety of retroviral vectors, leading to the design of safer constructs and the development of robust *in vitro* and *in vivo* genotoxic assays to predict the potential risk associated with their integration into the genome [10–12]. High-definition mapping of integration sites of vectors derived from the Moloney murine leukemia virus (MLV) and human immunodeficiency virus (HIV) in murine and human cells revealed non-random profiles with a strong tendency to target active regulatory regions for MLV-derived gamma-retroviral vectors [13,14] and transcribed regions for HIV-derived lentiviral vectors [15,16]. These integration patterns explain the relatively high risk to deregulate gene expression at the transcriptional or post-transcriptional level observed in pre-clinical, as well as in clinical studies (reviewed in [9]).

Small-scale surveys of integration sites of vectors derived from alpha-retroviruses, such as the avian sarcoma-leukosis virus (ASLV), in different cell types indicated a more random pattern compared to other retroviruses, with a slight preference for transcription units, but no apparent preference for promoters and transcription start sites (TSSs) [17–20]. This potentially more favorable integration profile prompted the development of a replication-deficient, self-inactivating (SIN) ASLV-derived vector capable of efficiently transducing murine and human cells [21]. This vector was able to sustain long-term transgene expression in murine and human hematopoietic progenitors at levels comparable to those obtained with SIN-MLV and SIN-HIV vectors and to correct the X-linked chronic granulomatous disease (X-CGD) phenotype in a mouse model of the disease [20,22].

We and others previously reported that MLV, SIN-MLV and SIN-HIV integrations are highly clustered in the human genome, with cell-specific patterns that correlate with the transcriptional program and the epigenetic landscape of each cell type [14–16,19,23–26]. In this study, we report a high-definition analysis of the integration patterns of SIN-MLV, SIN-ASLV and SIN-HIV vectors in human CD34$^+$ hematopoietic stem/progenitor cells (HSPCs), which was carried out to evaluate their comparative genotoxic potential in a clinically relevant target cell. We show that the SIN-ASLV integration profile is close to random, with no preferential targeting of TSSs or transcribed genes compared to SIN-MLV and SIN-HIV. The SIN-ASLV vector does not target CpG islands, conserved non-coding regions (CNCs) or elements enriched in transcription factor binding sites (TFBS), is less frequently associated with epigenetically defined promoter and enhancer regions compared to SIN-MLV and is randomly associated with repetitive elements in the genome. Similarly, we observed no preference for transcribed regions compared to SIN-HIV. Heterochromatic regions are excluded by the integration pattern of all three vectors. Interestingly, the ASLV vector showed no apparent clustering in the genome and has no association with the typical integration hot spots observed for MLV- and HIV-based vectors. These results highlight a safer integration profile of alpha-retroviral vectors in human cells, supporting their development as a clinical gene transfer tool.

2. Experimental

2.1. Vectors and Cells

Human CD34$^+$ HSPCs were purified form umbilical cord blood, pre-stimulated for 48 h in serum-free Iscove's modified Dulbecco medium supplemented with 20% Fetal Calf Serum (FCS), 20 ng/mL human thrombopoietin, 100 ng/mL Flt-3 ligand, 20 ng/mL interleukin-6 and 100 ng/mL stem cell factor, as previously described [23]. HSPCs were transduced with the SIN-ASLV vector, pAlpha.SIN.EFS.EGFP.WPRE (noTATA), expressing GFP under the control of the elongation factor 1α promoter, pseudotyped in an amphotropic envelope by three-plasmid transfection in 293T cells, as previously described [20]. Cells were infected by 3 rounds of spinoculation (1500 rpm for 45 min) in the presence of 4 μg/mL polybrene. Transduction efficiency was evaluated by cytofluorimetric analysis of GFP expression 48 h after infection.

2.2. Amplification, Sequencing, and Analysis of Retroviral Integration Sites

Genomic DNA was extracted from a pool of 3.5 × 10^6 CD34$^+$/GFP$^+$ cells enriched by fluorescence-activated cell sorting, after a brief period in culture to dilute unintegrated vectors. 3'-LTR vector-genome junctions were amplified by LM-PCR adapted to the GS-FLX Genome Sequencer (Roche/454 Life Sciences) pyrosequencing platform, as previously described [14]. Raw sequence reads were processed by an automated bioinformatic pipeline that eliminated small and redundant sequences [14] and mapped on the University of California at Santa Cruz (UCSC) hg19 release of the human genome [14]. All UCSC RefSeq genes having their TSS at ±50 kb from an integration site were annotated as targets. Genomic features were annotated when their genomic coordinates overlapped for ≥1 nucleotide with a ±1 kb interval around each integration site. We used UCSC tracks for both CpG islands and conserved TFBSs, and the previously described genomic coordinates of 82,335 mammalian conserved non-coding sequences (CNCs) [27]. Raw sequences having a single or ambiguous match in the genome (the latter mapping in multiple genomic positions with a difference in the identity <2) were blasted on the UCSC RepeatMasker database. DNase I hypersensitive sites from publicly available data [28] were annotated when overlapping for at least 1 bp with a ±1-kb interval around an integration. Repetitive elements were annotated when directly targeted by each integration site. Sequences having multiple matches were collapsed and counted as one when matching in the same genomic positions and were univocally associated with the single type of repetitive element they targeted.

For the association of the integrations with epigenetically defined chromatin states, we used publicly available ChIP-Seq data (NIH Roadmap Epigenomics Mapping Consortium database) that we re-annotated in the UCSC hg19 release of the human genome. We analyzed the distribution of integration sites around histone modifications (H3K4me1, H3K4me3, H3K36me3, H3K27me3) using the seqMINER platform [29]. Previously generated SIN-MLV, SIN-HIV integrations and random control sequences datasets [14] were also re-annotated on the UCSC hg19 genome. For all pairwise comparisons, we applied a 2-sided Fisher's exact test. The threshold for statistical significance was set at a p-value < 0.01.

3. Results and Discussion

3.1. SIN-ASLV Vectors Exhibit an Almost Random Integration Profile in the Genome of Human CD34⁺ HSPCs

To generate a high-definition alpha-retroviral integration profile in human HSPCs, we transduced umbilical cord blood-derived $CD34^+$ cells with a previously described SIN-ASLV vector carrying a GFP expression cassette under the control of the intron-less, 240-bp version of the elongation factor-1α (EFS) promoter [20]. Cells were transduced at 10% to 20% efficiency and were selected for GFP expression by cell sorting 10 days after infection, to dilute unintegrated vectors. Vector-genome junctions were amplified from genomic DNA by ligation-mediated (LM)-PCR and pyrosequenced, as previously described [14]. Raw sequences (available at GenBank with the accession number SRR1282019) were processed by a previously described bioinformatic pipeline [14] and mapped on the UCSC hg19 release of the human genome, to obtain 8250 unique insertion sites. Two datasets of SIN-MLV (13,097) and SIN-HIV (31,827) vector integrations, previously generated in human umbilical cord blood-derived $CD34^+$ cells in comparable experimental conditions, and a set of in-silico generated normalized random sites (40,000) [14,26] were re-annotated on the hg19 genome and used for comparison. To identify differences in the integration preferences of SIN-ASLV compared to SIN-MLV and SIN-HIV in human HSPCs, we first analyzed the distribution of integration sites around RefSeq genes in the human genome: integration was annotated as TSS-proximal when occurring in an interval of ±2.5 kb from the TSS of any RefSeq gene, intragenic when occurring inside a RefSeq gene >2.5 kb from the TSS and intergenic in all other cases.

The high-definition profile of SIN-ASLV integration showed only a modest preference for TSSs (6.97% of the integration sites were annotated as TSS-proximal) compared to SIN-HIV and random sites (3.45% and 3.16%, respectively), which was significantly lower than that observed for the SIN-MLV vector (23.38%, $p < 0.01$). Similarly, SIN-ASLV showed only a slight tendency to integrate into genes (49.48% *vs.* 40.58% of random sites), significantly lower than that observed for SIN-HIV vectors (76.77%, $p < 0.01$). As a consequence, the frequency of SIN-ASLV integration outside transcription units was only slightly lower than random (43.55% *vs.* 56.26%) and significantly higher than those observed for the other two vectors (34.36% and 19.78%, respectively, $p < 0.01$) (Table 1). A plot of the relative distance of SIN-ASLV integration sites in an interval of ±50 kb from any TSS revealed a spread distribution with only a modest accumulation in the ±2.5 kb interval around TSS compared to the SIN-MLV vector. A higher definition map (100-bp intervals) showed the absence of integrations in the basal promoter region, most likely occupied by the RNA PolII basal transcriptional machinery. Integrations of the SIN-HIV vector were under-represented in a much wider interval of ±2.5-kb around the TSS (Figure 1).

Table 1. Integration distribution around RefSeq genes and genomic features in the genome of human hematopoietic stem/progenitor cells (HSPCs).

Vector	Intergenic (%)	TSS-proximal (%)	Intragenic (%)	CpG islands (%)	CNCs (%)	TFBS (%)	Total integrations
SIN-ASLV	43.55 *	6.97	49.48	2.84	5.49	55.70	8250
SIN-MLV	34.36 *	23.38 *	42.26	17.68 *	8.42	69.95 *	13,097
SIN-HIV	19.78 *	3.45	76.77 *	1.23	4.58	54.61	31,827
Random	56.26	3.16	40.58	1.76	6.05	51.01	40,000

Percentage of self-inactivating (SIN)-Moloney murine leukemia virus (MLV), SIN-avian sarcoma-leukosis alpha-retrovirus (ASLV) and SIN-HIV integrations and random sequences targeting intergenic, transcription start sites (TSS)-proximal and intragenic regions, regions annotated as CpG islands, conserved non-coding (CNC) regions and transcription factor binding sites (TFBS). For all the comparison with random sites, we applied a two-sided Fisher's exact test. * $p < 0.01$.

This analysis indicates that SIN-ASLV vector integrations have an almost random distribution in the human genome, with only a modest preference for genes and promoter regions compared to SIN-HIV and SIN-MLV vectors, suggesting entirely different modalities of target site selection.

We previously reported that SIN-MLV integrations are enriched around annotated CpG islands and conserved TFBSs and moderately enriched around mammalian, evolutionarily conserved non-coding sequences (CNCs) [14,25,26]. SIN-ASLV integrations were found associated with these genomic features at almost a random frequency, as observed for SIN-HIV integrations, and at a much lower frequency compared to SIN-MLV integrations (CpGs: 2.84% *vs.* 17.68%; TFBSs: 55.70% *vs.* 69.95%; CNCs: 5.49% *vs.* 8.42%, $p < 0.01$ in all cases) (Table 1), suggesting again that SIN-ASLV integrations have no obvious association with functional genomic elements.

We then looked at the tendency of the three types of retroviral vectors to target repetitive elements, by blasting both single- and multiple-match sequences to the UCSC RepeatMasker database and by annotating repetitive elements directly targeted by each integration site. Interestingly, only SIN-ASLV integrations were associated with repetitive elements with an almost random frequency (51% *vs.* 50%), while both SIN-MLV and SIN-HIV integrations were significantly under-represented in repetitive regions (37% and 45%, respectively, $p < 0.01$) (Table 2). By looking at the different classes of repetitive elements, we found that all three vectors have a slightly higher preference to integrate in short interspersed nuclear elements (SINEs) compared to random controls (17% to 20% *vs.* 15%), probably as a consequence of the fact that SINEs are often located in transcribed regions and contain PolII promoters [30,31]. On the contrary, integrations in long interspersed nuclear elements (LINEs), long terminal repeats (LTRs) and other repetitive elements were under-represented or close to random (Table 2). Finally, integration in satellite elements was observed at a random frequency only for SIN-ASLV vectors (0.43% *vs.* 0.35%), while both SIN-MLV and SIN-HIV integrations were significantly under-represented in these regions (0.01% and 0.05%, respectively, $p < 0.01$) (Table 2).

Figure 1. Genomic distribution of SIN-MLV, SIN-ASLV and SIN-HIV integrations in human HSPCs. The distribution of the distance of SIN-MLV (red bars), SIN-ASLV (yellow bars) and SIN-HIV (blue bars) integration sites from the TSS of targeted genes at 2500-bp (**a**) or 50-bp (**b**) resolution. The percentage of genes targeted at each position is plotted on the y-axis. The black line indicates the distribution of random control sites.

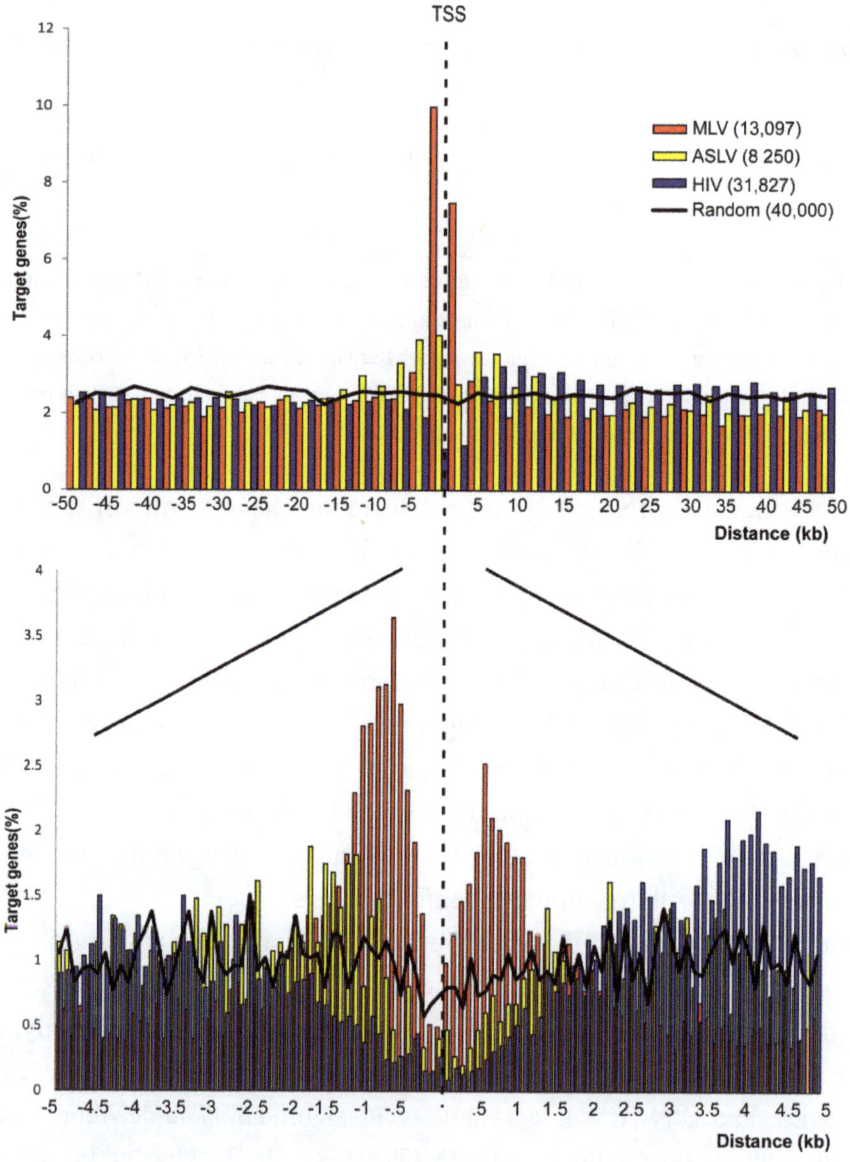

Table 2. Integrations targeting repetitive elements in the genome.

Vector	Repetitive elements (%)	LINEs (%)	SINEs (%)	Satellites (%)	LTRs (%)	Others (%)
SIN-ASLV (8,899)	51.29	19.50	19.77	0.43	6.26	5.35
SIN-MLV (13,606)	37.75	10.96	17.26	0.01	5.09	4.42
SIN-HIV (32,964)	45.78	19.45	16.86	0.05	4.08	5.35
Random (40,000)	50.96	21.27	14.69	0.35	9.57	5.10

Percentage of SIN-MLV, SIN-ASLV, SIN-HIV integrations and random sequences targeting repetitive elements and the percentage targeting each specific element: LINEs, short interspersed nuclear elements (SINEs), satellites, LTRs and all the other elements.

Overall, these data indicate a remarkably random pattern of integration for the ASLV-derived vector, which shows none of the characteristic preferences of gamma-retroviruses and lentiviruses for genes and genetic elements associated with gene function and regulation.

3.2. SIN-ASLV Integration Is Not Associated with Epigenetically-Defined Functional Genomic Regions

Many studies have reported a strong correlation between MLV and HIV integration sites and distinct epigenetic markers in different cell types (reviewed in [9]). In human $CD34^{+}$ HSPCs, MLV integrations are strongly associated with histone modifications marking transcriptionally active PolII promoters and enhancers, while HIV integrations correlate with epigenetic markers of active PolII elongation within transcription units [14,32]. We therefore investigated the association of SIN-ASLV integrations with defined epigenetic markers of functional genomic elements. Taking advantage of publicly available ChIP-Seq data in the genome of human $CD34^{+}$ HSPCs, we analyzed the association of SIN-MLV, SIN-ASLV and SIN-HIV integrations with specific histone modifications defining active or poised PolII promoters, enhancers, transcribed regions and heterochromatin (H3K4me1, H3K4me3, H3K36me3, H3K27me3 and H3K27ac).

More than 60% and 70% of SIN-MLV and SIN-HIV integrations sites, respectively, were univocally associated with a defined chromatin state, compared to only 40% of the SIN-ASLV integration sites, a frequency very close to the 30% observed for random sequences. In particular, SIN-ASLV integrations were found around regulatory regions, *i.e.*, enhancer ($H3K4me1^{+}$) and promoters ($H3K4me3^{+}$), at a much lower frequency compared to SIN-MLV (10% *vs.* 37% in enhancers and 6% *vs.* 26% in promoters, respectively, $p < 0.01$), a tendency comparable to that observed for SIN-HIV (10% in enhancers and 4% in promoters) and slightly higher than that of the random sample (3% and 2%, respectively). Moreover, SIN-ASLV integrations were poorly associated with a marker of transcribed gene bodies (H3K36me3) compared to SIN-HIV (15% *vs.* 38%, $p < 0.01$). All three vectors were under-represented in heterochromatic regions marked by H3K27me3 compared to random sites, with the SIN-ASLV vector showing the highest association (Figure 2A). This analysis is in agreement with the associations observed at the level of DNA sequence and genomic annotations, and confirms the preference of SIN-MLV and SIN-HIV vectors for, respectively, regulatory sequences and transcribed regions and an almost random integration pattern for SIN-ASLV. The modest bias observed for SIN-ASLV integrations in DNase I hypersensitive regions compared to the random sample (Table S1) can be explained by a certain tendency to integrate in "open" chromatin regions, as observed for most retroviruses [14,16,33,34].

The differences between SIN-ASLV and the other two vectors in targeting defined chromatin regions are highlighted by plotting the average integration densities of each vector type around each histone modification. Indeed, we clearly observe a peak of SIN-MLV integration sites in a ±2.5-kb interval from epigenetically-defined enhancers and promoters, while the distribution of the SIN-ASLV and SIN-HIV integrations around these elements is similar to that observed for random sequences (Figure 2B). The quasi-random association of ASLV integrations in regulatory element predicts a much lower genotoxic risk compared to MLV-derived vectors, whose tendency to target active regulatory elements is at the basis of their propensity to cause insertional deregulation of gene expression [9]. Most (>70%) of the genes targeted by all three vectors in HSPCs are actively expressed

(Figure S1), an expected finding, considering that retroviral target site selection is highly favored by an open chromatin state [14,16,33,34]. However, the SIN-ASLV vector targets the transcribed portion of active genes at a much lower frequency compared to HIV and is devoid of splicing signals, thus predicting a much lower risk to interfere with gene regulation at the post-transcriptional level [22,35,36].

Figure 2. Association of vector integration sites with different epigenetically-defined chromatin states. (**a**) The percentage of integration sites associated with specific, epigenetically defined genomic regions for each vector type. Chromatin states are categorized on the basis of the combination of different epigenetic marks mapped by ChIP-seq in human HSPCs. Only integration sites that are unambiguously associated with one chromatin state were used for the analysis. (**b**) The mean densities of H3K4me1, H3K4me3, H3K36me3 and H3K27me3 ChIP-seq fragments in a 5-kb window around all SIN-MLV (red), SIN-ASLV (yellow) and SIN-HIV (light blue) integration sites and random sequences (black). ac: H3K27ac.

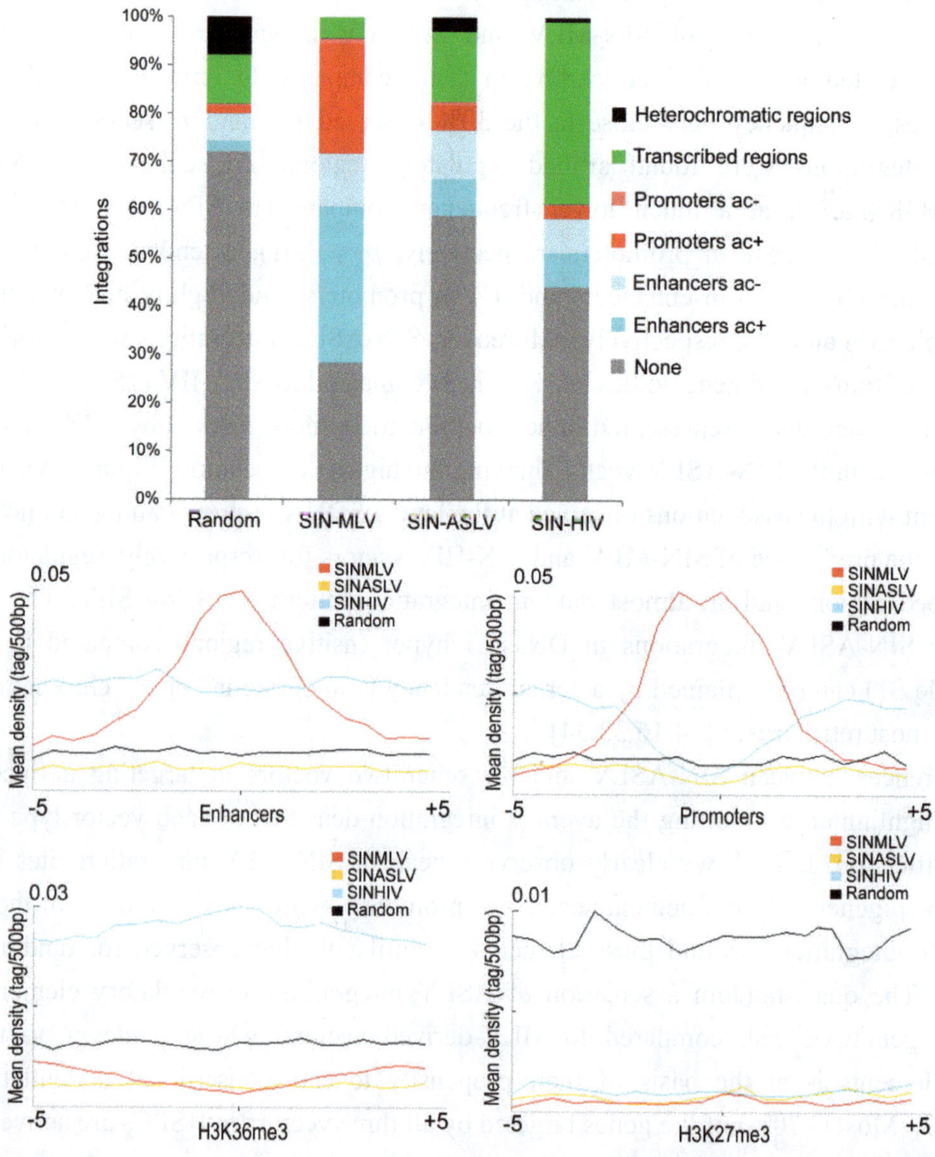

3.3. SIN-ASLV Vector Integrations Are Not Clustered in the Human Genome

The integration profile of MLV- and HIV-derived vectors in the human genome is characterized by heavy clustering into integration hot spots, where MLV forms narrow clusters overlapping active regulatory elements and HIV larger clusters targeting a subset of transcribed genes, in both cases in a cell-specific fashion [14–16,24,37]. On the contrary, the SIN-ASLV vector showed no significant clustering when we applied a statistical definition of clusters adjusted to the numerosity of the sample [14], which for the SIN-ASLV dataset was three integrations in 53,920 bp. By this threshold, we were able to identify only 484 clusters, a significantly lower frequency compared to the 1,415 and 2,724 identified for the SIN-MLV and SIN-HIV vectors, respectively ($p < 0.01$). Only 21% of all SIN-ASLV integrations are clustered, compared to 56% and 51% of SIN-MLV and SIN-HIV integrations, respectively ($p < 0.01$) (Table 3). Moreover, SIN-ASLV clusters are mostly made of few (three or four) integrations with few clusters containing up to nine integrations, while SIN-MLV and SIN-HIV clusters contain up to 37 and 122 integrations, respectively. From these data, it appears that, contrary to other retroviral vectors, SIN-ASLV integrations do not form hot spots of integrations in the human genome. Interestingly, when we looked at integration clusters at single genomic loci, we observed that the frequency of SIN-ASLV integrations at the typical MLV or HIV hot spots is very low and comparable to the frequency observed for random sequences. Figure 3 shows a comparison of the integration pattern of the three vectors in the NF1-EVI2A/B, RUNX1, LMO2 and PACS1 loci, four known hot spots for MLV or HIV integration. The same scenario is true for the EVI1/MDS1 (MECOM) locus and all other MLV or HIV integration hot spots (data not shown). The low frequency of SIN-ASLV integration in proto-oncogenes responsible for the severe adverse events observed in clinical gene therapy trials, such as LMO2, provide a further indication of its lower genotoxic profile.

Table 3. Clusters of integration sites in the genome of human HSPCs.

	SIN-MLV (13,097)	SIN-ASLV (8250)	SIN-HIV (31,827)
Clusters	1415	484	2724
Integrations in clusters (%)	56	21	51
Average cluster dimension	5.1	3.6	5.9

The number of SIN-MLV, SIN-ASLV and SIN-HIV clusters of integrations, the percentage of integrations in clusters and the average cluster dimension, calculated based on random sequences distribution in the genome. The threshold for cluster definition was defined at a p-value of <0.01 by a statistical algorithm that adjusts for the numerosity of the sample [14].

Although the SIN-ASLV integration profile shows none of the features typical of MLV- and HIV-derived vectors, it is not completely random and shows a general preference for euchromatic regions. It is now know that both MLV and HIV pre-integration complexes (PICs) are targeted to chromatin by a tethering mechanism involving the interaction of the viral integrase with host cell factors: the LEDGF/p75 chromatin component interacts with the HIV integrase and directs its integration into transcribed gene bodies [38,39], while the MLV integrase appears to bind to bromodomain-containing BET proteins specifically associated with acetylated histones around TSSs and active regulatory elements [40–42]. Although it is likely that ASLV also may adopt a tethering mechanism to direct its integration in favorable genomic regions, the details are unknown. The integration preferences uncovered by our analysis predict an interaction with a broader range of host cell factors, which tether the PICs to open chromatin regions with unspecified or very subtle functional characteristics, thus leading to a more random profile characterized by the absence of hot spots.

Figure 3. SIN-MLV, SIN-ASLV and SIN-HIV integration sites and clusters in CD34^{+} HSPC-specific loci. Distribution of SIN-MLV (red), SIN-ASLV (green) and SIN-HIV (blue) integration clusters (horizontal solid bars) and integrations (vertical marks) in the NF1-EVI2A/B, RUNX1, LMO2 and PACS1 loci, as displayed by the UCSC Genome Browser.

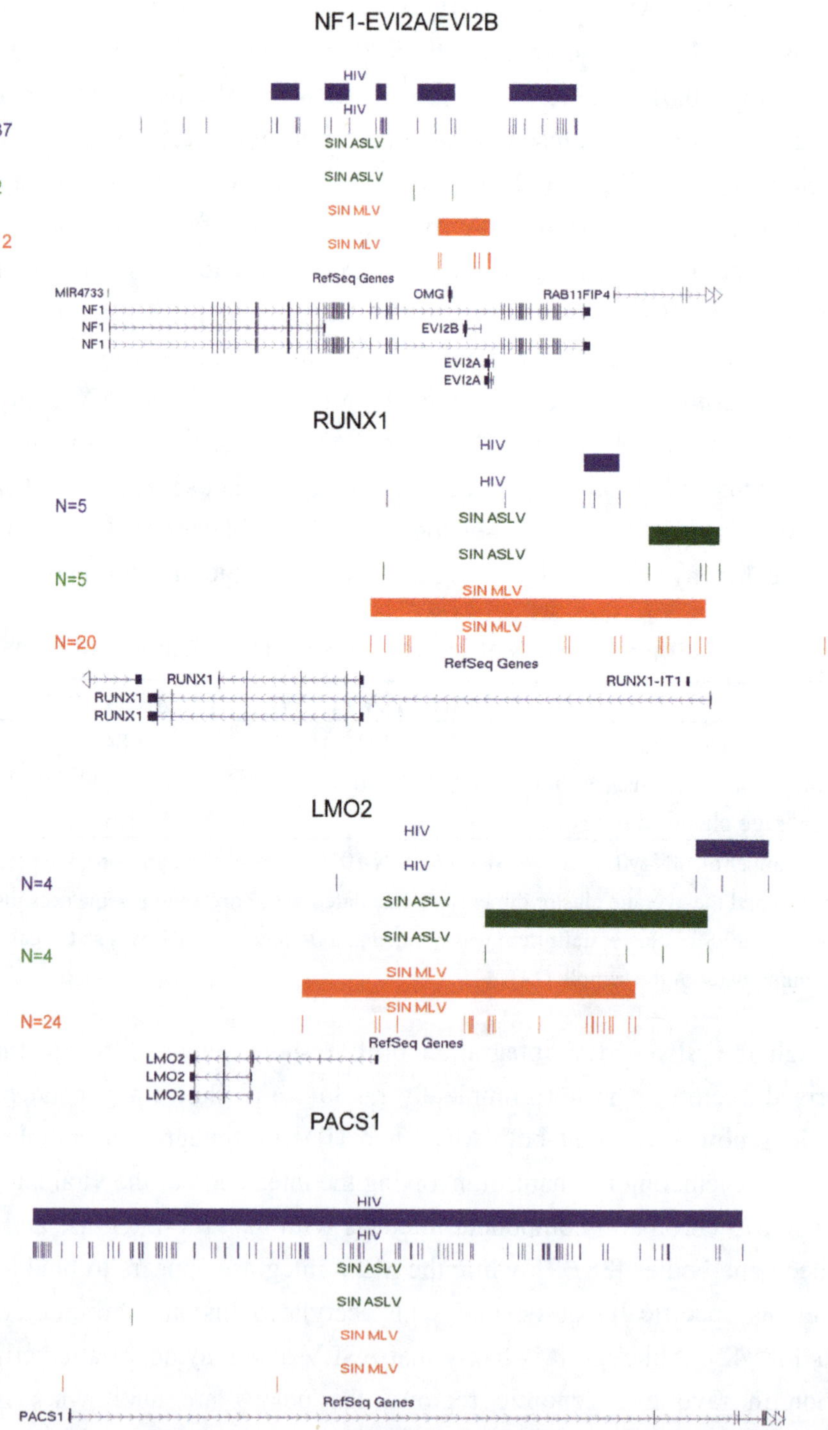

4. Conclusions

Overcoming the genotoxic consequences of retroviral vector integration in the host cell genome is one of the major issues for the application of retroviral-based gene transfer in clinical trials. The strong preference to target TSSs, active regulatory elements or transcribed genes, together with the high frequency of clustering around hot spots, is a characteristics shared by all retroviral vectors currently used in clinical gene therapy. These characteristics are at the basis of the potential of retroviral insertion to deregulate gene expression at the transcriptional or post-transcriptional level, which has been observed to cause clonal expansion and contribute to neoplastic transformation in a number of cases (reviewed by [9]). Many efforts are being made to improve the safety of currently available retroviral vectors by removing the viral transcriptional control element and avoiding dominant, long-range acting enhancers in the transgene expression cassette. Retargeting vector integration has proven more difficult and was so far unsuccessful. No strategy is obviously perfect, and even a completely random integration machinery would not abolish the risk of inducing an insertional oncogenic mutation in the host cell genome.

Based on a genome-wide analysis of >8000 integration sites in human HSPCs, we show that a SIN-ASLV vector has a quasi-random integration pattern that privileges active chromatin regions, but is not associated with active regulatory elements, like MLV, or with transcribed genes, like HIV. More importantly, the SIN-ASLV vector showed no integration hot spots and no preferences for subsets of genes with a defined ontology or genes that were previously identified as being activated by retroviral insertion into tumors. Previous evaluations of ASLV-derived vectors in pre-clinical models proved its ability to sustain long-term transgene expression in murine and human hematopoietic progenitors and to correct the pathology in a mouse model of X-linked Chronic Granulomatous Disease (X-CGD), with no evidence of post-transcriptional interference [20,22]. Combined with the use of short-range or cell-specific transcriptional regulatory elements, an ASLV vector appears to offer a very safe profile and to be an ideal candidate for *ex vivo* gene therapy applications.

Acknowledgments

This work was supported by grants from the European Research Council and the Italian National Research Council (EPIGEN). Ermanno Rizzi is financially supported by the Italian Research Ministry grants "Futuro in Ricerca" RBFR08U07M_003 and RBFR126B8I_003.

Author Contributions

Conceived and designed the experiments: Arianna Moiani, Fulvio Mavilio. Performed the experiments: Arianna Moiani, Ermanno Rizzi, and Julia Debora Suerth. Analyzed the data: Arianna Moiani, Francesco Gandolfi, Marco Severgnini, and Gianluca De Bellis. Contributed reagents/materials/analysis tools: Julia Debora Suerth and Axel Schambach. Wrote the paper: Arianna Moiani and Fulvio Mavilio.

References

1. Hacein-Bey-Abina, S.; Le Deist, F.; Carlier, F.; Bouneaud, C.; Hue, C.; de Villartay, J.P.; Thrasher, A.J.; Wulffraat, N.; Sorensen, R.; Dupuis-Girod, S.; *et al.* Sustained correction of X-linked severe combined immunodeficiency by *ex vivo* gene therapy. *N. Engl. J. Med.* **2002**, *346*, 1185–1193.
2. Ott, M.G.; Schmidt, M.; Schwarzwaelder, K.; Stein, S.; Siler, U.; Koehl, U.; Glimm, H.; Kuhlcke, K.; Schilz, A.; Kunkel, H.; *et al.* Correction of X-linked chronic granulomatous disease by gene therapy, augmented by insertional activation of MDS1-EVI1, PRMD16 or SETBP1. *Nat. Med.* **2006**, *12*, 401–409.
3. Mavilio, F.; Pellegrini, G.; Ferrari, S.; di Nunzio, F.; di Iorio, E.; Recchia, A.; Maruggi, G.; Ferrari, G.; Provasi, E.; Bonini, C.; *et al.* Correction of junctional epidermolysis bullosa by transplantation of genetically modified epidermal stem cells. *Nat. Med.* **2006**, *12*, 1397–1402.
4. Aiuti, A.; Cattaneo, F.; Galimberti, S.; Benninghoff, U.; Cassani, B.; Callegaro, L.; Scaramuzza, S.; Andolfi, G.; Mirolo, M.; Brigida, I.; *et al.* Gene therapy for immunodeficiency due to adenosine deaminase deficiency. *N. Engl. J. Med.* **2009**, *360*, 447–458.
5. Cartier, N.; Hacein-Bey-Abina, S.; Bartholomae, C.C.; Veres, G.; Schmidt, M.; Kutschera, I.; Vidaud, M.; Abel, U.; Dal-Cortivo, L.; Caccavelli, L.; *et al.* Hematopoietic stem cell gene therapy with a lentiviral vector in X-linked adrenoleukodystrophy. *Science* **2009**, *326*, 818–823.
6. Boztug, K.; Schmidt, M.; Schwarzer, A.; Banerjee, P.P.; Diez, I.A.; Dewey, R.A.; Bohm, M.; Nowrouzi, A.; Ball, C.R.; Glimm, H.; *et al.* Stem-cell gene therapy for the Wiskott-Aldrich syndrome. *N. Engl. J. Med.* **2010**, *363*, 1918–1927.
7. Biffi, A.; Montini, E.; Lorioli, L.; Cesani, M.; Fumagalli, F.; Plati, T.; Baldoli, C.; Martino, S.; Calabria, A.; Canale, S.; *et al.* Lentiviral hematopoietic stem cell gene therapy benefits metachromatic leukodystrophy. *Science* **2013**, doi:10.1126/science.1233158.
8. Aiuti, A.; Biasco, L.; Scaramuzza, S.; Ferrua, F.; Cicalese, M.P.; Baricordi, C.; Dionisio, F.; Calabria, A.; Giannelli, S.; Castiello, M.C.; *et al.* Lentiviral hematopoietic stem cell gene therapy in patients with Wiskott-Aldrich syndrome. *Science* **2013**, doi:10.1126/science.1233151.
9. Cavazza, A.; Moiani, A.; Mavilio, F. Mechanisms of retroviral integration and mutagenesis. *Human Gene Ther.* **2013**, *24*, 119–131.
10. Montini, E.; Cesana, D.; Schmidt, M.; Sanvito, F.; Ponzoni, M.; Bartholomae, C.; Sergi Sergi, L.; Benedicenti, F.; Ambrosi, A.; di Serio, C.; *et al.* Hematopoietic stem cell gene transfer in a tumor-prone mouse model uncovers low genotoxicity of lentiviral vector integration. *Nat. Biotechnol.* **2006**, *24*, 687–696.
11. Modlich, U.; Navarro, S.; Zychlinski, D.; Maetzig, T.; Knoess, S.; Brugman, M.H.; Schambach, A.; Charrier, S.; Galy, A.; Thrasher, A.J.; *et al.* Insertional transformation of hematopoietic cells by self-inactivating lentiviral and gammaretroviral vectors. *Mol. Ther.* **2009**, *17*, 1919–1928.
12. Montini, E.; Cesana, D.; Schmidt, M.; Sanvito, F.; Bartholomae, C.C.; Ranzani, M.; Benedicenti, F.; Sergi, L.S.; Ambrosi, A.; Ponzoni, M.; *et al.* The genotoxic potential of retroviral vectors is strongly modulated by vector design and integration site selection in a mouse model of hsc gene therapy. *J. Clin. Invest.* **2009**, *119*, 964–975.

13. Wu, X.; Li, Y.; Crise, B.; Burgess, S.M. Transcription start regions in the human genome are favored targets for mlv integration. *Science* **2003**, *300*, 1749–1751.
14. Cattoglio, C.; Pellin, D.; Rizzi, E.; Maruggi, G.; Corti, G.; Miselli, F.; Sartori, D.; Guffanti, A.; di Serio, C.; Ambrosi, A.; *et al.* High-definition mapping of retroviral integration sites identifies active regulatory elements in human multipotent hematopoietic progenitors. *Blood* **2010**, *116*, 5507–5517.
15. Schroder, A.R.; Shinn, P.; Chen, H.; Berry, C.; Ecker, J.R.; Bushman, F. HIV-1 integration in the human genome favors active genes and local hotspots. *Cell* **2002**, *110*, 521–529.
16. Wang, G.P.; Ciuffi, A.; Leipzig, J.; Berry, C.C.; Bushman, F.D. HIV integration site selection: Analysis by massively parallel pyrosequencing reveals association with epigenetic modifications. *Genome Res.* **2007**, *17*, 1186–1194.
17. Barr, S.D.; Leipzig, J.; Shinn, P.; Ecker, J.R.; Bushman, F.D. Integration targeting by avian sarcoma-leukosis virus and human immunodeficiency virus in the chicken genome. *J. Virol.* **2005**, *79*, 12035–12044.
18. Hu, J.; Renaud, G.; Gomes, T.J.; Ferris, A.; Hendrie, P.C.; Donahue, R.E.; Hughes, S.H.; Wolfsberg, T.G.; Russell, D.W.; Dunbar, C.E. Reduced genotoxicity of avian sarcoma leukosis virus vectors in rhesus long-term repopulating cells compared to standard murine retrovirus vectors. *Mol. Ther.* **2008**, *16*, 1617–1623.
19. Mitchell, R.S.; Beitzel, B.F.; Schroder, A.R.; Shinn, P.; Chen, H.; Berry, C.C.; Ecker, J.R.; Bushman, F.D. Retroviral DNA integration: ASLV, HIV, and MLV show distinct target site preferences. *PLoS Biol.* **2004**, *2*, e234.
20. Suerth, J.D.; Maetzig, T.; Brugman, M.H.; Heinz, N.; Appelt, J.U.; Kaufmann, K.B.; Schmidt, M.; Grez, M.; Modlich, U.; Baum, C.; *et al.* Alpharetroviral self-inactivating vectors: Long-term transgene expression in murine hematopoietic cells and low genotoxicity. *Mol. Ther.* **2012**, *20*, 1022–1032.
21. Suerth, J.D.; Maetzig, T.; Galla, M.; Baum, C.; Schambach, A. Self-inactivating alpharetroviral vectors with a split-packaging design. *J. Virol.* **2010**, *84*, 6626–6635.
22. Kaufmann, K.B.; Brendel, C.; Suerth, J.D.; Mueller-Kuller, U.; Chen-Wichmann, L.; Schwable, J.; Pahujani, S.; Kunkel, H.; Schambach, A.; Baum, C.; *et al.* Alpharetroviral vector-mediated gene therapy for X-CGD: Functional correction and lack of aberrant splicing. *Mol. Ther.* **2013**, *21*, 648–661.
23. Cattoglio, C.; Facchini, G.; Sartori, D.; Antonelli, A.; Miccio, A.; Cassani, B.; Schmidt, M.; von Kalle, C.; Howe, S.; Thrasher, A.J.; *et al.* Hot spots of retroviral integration in human $CD34^+$ hematopoietic cells. *Blood* **2007**, *110*, 1770–1778.
24. Cattoglio, C.; Maruggi, G.; Bartholomae, C.; Malani, N.; Pellin, D.; Cocchiarella, F.; Magnani, Z.; Ciceri, F.; Ambrosi, A.; von Kalle, C.; *et al.* High-definition mapping of retroviral integration sites defines the fate of allogeneic T cells after donor lymphocyte infusion. *PLoS One* **2010**, *5*, e15688.
25. Cavazza, A.; Cocchiarella, F.; Bartholomae, C.; Schmidt, M.; Pincelli, C.; Larcher, F.; Mavilio, F. Self-inactivating MLV vectors have a reduced genotoxic profile in human epidermal keratinocytes. *Gene Ther.* **2013**, doi:10.1038/gt.2013.18.

26. Moiani, A.; Miccio, A.; Rizzi, E.; Severgnini, M.; Pellin, D.; Suerth, J.D.; Baum, C.; de Bellis, G.; Mavilio, F. Deletion of the LTR enhancer/promoter has no impact on the integration profile of MLV vectors in human hematopoietic progenitors. *PLoS One* **2013**, *8*, e55721.
27. Kim, S.Y.; Pritchard, J.K. Adaptive evolution of conserved noncoding elements in mammals. *PLoS Genet.* **2007**, *3*, 1572–1586.
28. John, S.; Sabo, P.J.; Canfield, T.K.; Lee, K.; Vong, S.; Weaver, M.; Wang, H.; Vierstra, J.; Reynolds, A.P.; Thurman, R.E.; *et al.* Genome-scale mapping of DNase I hypersensitivity. *Curr. Protoc. Mol. Biol.* **2013**, doi:10.1002/0471142727.mb2127s103.
29. Ye, T.; Krebs, A.R.; Choukrallah, M.A.; Keime, C.; Plewniak, F.; Davidson, I.; Tora, L. Seqminer: An integrated chip-seq data interpretation platform. *Nucleic Acids Res.* **2010**, *39*, e35.
30. Ferrigno, O.; Virolle, T.; Djabari, Z.; Ortonne, J.P.; White, R.J.; Aberdam, D. Transposable b2 sine elements can provide mobile rna polymerase ii promoters. *Nat. Genet.* **2001**, *28*, 77–81.
31. Lunyak, V.V.; Prefontaine, G.G.; Nunez, E.; Cramer, T.; Ju, B.G.; Ohgi, K.A.; Hutt, K.; Roy, R.; Garcia-Diaz, A.; Zhu, X.; *et al.* Developmentally regulated activation of a SINE b2 repeat as a domain boundary in organogenesis. *Science* **2007**, *317*, 248–251.
32. Biasco, L.; Ambrosi, A.; Pellin, D.; Bartholomae, C.; Brigida, I.; Roncarolo, M.G.; di Serio, C.; von Kalle, C.; Schmidt, M.; Aiuti, A. Integration profile of retroviral vector in gene therapy treated patients is cell-specific according to gene expression and chromatin conformation of target cell. *EMBO Mol. Med.* **2011**, *3*, 89–101.
33. Berry, C.; Hannenhalli, S.; Leipzig, J.; Bushman, F.D. Selection of target sites for mobile DNA integration in the human genome. *PLoS Comput. Biol.* **2006**, *2*, e157.
34. Lewinski, M.K.; Yamashita, M.; Emerman, M.; Ciuffi, A.; Marshall, H.; Crawford, G.; Collins, F.; Shinn, P.; Leipzig, J.; Hannenhalli, S.; *et al.* Retroviral DNA integration: Viral and cellular determinants of target-site selection. *PLoS Pathog.* **2006**, *2*, e60.
35. Cesana, D.; Sgualdino, J.; Rudilosso, L.; Merella, S.; Naldini, L.; Montini, E. Whole transcriptome characterization of aberrant splicing events induced by lentiviral vector integrations. *J. Clin. Invest.* **2012**, *122*, 1667–1676.
36. Moiani, A.; Paleari, Y.; Sartori, D.; Mezzadra, R.; Miccio, A.; Cattoglio, C.; Cocchiarella, F.; Lidonnici, M.R.; Ferrari, G.; Mavilio, F. Lentiviral vector integration in the human genome induces alternative splicing and generates aberrant transcripts. *J. Clin. Invest.* **2012**, *122*, 1653–1666.
37. Ambrosi, A.; Glad, I.K.; Pellin, D.; Cattoglio, C.; Mavilio, F.; di Serio, C.; Frigessi, A. Estimated comparative integration hotspots identify different behaviors of retroviral gene transfer vectors. *PLoS Comput. Biol.* **2011**, *7*, e1002292.
38. Ciuffi, A.; Llano, M.; Poeschla, E.; Hoffmann, C.; Leipzig, J.; Shinn, P.; Ecker, J.R.; Bushman, F. A role for LEDGF/p75 in targeting HIV DNA integration. *Nat. Med.* **2005**, *11*, 1287–1289.
39. Llano, M.; Vanegas, M.; Fregoso, O.; Saenz, D.; Chung, S.; Peretz, M.; Poeschla, E.M. LEDGF/p75 determines cellular trafficking of diverse lentiviral but not murine oncoretroviral integrase proteins and is a component of functional lentiviral preintegration complexes. *J. Virol.* **2004**, *78*, 9524–9537.

40. De Rijck, J.; de Kogel, C.; Demeulemeester, J.; Vets, S.; El Ashkar, S.; Malani, N.; Bushman, F.D.; Landuyt, B.; Husson, S.J.; Busschots, K.; *et al.* The BET family of proteins targets Moloney murine leukemia virus integration near transcription start sites. *Cell Rep.* **2013**, *5*, 886–894.
41. Gupta, S.S.; Maetzig, T.; Maertens, G.N.; Sharif, A.; Rothe, M.; Weidner-Glunde, M.; Galla, M.; Schambach, A.; Cherepanov, P.; Schulz, T.F. Bromo- and extraterminal domain chromatin regulators serve as cofactors for murine leukemia virus integration. *J. Virol.* **2013**, *87*, 12721–12736.
42. Sharma, A.; Larue, R.C.; Plumb, M.R.; Malani, N.; Male, F.; Slaughter, A.; Kessl, J.J.; Shkriabai, N.; Coward, E.; Aiyer, S.S.; *et al.* BET proteins promote efficient murine leukemia virus integration at transcription start sites. *Proc. Natl. Acad. Sci. USA* **2013**, *110*, 12036–12041.

2

Architecture of Inherited Susceptibility to Colorectal Cancer: A Voyage of Discovery

Nicola Whiffin * and Richard S. Houlston

Molecular and Population Genetics Team, Genetics and Epidemiology, The Institute of Cancer Research, Sutton, SM2 5NG, UK; E-Mail: richard.houlston@icr.ac.uk

* Author to whom correspondence should be addressed; E-Mail: nicola.whiffin@icr.ac.uk.

Abstract: This review looks back at five decades of research into genetic susceptibility to colorectal cancer (CRC) and the insights these studies have provided. Initial evidence of a genetic basis of CRC stems from epidemiological studies in the 1950s and is further provided by the existence of multiple dominant predisposition syndromes. Genetic linkage and positional cloning studies identified the first high-penetrance genes for CRC in the 1980s and 1990s. More recent genome-wide association studies have identified common low-penetrance susceptibility loci and provide support for a polygenic model of disease susceptibility. These observations suggest a high proportion of CRC may arise in a group of susceptible individuals as a consequence of the combined effects of common low-penetrance risk alleles and rare variants conferring moderate CRC risks. Despite these advances, however, currently identified loci explain only a small fraction of the estimated heritability to CRC. It is hoped that a new generation of sequencing projects will help explain this missing heritability.

Keywords: colorectal cancer, genetics, susceptibility

---

1. Introduction

Colorectal cancer (CRC) is the third most common cancer worldwide with half a million new individuals diagnosed annually [1]. In the UK CRC affects ~40,000 individuals and is responsible for ~16,000 deaths each year (Cancer Research UK, 2013) amounting to a life-time risk of 5%–6%. It is now an established fact that inherited susceptibility has an important role in predisposition to CRC.

The earliest evidence for this came from epidemiological studies in the 1950s which showed a two- to three-fold increased risk of CRC in first degree relatives of patients [2]. Subsequent studies have identified a number of CRC susceptibility genes. These discoveries have greatly increased our understanding of the mechanisms underlying CRC biology and have provided promising targets for therapeutic intervention. Moreover, the ability to identify individuals at increased risk of CRC is of important clinical relevance.

2. Early Models of Genetic Susceptibility

Large families with multiple individuals affected by CRC have long been reported in the clinic. It was not until the late 1950s, however, that epidemiological studies attempted to quantify this familial clustering by comparing the incidence of CRC in first degree relatives (FDRs) of cases to control groups [3–5]. Recent analysis estimates an approximate two-fold increase in risk in FDRs [2]. This risk increases further to four-fold when the relative is diagnosed with early-onset CRC (<45 years of age), indicative of colorectal tumours developing in genetically susceptibly individuals at an earlier age.

In 1969 Ashley [6] proposed that colonic cancer development could be ascribed to a series of carcinogenic "hits" on normal intestinal mucosa cells. He further noted that the number of necessary "hits" was lower for genetically susceptible individuals with the Mendelian predisposition syndrome familial adenomatous polyposis (FAP). In the same year, DeMars [7] suggested that FAP, along with other apparently autosomal dominant syndromes, is autosomally recessive at the cellular level; individuals with a germline mutation in one allele of a tumour suppressor gene develop cancer as a result of subsequent somatic mutations in the other gene copy.

Anderson [8], in 1974, made the first argument for a polygenic mechanism to cancer susceptibility based on the increased risk in FDRs of cancer patients being limited to two- to three-fold. He stated these results were "not indicative of strong genetic effects" and rather suggested a mechanism involving many genes with small effects acting in concert with environmental factors with larger effects. Whilst there is growing evidence to suggest that his conclusion is, at least in part, correct, the reasoning behind this statement is flawed as the relative risks associated with FDRs are likely to be underestimated. This is because calculations of relative risks typically include both sporadic and genetically susceptible cases that are then compared to the general population which, to compound the problem, also contains individuals that are genetically susceptible to CRC.

3. Identification of Rare High-Penetrance Susceptibility Alleles

Evidence for Mendelian transmission of CRC was first provided by reports of large families with CRC segregating in a dominant fashion. Perhaps the most notable case report is "family G" first described in 1913 by Warthin and subsequently revisited by Lynch and Krush in 1971 [9]. This family of over 650 blood relatives provided scientists with one of the longest, most detailed cancer genealogies in the world and was instrumental in establishing the syndrome of hereditary non-polyposis colorectal cancer (HNPCC or Lynch syndrome).

Family based genetic linkage and positional cloning studies in the late 1980s and early 1990s led to the identification of numerous CRC susceptibility genes (Table 1). The *APC* gene on chromosome 5 was the first gene to be shown to be associated with CRC susceptibility when it was identified as mutated in FAP patients [10–14]. Subsequently, mutations in genes of the mismatch repair (MMR) pathway, particularly *MSH2*, *MLH1* and MSH6, the TGF-β signalling pathway genes, *SMAD4* and *BMPR1A*, and the serine/threonine kinase gene STK11, were revealed as the causes of HNPCC [15–26], Juvenile Polyposis syndrome (JPS) [27,28] and Peutz-Jeghers syndrome (PJS) [29] respectively. Risk alleles in these genes are rare (<0.1%) and confer a >10 fold increase in risk of CRC. These genes are tumour suppressors conforming to DeMars' "two-hit" model of cancer susceptibility through an apparently dominant mode of inheritance. The clinical utility of testing for high penetrance mutations in these genes has long been established and identification of individuals with such mutations has been shown reduce CRC incidence through prevention strategies and screening [30,31].

Table 1. Colorectal cancer predisposition syndromes and associated high-penetrance mutations.

Gene(s)	Syndrome	Risk in mutation carriers	Mode of inheritance	References
APC	FAP	90% by age 45	Dominant	[10–14]
Mismatch repair (MLH1/MSH2/MSH6/PMS2)	HNPCC/Lynch syndrome	40%–80% by age 75	Dominant	[15–26,32,33]
SMAD4/BMPR1A	JPS	17%–68% by age 60	Dominant	[26,27]
STK11	PJS	39% by age 70	Dominant	[28]
MUTYH	MYH-associated polyposis	35%–53%	Recessive	[34,35]
POLD1/POLE	Oligopolyposis		Dominant	[36]

4. More Recent Models of Genetic Susceptibility

Studies examining the difference in CRC development between monozygotic and dizygotic twins estimated that ~35% of CRC could be ascribed to a genetic predisposition [37]. However, <10% of all CRC can be accounted for by germline mutations in *APC* and the MMR genes and crucially ~70% of the familial risk of CRC remains unexplained [38].

Over the past 20 years, extensive efforts to identify additional, highly penetrant cancer susceptibility genes for CRC through conventional linkage scans have met with limited success [39,40]. This strongly implies that any additional high penetrance CRC gene will individually account for only a small proportion of the familial risk. Statistical modelling of the pattern of familial occurrence of CRC after exclusion of known high-risk genes suggests that much of the inherited susceptibility is likely to be polygenic with the co-inheritance of multiple genetic variants, each with a modest individual effect, causing a wide range of risk in the population (Figure 1).

Over the past two decades candidate gene studies have identified rare moderately-penetrant risk alleles (minor allele frequency (MAF) < 2%; relative risks (RRs) > 2.0) and more recent genome-wide association studies (GWAS) have identified common, low-penetrance alleles (MAF > 10%; OR < 1.5). In reality, these variants are likely to occur as a continuum and the separate classes of risk alleles merely reflect the subgroups detectable using current methodologies.

5. Rare, Moderately-Penetrant Disease-Causing Variants

The "rare-variant" hypothesis suggests that much of the remaining heritability could be due to the combined effect of rare, moderately-penetrant risk alleles [41]. These variants are hypothesised to act independently and to confer modest, but detectable, increases in risk.

Figure 1. Polygenic model of disease susceptibility. The distribution of risk alleles in both cases and controls follows a normal distribution. However, cases have a shift towards a higher number of high risk alleles.

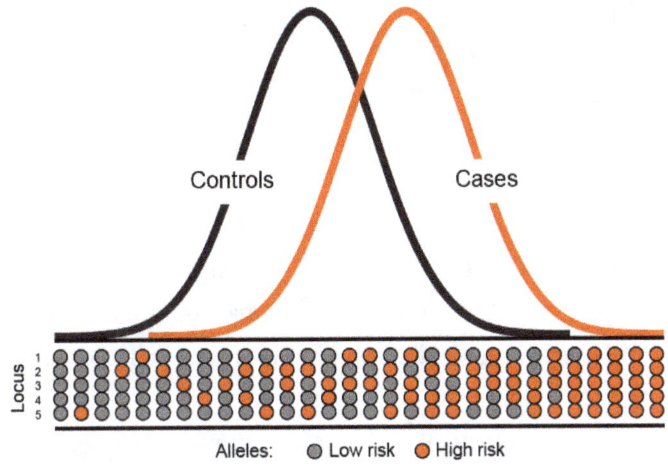

Attempts to identify this class of disease allele have mainly been through resequencing candidate genes in affected families, the success of which has been hampered by our limited knowledge of tumour biology. The identification of the missense variant, APC I1307K, carried by ~6% of Ashkenazi Jews and conferring around a two-fold increase in risk of CRC [42] and the more recent discovery of the functional promoter variant -93G>A of MLH1, shown to predispose to microsatellite unstable CRC [43], represent rare successes of this approach.

A priori rare disease-causing alleles are likely to act in a dominant fashion; however, functional variants in the base-excision repair gene *MUTYH* provide an example of a recessive model of inheritance [34]. Biallelic or compound heterozygosity of the G396D and Y179C mutations in *MUTYH*, which are carried by around 1%–2% of the UK population, confers a CRC risk comparable to that seen in carriers of germline MMR mutations [35].

Mechanistically these variants are likely to be directly causal. For the variants in *MUTYH*, insights into biological basis of susceptibility came from the method in which they were discovered; a FAP family with no apparent *APC* mutation was found to possess a mutator phenotype reflective of defective base excision, resulting in somatic mutation of *APC* and other genes [34,44]. In contrast, the *APC* I1307K T>A variant appears to increase replication errors in *APC* through generation of a run of eight adenines [42].

6. Identification of Common Low-Penetrance Alleles

The "common-variant, common-disease" hypothesis states that a substantial proportion of the remaining risk is likely to be accounted for by the summation of numerous low-penetrant genetic variants, each with a relatively high frequency in the population [45]. These variants have more subtle effects on gene regulation and predominantly reside within non-coding regions of the genome. Each individual variant is associated with only a modest increase in risk; however, collectively they may confer a substantial increase in disease susceptibility. These alleles rarely cause multiple cases in families and therefore cannot be detected through genetic linkage studies [46]. Initial attempts to identify this class of allele through candidate gene association studies were based on small case-control series and had little success; any proposed variants were not successfully validated in subsequent studies.

Genome-wide association studies (GWAS), typically based on genotyping of 300,000 to over 1 million SNPs, have proved to be a powerful approach in identifying common, low penetrance susceptibility loci for CRC without prior knowledge of location and function.

Since the first CRC GWAS in 2007, 18 CRC susceptibility loci have been identified in European populations (Table 2) [47–53]. While each individual risk allele confers only a small relative risk (1.06 < OR < 1.26), the SNPs are common (MAF > 10%) and hence contribute significantly to the overall incidence of CRC. Moreover, by acting in concert they can impact significantly on an individual's risk of developing CRC (Figure 2) [54]. The design of association studies is advantageous as large numbers of unrelated case and control samples may be readily obtained, providing adequate power to detect loci with relatively small effects. This is in contrast to the difficulty in recruiting the extensive pedigrees required for linkage studies.

Importantly, few genes implicated in GWAS were previously evaluated in candidate gene studies, highlighting the importance of such an agnostic approach for gene discovery and understanding of CRC aetiology. None of the currently identified loci, for example, are involved in DNA repair, which is the principle pathway underscoring high-penetrance CRC susceptibility. Interestingly, five of the loci discovered to date are within or near to genes involved in the TGF-β signalling pathway [47,50]. This pathway has already been implicated in pathogenesis of CRC, as the dominant CRC predisposition syndrome JPS is caused by high penetrance mutations in the TGF-β family genes *SMAD4* and *BMPR1A* [13,14].

Table 2. Loci identified as associated with colorectal cancer through genome-wide association studies and meta-analyses.

Locus	Nearest Gene(s)	GWAS tagSNP	Location	Risk Allele	Alt Allele	RAF
1q41	*DUSP10*	rs6691170	222,045,446	T	G	0.40
3q26.2	*TERC, MYNN*	rs10936599	169,492,101	C	T	0.75
6p21.2	*CDKN1A*	rs1321311	36,622,900	T	G	0.21
8q23.3	*EIF3H*	rs16892766	117,630,683	C	A	0.09
8q24.21	*MYC*	rs6983267	128,413,305	G	T	0.52
10p14	*GATA3*	rs10795668	8,701,219	G	A	0.67
11q13.4	*POLD3*	rs3824999	74,345,550	C	A	0.47
11q23.1	*FLJ45803*	rs3802842	111,171,709	C	A	0.27
12q13	*DIP2B, ATF1*	rs11169552	51,155,663	C	T	0.75
14q22.2	*BMP4*	rs4444235	54,410,919	C	T	0.48
15q13.3	*SCG5, GREM1*	rs4779584	32,994,756	T	C	0.19
16q22.1	*CDH1*	rs9929218	68,820,946	G	A	0.71
18q21.2	*SMAD7*	rs4939827	46,453,463	T	C	0.53
19q13.11	*RHPN2, GPATCH1*	rs10411210	33,532,300	C	T	0.90
20p12.3	*BMP2*	rs961253	6,404,281	A	C	0.37
		rs4813802	6,699,595	G	T	0.34
20q13.33	*LAMA5*	rs4925386	60,921,044	C	T	0.68
Xp22.2	SHROOM2	rs5934683	9,751,474	T	C	0.56

Figure 2. Plot showing the increase in odds ratio for colorectal cancer with an increasing number of risk alleles.

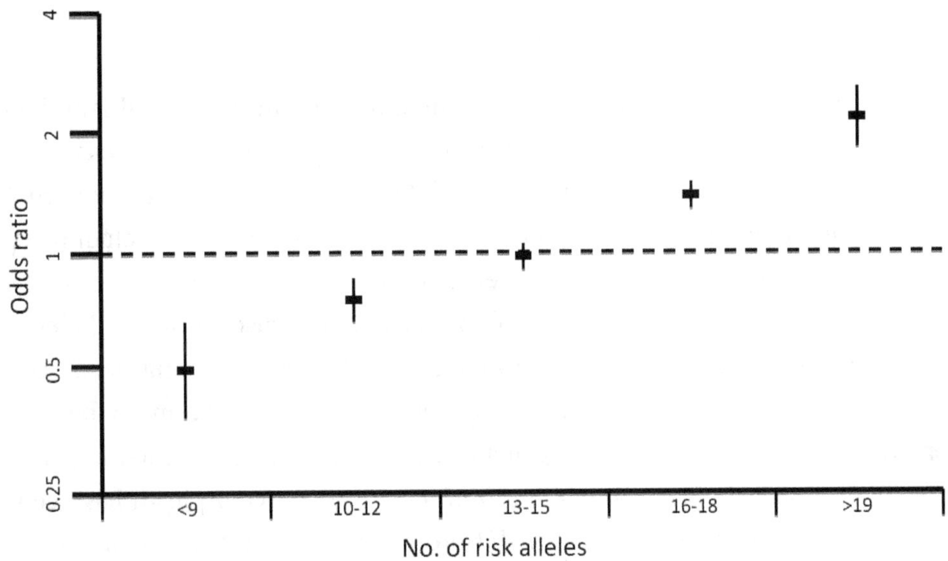

7. Functional Effects of GWAS Loci

Elucidating the basis of association at common low-penetrance loci represents a significant challenge. The tagging SNPs (tagSNPs) used in GWAS are not necessarily strong candidates for being causal and were instead chosen to capture variation across large genomic regions. Hence, establishing which of a set of highly correlated variants in linkage disequilibrium (LD) with the tagSNP is the true causal variant is a challenge. In addition, many GWAS loci map to non-coding regions or gene deserts, suggesting the true cause of the association at these regions is through subtle effects on gene expression rather than changes to protein coding sequence.

Fine mapping of CRC loci is still in its infancy with most studies attempting only to narrow down the location of a likely functional variant through imputation or re-sequencing [55–57]. These studies suggest candidate variants but very few functional studies have been carried out to test these assertions. To date, in only four regions has a SNP been proposed as the likely functional candidate and a mechanism of action suggested (8q23.3 [58], 8q24.21 [59,60], 11q23.1 [61], 18q21.1 [62]). The most intriguing of these regions is 8q24.21, which has pleiotropic effects on cancer susceptibility, also harbouring risk loci for breast [57], ovarian [63], bladder [64], CLL [65] and multiple independent loci for prostate cancer [66–69]. This is in contrast to most associations found to date which appear to be disease specific. The rs6983267 risk SNP is associated with both prostate and colorectal cancers and lies within an evolutionarily conserved region. The two alleles of rs6983267 show differential binding of the TCF4 transcription factor [59] to an enhancer element that has been shown to physically interact with the *MYC* gene promoter [60]. *MYC* is amplified or over-expressed in multiple cancer types leading to up-regulation of many genes controlling cell proliferation, so it is predicted that variation at this locus acts through subtle effects on *MYC* gene expression. The risk allele of rs6983267 has also been suggested as a marker of worse prognosis in CRC patients [70]. The strongest CRC GWAS association is at 18q21.1 (RR = 1.26) within the *SMAD7* gene, which acts as an antagonist of the TGF-β signalling pathway. Resequencing of the 17 kb region of linkage disequilibrium surrounding the GWAS tagSNP rs4939827 identified a novel variant, termed Novel1 (rs58920878), which was shown to affect *SMAD7* expression [62].

8. Impact of Common Variation on CRC Risk

Collectively, the currently identified risk loci explain only ~6% of the overall familial risk of CRC. This estimate is likely conservative as the effect of the causal variant at each locus is expected to be greater than the association detected through a GWAS tagSNP. In addition, as evinced by the 14q22 (*BMP4*) [52] association multiple risk variants may exist at each locus, including low-frequency variants with significantly larger effects. Moreover, epistatic interactions may exist between common risk loci, which could make the contribution of low-penetrance susceptibility alleles much higher. Such interactions remain difficult to detect due to substantial multiple testing penalties, and existing studies suggest the effects of most common low-penetrance alleles seem to be independent. In addition, interactions of these alleles with epigenetic regulation or environmental factors may lead to a greater increase in disease risk. For example the *MLH1*-93G>A polymorphism has been shown to be associated with increased methylation of the *MLH1* promoter [71]. Another important consideration is the possible modification of the effect of a low-penetrance allele by the presence of a high-penetrance mutation. The only evidence to support this assertion in CRC is a small study implicating the 8q23.3 and 11q23.1 CRC SNPs as modifiers of CRC risk in MMR mutation carriers [72]. Although there was initial hope for the use of low penetrance variants in the clinic, the small proportion of familial risk explained and the apparent lack of epistatic interactions between the variants leads to them being of low predictive power. The increased risk associated with having a high number of risk alleles (Figure 2) has the prospect of identifying individuals in the general population who might benefit from earlier screening [54].

9. Identifying Novel CRC Susceptibility Loci

It is unlikely that there are any common CRC risk loci with appreciable MAF (>10%) and with relative risks >1.1 that remain to be uncovered given the size of existing GWAS studies. Small variant effect sizes combined with stringent thresholds for establishing statistical significance and financial constraints on the number of variants which can be followed up constrain GWAS study power. However, many GWAS have long tails of association with alleles of increasingly small effect, suggesting much of the remaining susceptibility may be embodied in a multitude of common risk alleles. Larger GWAS studies combining multiple phases and tens of thousands of cases may identify many more of these susceptibility loci, although the effect of these on CRC risk is likely to be minimal. Such studies have been conducted in both breast [73] and prostate [74] cancer identifying 41 and 23 novel risk loci respectively.

Commercial arrays used for GWAS capture a large proportion of common SNPs with minor allele frequencies (MAF) > 10%. However, a much lower fraction of less common (5%–10% MAF) and rare (MAF < 5%) SNPs are captured by these arrays. The power of GWAS to detect variants with MAF < 10% is therefore limited. Additionally, GWAS arrays are not optimally formatted to capture indels and copy number variants, both of which are likely to have roles in disease susceptibility. New "exome chips" have recently been released that aim to address some of these limitations. However the success of these new arrays remains to be evaluated.

Since the completion of the human genome project in 2003, the utilisation of massively parallel sequencing technologies to identify variants has become feasible. Moreover, these methods can be used to detect small indels, substitutions and structural variants. Although in their infancy, such studies are beginning to identify additional variants. For example, highly-penetrant mutations have recently been identified in the proof reading domains of POLD1 (S478N) and POLE (L424V) in CRC families [36]. Another recent study identified 11 candidate CRC susceptibility genes with truncating mutations in two or more of 96 familial CRC cases [75]. To maximally utilise sequencing data, new bioinformatic techniques are required to remove sequencing errors and prioritise variants. Additionally, due to financial considerations such studies remain small and not powered to detect less common variants with moderate effects risk of disease. Using cases enriched for genetic susceptibility evidenced by a strong family history or early disease onset is a useful technique to increase the efficiency of these studies. In addition, utilising whole exome sequencing, as performed by Palles *et al.* [36], can dramatically reduce the costs associated with such projects. Coding variants are also much easier to interpret than those in non-coding regions. However, with every individual's genome harboring 250–300 putative loss of function variants [76] and many missense variants being of unknown effect, identification of the disease causing variant presents a significant challenge. Recent studies are working to interpret this class of variants [77] and algorithms such as SIFT [78], PolyPhen2 [79] and CONDEL [80] aim to guide researchers by predicting the functional effects of coding mutations. Work to develop similar methods to deal with non-coding regions is still in its infancy [81,82], however, recent evidence [83,84] suggests that these regions are also *a priori* likely to contain variants with a large effect on CRC risk.

10. Conclusions

Much has been achieved in the study of genetic susceptibility to CRC in the last five decades. The architecture underlying this susceptibility is now recognised to be defined by a spectrum of predisposition alleles with different effect sizes and frequencies in the population. GWAS has proved a successful approach for identification of novel low-penetrance CRC risk alleles, improving our understanding of disease aetiology and providing novel therapeutic targets. Determining the biological processes underlying the associations at these loci presents a significant challenge and will likely require large collaborations between genetic researchers and functional biologists. Despite these advances, a large proportion of the heritability to CRC remains unaccounted for. It is hoped that the new generation of sequencing projects will help to uncover this missing heritability.

Acknowledgements

Principle funding was provided to RSH by Cancer Research UK (C1298/A8362-Bobby Moore Fund for Cancer Research UK). NW is in receipt of a PhD studentship from the Institute of Cancer Research.

Author Contributions

Manuscript was conceived and written by NW and RSH.

References

1. Jemal, A.; Siegel, R.; Ward, E.; Murray, T.; Xu, J.; Smigal, C.; Thun, M.J. Cancer statistics, 2006. *CA Cancer J. Clin.* **2006**, *56*, 106–130.
2. Johns, L.E.; Houlston, R.S. A systematic review and meta-analysis of familial colorectal cancer risk. *Am. J. Gastroenterol.* **2001**, *96*, 2992–3003.
3. Woolf, C.M. A genetic study of carcinoma of the large intestine. *Am. J. Hum. Genet.* **1958**, *10*, 42–47.
4. Macklin, M.T. Inheritance of cancer of the stomach and large intestine in man. *J. Natl. Canc. Inst.* **1960**, *24*, 551–571.
5. Lovett, E. Familial factors in the etiology of carcinoma of the large bowel. *Proc. R. Soc. Med.* **1974**, *67*, 751–752.
6. Ashley, D.J. Colonic cancer arising in polyposis coli. *J. Med. Genet.* **1969**, *6*, 376–378.
7. DeMars, R. *23rd Annual Symposium on Fundamental Cancer Research*; Williams and Wilkins: Baltimore, MD, USA, 1969; pp. 105–106.
8. Anderson, D.E. Genetic study of breast cancer: Identification of a high risk group. *Cancer* **1974**, *34*, 1090–1097.
9. Lynch, H.T.; Krush, A.J. Cancer family "G" Revisited: 1895–1970. *Cancer* **1971**, *27*, 1505–1511.
10. Bodmer, W.F.; Bailey, C.J.; Bodmer, J.; Bussey, H.J.; Ellis, A.; Gorman, P.; Lucibello, F.C.; Murday, V.A.; Rider, S.H.; Scambler, P.; *et al.* Localization of the gene for familial adenomatous polyposis on chromosome 5. *Nature* **1987**, *328*, 614–616.
11. Leppert, M.; Dobbs, M.; Scambler, P.; O'Connell, P.; Nakamura, Y.; Stauffer, D.; Woodward, S.; Burt, R.; Hughes, J.; Gardner, E.; *et al.* The gene for familial polyposis coli maps to the long arm of chromosome 5. *Science* **1987**, *238*, 1411–1413.
12. Herrera, L.; Kakati, S.; Gibas, L.; Pietrzak, E.; Sandberg, A.A. Gardner syndrome in a man with an interstitial deletion of 5q. *Am. J. Med. Genet.* **1986**, *25*, 473–476.
13. Groden, J.; Thliveris, A.; Samowitz, W.; Carlson, M.; Gelbert, L.; Albertsen, H.; Joslyn, G.; Stevens, J.; Spirio, L.; Robertson, M.; *et al.* Identification and characterization of the familial adenomatous polyposis coli gene. *Cell* **1991**, *66*, 589–600.
14. Kinzler, K.W.; Nilbert, M.C.; Su, L.K.; Vogelstein, B.; Bryan, T.M.; Levy, D.B.; Smith, K.J.; Preisinger, A.C.; Hedge, P.; McKechnie, D.; *et al.* Identification of fap locus genes from chromosome 5q21. *Science* **1991**, *253*, 661–665.
15. Miyaki, M.; Konishi, M.; Tanaka, K.; Kikuchi-Yanoshita, R.; Muraoka, M.; Yasuno, M.; Igari, T.; Koike, M.; Chiba, M.; Mori, T. Germline mutation of msh6 as the cause of hereditary nonpolyposis colorectal cancer. *Nat. Genet.* **1997**, *17*, 271–272.
16. Papadopoulos, N.; Nicolaides, N.C.; Wei, Y.F.; Ruben, S.M.; Carter, K.C.; Rosen, C.A.; Haseltine, W.A.; Fleischmann, R.D.; Fraser, C.M.; Adams, M.D.; *et al.* Mutation of a mutl homolog in hereditary colon cancer. *Science* **1994**, *263*, 1625–1629.

17. Bronner, C.E.; Baker, S.M.; Morrison, P.T.; Warren, G.; Smith, L.G.; Lescoe, M.K.; Kane, M.; Earabino, C.; Lipford, J.; Lindblom, A.; *et al.* Mutation in the DNA mismatch repair gene homologue hmlh1 is associated with hereditary non-polyposis colon cancer. *Nature* **1994**, *368*, 258–261.
18. Strand, M.; Prolla, T.A.; Liskay, R.M.; Petes, T.D. Destabilization of tracts of simple repetitive DNA in yeast by mutations affecting DNA mismatch repair. *Nature* **1993**, *365*, 274–276.
19. Peltomaki, P.; Aaltonen, L.A.; Sistonen, P.; Pylkkanen, L.; Mecklin, J.P.; Jarvinen, H.; Green, J.S.; Jass, J.R.; Weber, J.L.; Leach, F.S.; *et al.* Genetic mapping of a locus predisposing to human colorectal cancer. *Science* **1993**, *260*, 810–812.
20. Aaltonen, L.A.; Peltomaki, P.; Leach, F.S.; Sistonen, P.; Pylkkanen, L.; Mecklin, J.P.; Jarvinen, H.; Powell, S.M.; Jen, J.; Hamilton, S.R.; *et al.* Clues to the pathogenesis of familial colorectal cancer. *Science* **1993**, *260*, 812–816.
21. Lindblom, A.; Tannergard, P.; Werelius, B.; Nordenskjold, M. Genetic mapping of a second locus predisposing to hereditary non-polyposis colon cancer. *Nat. Genet.* **1993**, *5*, 279–282.
22. Leach, F.S.; Nicolaides, N.C.; Papadopoulos, N.; Liu, B.; Jen, J.; Parsons, R.; Peltomaki, P.; Sistonen, P.; Aaltonen, L.A.; Nystrom-Lahti, M.; *et al.* Mutations of a muts homolog in hereditary nonpolyposis colorectal cancer. *Cell* **1993**, *75*, 1215–1225.
23. Drummond, J.T.; Li, G.M.; Longley, M.J.; Modrich, P. Isolation of an hmsh2-p160 heterodimer that restores DNA mismatch repair to tumor cells. *Science* **1995**, *268*, 1909–1912.
24. Palombo, F.; Gallinari, P.; Iaccarino, I.; Lettieri, T.; Hughes, M.; D'Arrigo, A.; Truong, O.; Hsuan, J.J.; Jiricny, J. Gtbp, a 160-kilodalton protein essential for mismatch-binding activity in human cells. *Science* **1995**, *268*, 1912–1914.
25. Akiyama, Y.; Sato, H.; Yamada, T.; Nagasaki, H.; Tsuchiya, A.; Abe, R.; Yuasa, Y. Germ-line mutation of the hmsh6/gtbp gene in an atypical hereditary nonpolyposis colorectal cancer kindred. *Canc. Res.* **1997**, *57*, 3920–3923.
26. Shibata, D.; Peinado, M.A.; Ionov, Y.; Malkhosyan, S.; Perucho, M. Genomic instability in repeated sequences is an early somatic event in colorectal tumorigenesis that persists after transformation. *Nat. Genet.* **1994**, *6*, 273–281.
27. Howe, J.R.; Roth, S.; Ringold, J.C.; Summers, R.W.; Jarvinen, H.J.; Sistonen, P.; Tomlinson, I.P.; Houlston, R.S.; Bevan, S.; Mitros, F.A.; *et al.* Mutations in the smad4/dpc4 gene in juvenile polyposis. *Science* **1998**, *280*, 1086–1088.
28. Howe, J.R.; Bair, J.L.; Sayed, M.G.; Anderson, M.E.; Mitros, F.A.; Petersen, G.M.; Velculescu, V.E.; Traverso, G.; Vogelstein, B. Germline mutations of the gene encoding bone morphogenetic protein receptor 1a in juvenile polyposis. *Nat. Genet.* **2001**, *28*, 184–187.
29. Hemminki, A.; Markie, D.; Tomlinson, I.; Avizienyte, E.; Roth, S.; Loukola, A.; Bignell, G.; Warren, W.; Aminoff, M.; Hoglund, P.; *et al.* A serine/threonine kinase gene defective in peutz-jeghers syndrome. *Nature* **1998**, *391*, 184–187.
30. Jarvinen, H.J.; Aarnio, M.; Mustonen, H.; Aktan-Collan, K.; Aaltonen, L.A.; Peltomaki, P.; de La Chapelle, A.; Mecklin, J.P. Controlled 15-year trial on screening for colorectal cancer in families with hereditary nonpolyposis colorectal cancer. *Gastroenterology* **2000**, *118*, 829–834.
31. Burn, J.; Gerdes, A.M.; Macrae, F.; Mecklin, J.P.; Moeslein, G.; Olschwang, S.; Eccles, D.;

Evans, D.G.; Maher, E.R.; Bertario, L.; *et al.* Long-term effect of aspirin on cancer risk in carriers of hereditary colorectal cancer: An analysis from the capp2 randomised controlled trial. *Lancet* **2011**, *378*, 2081–2087.

32. Nicolaides, N.C.; Papadopoulos, N.; Liu, B.; Wei, Y.F.; Carter, K.C.; Ruben, S.M.; Rosen, C.A.; Haseltine, W.A.; Fleischmann, R.D.; Fraser, C.M.; *et al.* Mutations of two pms homologues in hereditary nonpolyposis colon cancer. *Nature* **1994**, *371*, 75–80.
33. Hendriks, Y.M.; Jagmohan-Changur, S.; van der Klift, H.M.; Morreau, H.; van Puijenbroek, M.; Tops, C.; van Os, T.; Wagner, A.; Ausems, M.G.; Gomez, E.; *et al.* Heterozygous mutations in pms2 cause hereditary nonpolyposis colorectal carcinoma (lynch syndrome). *Gastroenterology* **2006**, *130*, 312–322.
34. Al-Tassan, N.; Chmiel, N.H.; Maynard, J.; Fleming, N.; Livingston, A.L.; Williams, G.T.; Hodges, A.K.; Davies, D.R.; David, S.S.; Sampson, J.R.; *et al.* Inherited variants of myh associated with somatic g:C-->t:A mutations in colorectal tumors. *Nat. Genet.* **2002**, *30*, 227–232.
35. Lubbe, S.J.; di Bernardo, M.C.; Chandler, I.P.; Houlston, R.S. Clinical implications of the colorectal cancer risk associated with mutyh mutation. *J. Clin. Oncol.* **2009**, *27*, 3975–3980.
36. Palles, C.; Cazier, J.B.; Howarth, K.M.; Domingo, E.; Jones, A.M.; Broderick, P.; Kemp, Z.; Spain, S.L.; Guarino, E.; Salguero, I.; *et al.* Germline mutations affecting the proofreading domains of pole and pold1 predispose to colorectal adenomas and carcinomas. *Nat. Genet.* **2013**, *45*, 136–144.
37. Lichtenstein, P.; Holm, N.V.; Verkasalo, P.K.; Iliadou, A.; Kaprio, J.; Koskenvuo, M.; Pukkala, E.; Skytthe, A.; Hemminki, K. Environmental and heritable factors in the causation of cancer—Analyses of cohorts of twins from sweden, denmark, and finland. *N. Engl. J. Med.* **2000**, *343*, 78–85.
38. Lubbe, S.J.; Webb, E.L.; Chandler, I.P.; Houlston, R.S. Implications of familial colorectal cancer risk profiles and microsatellite instability status. *J. Clin. Oncol.* **2009**, *27*, 2238–2244.
39. Papaemmanuil, E.; Carvajal-Carmona, L.; Sellick, G.S.; Kemp, Z.; Webb, E.; Spain, S.; Sullivan, K.; Barclay, E.; Lubbe, S.; Jaeger, E.; *et al.* Deciphering the genetics of hereditary non-syndromic colorectal cancer. *Eur. J. Hum. Genet.* **2008**, *16*, 1477–1486.
40. Wiesner, G.L.; Daley, D.; Lewis, S.; Ticknor, C.; Platzer, P.; Lutterbaugh, J.; MacMillen, M.; Baliner, B.; Willis, J.; Elston, R.C.; *et al.* A subset of familial colorectal neoplasia kindreds linked to chromosome 9q22.2–31.2. *Proc. Natl. Acad. Sci. USA* **2003**, *100*, 12961–12965.
41. Bodmer, W.; Bonilla, C. Common and rare variants in multifactorial susceptibility to common diseases. *Nat. Genet.* **2008**, *40*, 695–701.
42. Laken, S.J.; Petersen, G.M.; Gruber, S.B.; Oddoux, C.; Ostrer, H.; Giardiello, F.M.; Hamilton, S.R.; Hampel, H.; Markowitz, A.; Klimstra, D.; *et al.* Familial colorectal cancer in ashkenazim due to a hypermutable tract in apc. *Nat. Genet.* **1997**, *17*, 79–83.
43. Whiffin, N.; Broderick, P.; Lubbe, S.J.; Pittman, A.M.; Penegar, S.; Chandler, I.; Houlston, R.S. Mlh1-93G>A is a risk factor for msi colorectal cancer. *Carcinogenesis* **2011**, *32*, 1157–1161.
44. Lipton, L.; Halford, S.E.; Johnson, V.; Novelli, M.R.; Jones, A.; Cummings, C.; Barclay, E.; Sieber, O.; Sadat, A.; Bisgaard, M.L.; *et al.* Carcinogenesis in myh-associated polyposis follows a distinct genetic pathway. *Canc. Res.* **2003**, *63*, 7595–7599.
45. Reich, D.E.; Lander, E.S. On the allelic spectrum of human disease. *Trends Genet.* **2001**, *17*,

502–510.
46. Risch, N.; Merikangas, K. The future of genetic studies of complex human diseases. *Science* **1996**, *273*, 1516–1517.
47. Broderick, P.; Carvajal-Carmona, L.; Pittman, A.M.; Webb, E.; Howarth, K.; Rowan, A.; Lubbe, S.; Spain, S.; Sullivan, K.; Fielding, S.; *et al.* A genome-wide association study shows that common alleles of smad7 influence colorectal cancer risk. *Nat. Genet.* **2007**, *39*, 1315–1317.
48. Dunlop, M.G.; Dobbins, S.E.; Farrington, S.M.; Jones, A.M.; Palles, C.; Whiffin, N.; Tenesa, A.; Spain, S.; Broderick, P.; Ooi, L.Y.; *et al.* Common variation near cdkn1a, pold3 and shroom2 influences colorectal cancer risk. *Nat. Genet.* **2012**, *44*, 770–776.
49. Houlston, R.S.; Cheadle, J.; Dobbins, S.E.; Tenesa, A.; Jones, A.M.; Howarth, K.; Spain, S.L.; Broderick, P.; Domingo, E.; Farrington, S.; *et al.* Meta-analysis of three genome-wide association studies identifies susceptibility loci for colorectal cancer at 1q41, 3q26.2, 12q13.13 and 20q13.33. *Nat. Genet.* **2010**, *42*, 973–977.
50. Houlston, R.S.; Webb, E.; Broderick, P.; Pittman, A.M.; di Bernardo, M.C.; Lubbe, S.; Chandler, I.; Vijayakrishnan, J.; Sullivan, K.; Penegar, S.; *et al.* Meta-analysis of genome-wide association data identifies four new susceptibility loci for colorectal cancer. *Nat. Genet.* **2008**, *40*, 1426–1435.
51. Tenesa, A.; Farrington, S.M.; Prendergast, J.G.; Porteous, M.E.; Walker, M.; Haq, N.; Barnetson, R.A.; Theodoratou, E.; Cetnarskyj, R.; Cartwright, N.; *et al.* Genome-wide association scan identifies a colorectal cancer susceptibility locus on 11q23 and replicates risk loci at 8q24 and 18q21. *Nat. Genet.* **2008**, *40*, 631–637.
52. Tomlinson, I.P.; Carvajal-Carmona, L.G.; Dobbins, S.E.; Tenesa, A.; Jones, A.M.; Howarth, K.; Palles, C.; Broderick, P.; Jaeger, E.E.; Farrington, S.; *et al.* Multiple common susceptibility variants near bmp pathway loci grem1, bmp4, and bmp2 explain part of the missing heritability of colorectal cancer. *PLoS Genet.* **2011**, *7*, e1002105.
53. Tomlinson, I.P.; Webb, E.; Carvajal-Carmona, L.; Broderick, P.; Howarth, K.; Pittman, A.M.; Spain, S.; Lubbe, S.; Walther, A.; Sullivan, K.; *et al.* A genome-wide association study identifies colorectal cancer susceptibility loci on chromosomes 10p14 and 8q23.3. *Nat. Genet.* **2008**, *40*, 623–630.
54. Lubbe, S.J.; di Bernardo, M.C.; Broderick, P.; Chandler, I.; Houlston, R.S. Comprehensive evaluation of the impact of 14 genetic variants on colorectal cancer phenotype and risk. *Am. J. Epidemiol.* **2012**, *175*, 1–10.
55. Whiffin, N.; Dobbins, S.E.; Hosking, F.J.; Palles, C.; Tenesa, A.; Wang, Y.; Farrington, S.M.; Jones, A.M.; Broderick, P.; Campbell, H.; *et al.* Deciphering the genetic architecture of low-penetrance susceptibility to colorectal cancer. *Hum. Mol. Genet.* **2013**, *22*, 5075–5082.
56. Carvajal-Carmona, L.G.; Cazier, J.B.; Jones, A.M.; Howarth, K.; Broderick, P.; Pittman, A.; Dobbins, S.; Tenesa, A.; Farrington, S.; Prendergast, J.; *et al.* Fine-mapping of colorectal cancer susceptibility loci at 8q23.3, 16q22.1 and 19q13.11: Refinement of association signals and use of in silico analysis to suggest functional variation and unexpected candidate target genes. *Hum. Mol. Genet.* **2011**, *20*, 2879–2888.
57. Easton, D.F.; Pooley, K.A.; Dunning, A.M.; Pharoah, P.D.; Thompson, D.; Ballinger, D.G.; Struewing, J.P.; Morrison, J.; Field, H.; Luben, R.; *et al.* Genome-wide association study identifies novel breast cancer susceptibility loci. *Nature* **2007**, *447*, 1087–1093.

58. Pittman, A.M.; Naranjo, S.; Jalava, S.E.; Twiss, P.; Ma, Y.; Olver, B.; Lloyd, A.; Vijayakrishnan, J.; Qureshi, M.; Broderick, P.; *et al.* Allelic variation at the 8q23.3 colorectal cancer risk locus functions as a cis-acting regulator of eif3h. *PLoS Genet.* **2010**, *6*, e1001126.
59. Tuupanen, S.; Turunen, M.; Lehtonen, R.; Hallikas, O.; Vanharanta, S.; Kivioja, T.; Bjorklund, M.; Wei, G.; Yan, J.; Niittymaki, I.; *et al.* The common colorectal cancer predisposition snp rs6983267 at chromosome 8q24 confers potential to enhanced wnt signaling. *Nat. Genet.* **2009**, *41*, 885–890.
60. Pomerantz, M.M.; Ahmadiyeh, N.; Jia, L.; Herman, P.; Verzi, M.P.; Doddapaneni, H.; Beckwith, C.A.; Chan, J.A.; Hills, A.; Davis, M.; *et al.* The 8q24 cancer risk variant rs6983267 shows long-range interaction with myc in colorectal cancer. *Nat. Genet.* **2009**, *41*, 882–884.
61. Pittman, A.M.; Webb, E.; Carvajal-Carmona, L.; Howarth, K.; di Bernardo, M.C.; Broderick, P.; Spain, S.; Walther, A.; Price, A.; Sullivan, K.; *et al.* Refinement of the basis and impact of common 11q23.1 variation to the risk of developing colorectal cancer. *Hum. Mol. Genet.* **2008**, *17*, 3720–3727.
62. Pittman, A.M.; Naranjo, S.; Webb, E.; Broderick, P.; Lips, E.H.; van Wezel, T.; Morreau, H.; Sullivan, K.; Fielding, S.; Twiss, P.; *et al.* The colorectal cancer risk at 18q21 is caused by a novel variant altering smad7 expression. *Genome Res.* **2009**, *19*, 987–993.
63. Ghoussaini, M.; Song, H.; Koessler, T.; Al Olama, A.A.; Kote-Jarai, Z.; Driver, K.E.; Pooley, K.A.; Ramus, S.J.; Kjaer, S.K.; Hogdall, E.; *et al.* Multiple loci with different cancer specificities within the 8q24 gene desert. *J. Natl. Canc. Inst.* **2008**, *100*, 962–966.
64. Kiemeney, L.A.; Thorlacius, S.; Sulem, P.; Geller, F.; Aben, K.K.; Stacey, S.N.; Gudmundsson, J.; Jakobsdottir, M.; Bergthorsson, J.T.; Sigurdsson, A.; *et al.* Sequence variant on 8q24 confers susceptibility to urinary bladder cancer. *Nat. Genet.* **2008**, *40*, 1307–1312.
65. Crowther-Swanepoel, D.; Broderick, P.; di Bernardo, M.C.; Dobbins, S.E.; Torres, M.; Mansouri, M.; Ruiz-Ponte, C.; Enjuanes, A.; Rosenquist, R.; Carracedo, A.; *et al.* Common variants at 2q37.3, 8q24.21, 15q21.3 and 16q24.1 influence chronic lymphocytic leukemia risk. *Nat. Genet.* **2010**, *42*, 132–136.
66. Al Olama, A.A.; Kote-Jarai, Z.; Giles, G.G.; Guy, M.; Morrison, J.; Severi, G.; Leongamornlert, D.A.; Tymrakiewicz, M.; Jhavar, S.; Saunders, E.; *et al.* Multiple loci on 8q24 associated with prostate cancer susceptibility. *Nat. Genet.* **2009**, *41*, 1058–1060.
67. Amundadottir, L.T.; Sulem, P.; Gudmundsson, J.; Helgason, A.; Baker, A.; Agnarsson, B.A.; Sigurdsson, A.; Benediktsdottir, K.R.; Cazier, J.B.; Sainz, J.; *et al.* A common variant associated with prostate cancer in european and african populations. *Nat. Genet.* **2006**, *38*, 652–658.
68. Yeager, M.; Orr, N.; Hayes, R.B.; Jacobs, K.B.; Kraft, P.; Wacholder, S.; Minichiello, M.J.; Fearnhead, P.; Yu, K.; Chatterjee, N.; *et al.* Genome-wide association study of prostate cancer identifies a second risk locus at 8q24. *Nat. Genet.* **2007**, *39*, 645–649.
69. Gudmundsson, J.; Sulem, P.; Manolescu, A.; Amundadottir, L.T.; Gudbjartsson, D.; Helgason, A.; Rafnar, T.; Bergthorsson, J.T.; Agnarsson, B.A.; Baker, A.; *et al.* Genome-wide association study identifies a second prostate cancer susceptibility variant at 8q24. *Nat. Genet.* **2007**, *39*, 631–637.
70. Takatsuno, Y.; Mimori, K.; Yamamoto, K.; Sato, T.; Niida, A.; Inoue, H.; Imoto, S.; Kawano, S.; Yamaguchi, R.; Toh, H.; *et al.* The rs6983267 snp is associated with myc transcription efficiency,

which promotes progression and worsens prognosis of colorectal cancer. *Ann. Surg. Oncol.* **2013**, *20*, 1395–1402.

71. Mrkonjic, M.; Roslin, N.M.; Greenwood, C.M.; Raptis, S.; Pollett, A.; Laird, P.W.; Pethe, V.V.; Chiang, T.; Daftary, D.; Dicks, E.; *et al.* Specific variants in the mlh1 gene region may drive DNA methylation, loss of protein expression, and msi-h colorectal cancer. *PLoS One* **2010**, *5*, e13314.
72. Wijnen, J.T.; Brohet, R.M.; van Eijk, R.; Jagmohan-Changur, S.; Middeldorp, A.; Tops, C.M.; van Puijenbroek, M.; Ausems, M.G.; Gomez Garcia, E.; Hes, F.J.; *et al.* Chromosome 8q23.3 and 11q23.1 variants modify colorectal cancer risk in lynch syndrome. *Gastroenterology* **2009**, *136*, 131–137.
73. Michailidou, K.; Hall, P.; Gonzalez-Neira, A.; Ghoussaini, M.; Dennis, J.; Milne, R.L.; Schmidt, M.K.; Chang-Claude, J.; Bojesen, S.E.; Bolla, M.K.; *et al.* Large-scale genotyping identifies 41 new loci associated with breast cancer risk. *Nat. Genet.* **2013**, *45*, 353–361, 361e1–361e2.
74. Eeles, R.A.; Olama, A.A.; Benlloch, S.; Saunders, E.J.; Leongamornlert, D.A.; Tymrakiewicz, M.; Ghoussaini, M.; Luccarini, C.; Dennis, J.; Jugurnauth-Little, S.; *et al.* Identification of 23 new prostate cancer susceptibility loci using the icogs custom genotyping array. *Nat. Genet.* **2013**, *45*, 385–391, 391e1–391e2.
75. Gylfe, A.E.; Katainen, R.; Kondelin, J.; Tanskanen, T.; Cajuso, T.; Hanninen, U.; Taipale, J.; Taipale, M.; Renkonen-Sinisalo, L.; Jarvinen, H.; *et al.* Eleven candidate susceptibility genes for common familial colorectal cancer. *PLoS Genet.* **2013**, *9*, e1003876.
76. Abecasis, G.R.; Altshuler, D.; Auton, A.; Brooks, L.D.; Durbin, R.M.; Gibbs, R.A.; Hurles, M.E.; McVean, G.A. A map of human genome variation from population-scale sequencing. *Nature* **2010**, *467*, 1061–1073.
77. Thompson, B.A.; Spurdle, A.B.; Plazzer, J.-P.; Greenblatt, M.S.; Akagi, K.; Al-Mulla, F.; Bapat, B.; Bernstein, I.; Capellá, G.; den Dunnen, J.T.; *et al.* Application of a 5-tiered scheme for standardized classification of 2360 unique mismatch repair gene variants in the insight locus-specific database. *Nat. Genet.* **2014**, *46*, 107–115.
78. Kumar, P.; Henikoff, S.; Ng, P.C. Predicting the effects of coding non-synonymous variants on protein function using the sift algorithm. *Nat. Protoc.* **2009**, *4*, 1073–1081.
79. Adzhubei, I.A.; Schmidt, S.; Peshkin, L.; Ramensky, V.E.; Gerasimova, A.; Bork, P.; Kondrashov, A.S.; Sunyaev, S.R. A method and server for predicting damaging missense mutations. *Nat. Methods* **2010**, *7*, 248–249.
80. Gonzalez-Perez, A.; Lopez-Bigas, N. Improving the assessment of the outcome of nonsynonymous snvs with a consensus deleteriousness score, condel. *Am. J. Hum. Genet.* **2011**, *88*, 440–449.
81. Khurana, E.; Fu, Y.; Colonna, V.; Mu, X.J.; Kang, H.M.; Lappalainen, T.; Sboner, A.; Lochovsky, L.; Chen, J.; Harmanci, A.; *et al.* Integrative annotation of variants from 1092 humans: Application to cancer genomics. *Science* **2013**, *342*, 1235587.
82. Kircher, M.; Witten, D.M.; Jain, P.; O'Roak, B.J.; Cooper, G.M.; Shendure, J. A general framework for estimating the relative pathogenicity of human genetic variants. *Nat. Genet.* **2014**, *46*, 310–315.
83. Weedon, M.N.; Cebola, I.; Patch, A.M.; Flanagan, S.E.; de Franco, E.; Caswell, R.; Rodriguez-Segui, S.A.; Shaw-Smith, C.; Cho, C.H.; Lango Allen, H.; *et al.* Recessive mutations

in a distal ptf1a enhancer cause isolated pancreatic agenesis. *Nat. Genet.* **2014**, *46*, 61–64.

84. Horn, S.; Figl, A.; Rachakonda, P.S.; Fischer, C.; Sucker, A.; Gast, A.; Kadel, S.; Moll, I.; Nagore, E.; Hemminki, K.; *et al*. Tert promoter mutations in familial and sporadic melanoma. *Science* **2013**, *339*, 959–961.

3

Revisiting Respect for Persons in Genomic Research

Debra J. H. Mathews [1,*] **and Leila Jamal** [2,3]

[1] Johns Hopkins Berman Institute of Bioethics, 1809 Ashland Avenue, Baltimore, MD 21205, USA
[2] Department of Health Policy and Management, Johns Hopkins Bloomberg School of Health, 615 North Wolfe St., Baltimore, MD 21205, USA; E-Mail: ljamal@jhsph.edu
[3] Division of Neurogenetics, Kennedy Krieger Institute, 801 N. Broadway, Rm. 564, Baltimore, MD 21205, USA

* Author to whom correspondence should be addressed; E-Mail: dmathews@jhu.edu;

Abstract: The risks and benefits of research using large databases of personal information are evolving in an era of ubiquitous, internet-based data exchange. In addition, information technology has facilitated a shift in the relationship between individuals and their personal data, enabling increased individual control over how (and how much) personal data are used in research, and by whom. This shift in control has created new opportunities to engage members of the public as partners in the research enterprise on more equal and transparent terms. Here, we consider how some of the technological advances driving and paralleling developments in genomics can also be used to supplement the practice of informed consent with other strategies to ensure that the research process as a whole honors the notion of respect for persons upon which human research subjects protections are premised. Further, we suggest that technological advances can help the research enterprise achieve a more thoroughgoing respect for persons than was possible when current policies governing human subject research were developed. Questions remain about the best way to revise policy to accommodate these changes.

Keywords: respect for persons; genomics; genetics; informed consent

---

1. Introduction

The risks and benefits of research using large databases of personal information are evolving in an era of ubiquitous, internet-based data exchange. Here, we consider some of the technological advances driving and paralleling developments in genomics, and how they can be used to supplement the practice of informed consent to ensure that the research process as a whole honors the notion of respect for persons upon which human research subjects protections are premised.

The cost of next-generation sequencing has declined precipitously in recent years, increasing the potential of genomic research to expand knowledge of human biology and disease [1]. To render human genome data meaningful for individuals, investigators must collect and analyze information contributed by many individuals from diverse populations over long periods of time. To build large datasets, people are asked to donate biospecimens and personal data, including genomic data, to repositories of de-identified tissue and data used by many researchers [2]. Indeed, in an effort to harness the scientific potential of such large datasets, many of the world's leading research institutions recently announced ambitious plans to build a global, interoperable framework for sharing genomic and other research data more broadly in the future [3], and the NIH is currently developing a revised data-sharing policy [4]. As this new era of genomic research progresses, it is critical that we attend not only to the benefits that such broad sharing will have for science and medicine, but also to the proportionality of risks and benefits borne by contributors to biorepositories and genome databases.

The structures and norms guiding the development and use of such repositories were established at a time when the re-identification of individual data contributors was thought to be unlikely, and the anonymization of personal data was a reasonable strategy for mitigating risks to research subjects from loss of confidentiality and subsequent discrimination. As we have learned over the past five years, it is no longer possible to credibly guarantee that anonymized or de-identified samples and data will remain de-identified in large data repositories [5–7]. The increased technical capacity to reidentify individuals in databases can be addressed in a number of ways: (1) we can clamp down on sharing; (2) we can merely be transparent about the risks during the informed consent process and allow those individuals willing to assume the risks to do so [8]; or (3) we can shift our attention to increasing penalties for re-identification and misuse of identifiable data [9]. Limiting use would be an unfortunate and ill-considered outcome, reducing research and medical benefits to society and foiling the intentions of many individual contributors who are, after all, providing samples and data to further science and clinical innovation. Transparency and penalties for misuse may be necessary to address the increased risk of re-identification, but they are not sufficient. Here, we suggest that, where technological capacity exists, technological advances can help the research enterprise achieve a more thoroughgoing respect for persons than was possible when current policies governing human subject research were developed. Further, by restricting access to data and failing to recognize that some individuals may exercise their autonomy by enabling use of their genomic and personal data, researchers and regulators hobble science and fail to truly honor the notion of respect for persons that underlies the entire enterprise. That said, questions remain about the best way to revise policy to accommodate the changed landscape.

2. Background

Concerns about the ethical use of human genomic and other personal data in prospective cohort studies are longstanding [10]. However, the increased use of next-generation sequencing in research reanimates three challenges on an unprecedented scale. First, next-generation sequencing can generate data from every known disease-associated gene or DNA sample. As more is learned about the contribution of genomic factors to disease risk, an individual genome sequence will acquire new meaning to the person from whom it originated and will contribute to the interpretation of others' genomes.

Second, next-generation sequencing has co-evolved with powerful computing infrastructures for analyzing and exchanging enormous volumes of personal data. To facilitate the efficient use of resources, there has been a growing tendency to establish large databases and open-access policies to store and share human genomic and other research data. This trend favors the "emergence" of many hypotheses from large datasets long after a participant's initial informed consent to research, and facilitates the re-use and combining of datasets by multiple researchers. As a result, secondary and tertiary data users may be far removed from the original context in which research data were obtained, blurring the lines of accountability for responsible data use.

Third, it has become easier to re-identify individual contributors to databases based on publicly-available internet data, as the latter has grown more abundant [5–7]. Consequently, the privacy risks associated with contributing biospecimens and genomic data to research must now be assessed broadly, rather than in relation to the activities of any one project.

A current challenge facing policymakers is to develop standards for using not only archived tissues samples and data, but also newly generated genomic information in research to benefit society while respecting heterogeneous beliefs about privacy [11–14] and while safeguarding research participants from uncertain risks. This dilemma is often framed as a tension between serving individual autonomy interests by keeping data confidential on the one hand, and advancing public beneficence by sharing data liberally on the other. However, this polarized view may be oversimplified. Internet users have increasingly come to use social media—blogs, Facebook, Twitter, wikis, forums—to become content creators and sharers in their own right. While norms are still evolving, information technology (IT) has facilitated a shift in the relationship between individuals and their personal data, enabling increased individual control over how (and how much) personal data are used in research, and by whom. This shift in control has created new opportunities to engage members of the public as partners in the research enterprise on more equal and transparent terms. Conceptions of privacy—including what should remain private and what privacy means in various online spaces—and risks of breaching confidentiality are changing even as genomic data are accumulating rapidly.

3. The Rationale for Informed Consent

An ethical duty to secure the autonomous and voluntary informed consent of human research subjects emerged in response to specific and grave concerns—about physical harm, discrimination, stigma—that arose from inhumane and coercive research practices in the U.S., Europe and elsewhere during the 20th century [15,16]. Today, to uphold the bioethical principle of respect for persons, the

United States Federal Policy for the Protection of Human Subjects ("The Common Rule") requires investigators to obtain informed consent from prospective research subjects before collecting or using their individually identifiable biological materials or data in research studies [17]. The doctrine of informed consent was conceived to ensure respect for persons as autonomous agents in clinical care and research. Motivated to prevent further unethical research practices, the U.S. National Research Act of 1974 both mandated Institutional Review Board (IRB) review for research and convened a National Commission for the Protection of Human Subjects of Biomedical and Behavioral Research, which produced The Belmont Report, the foundation of much of the Common Rule.

The Belmont Report identifies three ethical principles: respect for persons, beneficence, and justice, which are paired with three corresponding means of translating principle into action: informed consent, assessing risks and benefits, and fair selection of subjects. The original Belmont concept of "autonomy" embedded in respect for persons is elaborated as follows:

An autonomous person is an individual capable of deliberation about personal goals and of acting under the direction of such deliberation. To respect autonomy is to give weight to an autonomous person's considered opinions and choices <u>while refraining from obstructing their actions unless they are clearly detrimental to others</u>. To show lack of respect for an autonomous agent is to repudiate that person's considered judgments or to withhold information necessary to make a considered judgment, when there are no compelling reasons to do so [18] [underlining added].

The Belmont Report formed the basis of the first formal research regulations adopted by the Department of Health and Human Services (HHS) in 1981, only slightly modified in the currently prevailing Common Rule.

4. The Changing Research Landscape

It is widely agreed that since the adoption of the Common Rule, the advent of genomic research has changed the research landscape, as have its risks and benefits, as a result of technological advances that make it cheaper and easier to generate, analyze, and share large volumes of data [19,20]. Just as significant, many technological advances in the same period have diversified the tools available to mitigate or offset the risks facing contributors to genomic research.

4.1. The Shifting Relationship between Identifiability and Ethics Review

Historically, the risks of genetic and genomic research have been mitigated by nondisclosure (e.g., of non-paternity), and sample and data anonymization or de-identification. Stripping identifiers or severing links between tissues and tissue donors were, justifiably, seen as effective measures to mitigate risks to individuals' privacy interests, by restricting access to their personal information. Yet privacy is a complex, variably defined concept encompassing a plurality of related issues; informational secrecy is merely one of its dimensions. Further, the practice of respecting privacy by restricting access to individual information undermines the pursuit of public benefit through aggregation of large amounts of personal data in research databases, and may not actually align with research subjects' values [21,22].

The concerns addressed by restricting access to personal information include threats to valued social and economic opportunities as a result of privacy breaches and threats to individual autonomy, including risk of social stigma and unwanted scrutiny, making it harder to exercise basic liberties in the course of daily life [23]. Further, some individuals simply do not want others (e.g., researchers) to know information about them that they do not know themselves, or that they do not wish to know about themselves.

The moral case for gaining access to personal information also varies. In science, the argument is often made that such access will advance scientific knowledge, leading to improved healthcare and other societal benefits [24,25]. Justifying the use of personal information to achieve ends like these is difficult when the contribution of individual information to these outcomes is unclear, and even harder when not all parties involved are in agreement about the desirability of the ends. The various interests protected and hindered by confidentiality provisions make it impossible to arrive at a consensus risk-benefit profile for a pool of research subjects that can be assessed each time personal information is transferred from one holder to another.

Given the choice, some individuals might decline to make their personally identifiable health information available to researchers; others might elect to share their data to enable scientists to develop new treatments, to help advance biomedical science, or to forge connections to other individuals with common diagnoses or health concerns; still others might choose to share with academic but not commercial researchers, or with breast cancer researchers, but not those who study psychiatric disease. Whether a person is motivated to enroll in research by personal history of illness, intellectual curiosity, or feelings of altruism or social responsibility, the tradeoffs involved in contributing personal information to a biorepository are dynamic and variable over time, and contributors' values and goals are diverse. Current policy that uniformly restricts access to data as a form of privacy protection both fails to respect those participants who would wish to have and share their data freely and limits the potential benefits to science and society that may accrue from the use of those data.

In recent years, it has become increasingly possible to re-identify individual data contributors to large electronic datasets [5–7]. This is significant because under the regulatory status quo, full ethics review is primarily reserved for projects using personal data considered "identifiable" under the Common Rule, meaning that the identity of the subject can be "readily ascertained" by the investigator from the information. Informed consent is not typically sought from individuals before their "de-identified" data are used in research. In human genomics, this policy is problematic due to the inherent identifiability of human sequence data and the need sometimes to interpret these data in the context of detailed phenotypic information.

The prevailing notion that investigators can balance the risk-benefit profile of genomic research by divorcing data from individual identifiers is also problematic because de-identification may actually impoverish the quality of research data to an extent that undermines scientific progress. De-identification might also preclude the return of individual research results to participants in instances when such results have implications for their well-being. Further, de-identification denies participants the opportunity to exercise their autonomy by managing the use of their data over time, as their circumstances and views change. From an individual's perspective, the foreclosure of these benefits and limitations on their autonomy might actually worsen the risk-benefit profile of participating in research.

4.2. Growth of Online Data-Sharing

Simultaneous with the emergence of next-gen sequencing technologies, there has been a profound shift in the nature of online information sharing in the course of daily life. Today's Internet contains vast quantities of user-volunteered, identifiable data disclosed for purposes as varied as commercial exchange, social networking, recreational gaming, and health support and promotion. Facebook, Pinterest, patient discussion boards, posted Fitbit reports and myriad other forms of Internet sharing have changed what, how and with whom we share. In many online health-related communities, members develop and test their own hypotheses, assuming roles typically reserved for "experts", and operating outside traditional human subjects protections frameworks (see Section 5.4 below). Further, some have begun to advocate not for the ability to keep one's data private, but rather for the ability to have and to share one's data freely [26]. Such calls for the freedom to share reflect the oft-ignored feature of autonomy as defined in the Belmont Report, respect for individuals' ability to pursue their interests so long as they do not harm others (see underling above).

Norms of information exchange are also changing. When investigators and institutions are trusted, research participants tend not to mind contributing identifiable data to multiple research projects provided that they are kept informed about the nature of the research to which they are contributing [27,28]. Furthermore, several studies have shown that individual concerns about privacy are highly variable and seem to be affected by the tradeoffs that individuals make among three considerations: their privacy concerns, their perceptions of the utility of study participation, and the degree of reciprocity they perceive from investigators using their data [29,30].

Taken together—the limitations of informed consent, the growing ease of re-identifying donors and the value of donor-associated data, the proliferation of new IT platforms, and evidence for a so-called "privacy-utility tradeoff" made by research participants—these new realities suggest it is time to revise how we configure an ethical relationship between donors and users of genomic research data. If we wish to uphold the notion of respect for persons on which we base human research subject protections, we must both "give weight to an autonomous person's considered opinions and choices" and refrain "from obstructing their actions unless they are clearly detrimental to others." Limiting autonomy by restricting individuals' access to and sharing of their own data, or ability to modify their preferences regarding data use over time fails to uphold the second requirement of respect for persons.

5. Application of IT to both Research and Research Subject Protections

The importance of trust and reciprocity to research participation suggests that revising the relationship between donors and users toward a more collaborative model might also encourage and support participation in genomic research, to the potential benefit of both parties and society as a whole. Many argue that research subjects must become more active partners in the research process itself: true participants, rather than mere subjects [10–12]. To realize this aim, and achieve the hoped for trust and reciprocity, new digital systems for collecting and curating research data (including genomic data) have been developed by innovators in both the for-profit and non-profit sectors. Below, we describe a heterogeneous group of evolving new approaches to collecting and using biospecimens and genomic data in research. Given their novelty and continuing evolution, it is not our aim to classify them prematurely or draw a false equivalence among them. Our goals are to draw attention to the innovative ways these approaches re-imagine the relationship between research participants and researchers, and to highlight some of the empirical questions that must be addressed, as we attempt to evaluate the ethical implications of the new research models.

5.1. The Personal Genome Project and Open Consent

The Harvard-based Personal Genome Project (PGP) [31] has abandoned the notion that de-identification of genomic research data and samples is plausible or even desirable, privileging the values of "veracity" and reciprocity in the conduct of research [32]. The PGP is a longitudinal genome research study enrolling participants through a detailed, web-based informed consent process (including a mandatory genetics exam) that secures "open consent" from participants for ongoing research use of their individual genomic and phenotypic data. PGP participants are free to upload as little or as much personal information as they wish to their online PGP profiles, within its defined parameters. Although these profiles do not display names, the PGP makes no promises that data contributed to the project will remain de-identified or anonymous. In return for assuming the risks of re-identification, the PGP offers participants individual research data and hosts an annual research meeting to which participants are invited, demonstrating the PGP's belief that reciprocity may play an important role in earning and securing the trust of their study participants.

5.2. Portable Legal Consent

The Portable Legal Consent (PLC), developed by the Consent to Research project, is designed to address the challenges of broad data sharing. The PLC gives participants who wish to donate data to research the opportunity to attach a single research consent to their health and genetic data, which they then upload to a secure website. These data can then be used for research purposes by any researcher who agrees to specific terms of data use including: an intent to publish research results in an open-access forum, a promise not to attempt to re-identify individual research participants, and a promise not to distribute data among third parties who do not agree to the PLC conditions. While participants may withdraw their data from the database at any time, they are clearly advised that once data are uploaded, it may not be possible to remove them from all sources (for example, from researchers who have already downloaded, shared, or used the data).

5.3. Registry for All Disease ("Reg4All")

In 2012, the umbrella disease advocacy organization Genetic Alliance created Reg4All [33] to collect information relevant to many health conditions. Using a "dynamic consent" platform, Reg4All participants select fine-grained consent rules to determine how their personal data are viewed, by whom, and for what purposes. The system's privacy settings include "deny the use of my data in any form for any purpose"; "allow discovery and retrieval of all of my data in the registry", and "make my data available to ONLY this research project". Preferences also allow varying degrees of contact between registry participants and investigators interested in using their data. Participants may make their data available to specific clinical trials and research studies, or they may allow their data to be used openly by all. For each decision about data use, a participant may choose to give consent, deny consent, or postpone the decision until later. A participant may choose to enter their preferences once and retain them, or they may choose to change their choices at a later date. The overall vision of Reg4All is to re-imagine the researcher-participant relationship as a reciprocal collaboration over time.

5.4. "Apomediated", Peer-Produced Research

The term "apomediation" describes the relatively non-hierarchical nature of information-sharing in some research communities [34,35]. Apomediated initiatives create virtual spaces in which individuals are encouraged to propose and carry out their own research studies using self-reported data. Examples include PatientsLikeMe (PLM), which provides self-tracking and social networking tools to its over 220,000 users in exchange for permission to share their data with researchers listed on the PLM website. Since 2012, PLM's peer-reviewed publications have covered measures of functional disability in multiple sclerosis, epilepsy care quality, and Parkinson's disease progression [36–38]. Other initiatives include DIYGenomics, which has hosted a crowd-sourced study of the relationship between polymorphisms in the Methylenetetrahydrofolate reductase (MTHFR) gene, homocysteine levels, and vitamin B deficiency, and Genomera, which in beta version allows members of online communities to initiate studies related to nutrition, sleep patterns, exercise, and genome variation [39].

6. Open Questions

The ability of IT and social media to change how genomic and other health data are shared and interpreted has generated excitement among health-oriented constituencies. Advocacy organizations have embraced social media's role in helping patients become more engaged in their own healthcare and in research [40–42]. That said, using social media to share personal information raises its own ethical issues, and robust, longitudinal studies examining the effectiveness or safety of using social media to manage health information are needed. Some question whether existing initiatives are as "participant-centric" as they claim, given that commercial incentives may generate conflicts of interest in some cases [43]. One obvious concern is that personal information may be acquired surreptitiously or abused [44]. Another concern is that "gamified" survey data may not always be contributed voluntarily by users, given the compulsive nature of some forms of internet gaming [45]. Yet other concerns focus on financial motivations of the entities controlling the data—will participant and researcher incentives always stay in alignment [43]?

Thus far, we have few data on basic questions about these new models for doing research, such as: do granular data-sharing choices unduly hinder or bias the collection of research data? Who, if anyone, is alienated or excluded by systems like those we have described above? It is important to acknowledge that many participants in genomics research will not have ready access to or experience with the kinds of technologies we discuss here—will variation in access to technology lead to or exacerbate existing disparities between different research populations? Which, and how many, data-sharing options are necessary to secure autonomous and respectful research participation? What happens when study participants assume roles traditionally held by researchers?

Interactive websites have been demonstrated to be effective at educating the public about genomics, and individual data-sharing attitudes have been found to be highly nuanced and variable. We believe that the approaches highlighted above are promising strategies for managing many of the challenges of modern genomic research, while fostering autonomy. However, to realize their full potential, they must be developed in parallel with empirical studies of their benefits and harms, both intended and unintended.

7. Conclusions

Current informed consent practices are unequal to the task of upholding authentic respect for persons in contemporary genomic research. New models that take advantage of advances in both genomic research and IT promise to address this shortfall, but require further study of their associated benefits and harms. Careful study will be necessary to guide the evolution of these new models, and to ensure that research both adequately balances protections and benefits against the burdens and uncertainties borne by participants in genomic studies, and does not unnecessarily limit participants' actions.

Prior work in bioethics has addressed privacy concerns narrowly, by focusing on privacy as a strict function of identifiability or a form of informational secrecy [46–48]. This focus misses other broad interests individuals may have in sharing their own health and genomic data and information. The conception of privacy as informational secrecy lends itself to a view of genomic information-sharing as a false dichotomy, in which information is either wholly private or wholly public. By restricting access to data and failing to recognize that some individuals may exercise their autonomy by enabling use of their genomic and personal data, researchers and regulators hobble science and fail to truly honor the notion of respect for persons that underlies the entire enterprise.

The scientific, bioethics, and research oversight communities frequently frame the debate as privacy *versus* public beneficence and equate respect for persons with informed consent. Such norms and practices impede meaningful reform of human subjects protections. Further, we lack the empirical evidence necessary to evaluate emerging models of engaging with research subjects and participants that more fully embody the original concept of respect for persons. The research enterprise as a whole must accommodate the cultural shift that is taking place in the relationship between individuals and their health information. Appreciating and understanding this transformation will be an indispensible step in adapting ethical guidelines to the realities of modern information use and patients who want and expect to be true participants in research.

Acknowledgments

The authors would like to thank Robert Cook-Deegan for his helpful comments on an earlier version of this manuscript.

Author Contributions

Conception and design: DJHM, LJ. Wrote the paper: DJHM, LJ.

References

1. Pasche, B.; Absher, D. Whole-genome sequencing: A step closer to personalized medicine. *JAMA* **2011**, *305*, 1596–1597.
2. Greenbaum, D.; Sboner, A.; Mu, X.J.; Gerstein, M. Genomics and privacy: Implications of the new reality of closed data for the field. *PLoS Comput. Biol.* **2011**, *7*, E1002278.
3. Hayden, E.C. Geneticists push for global data-sharing. *Nature* **2013**, *498*, 16–17.
4. Draft NIH Genomic Data Sharing Policy Request for Public Comments; 2013. Available online: http://www.bioethics.net/2013/10/nih-requests-comment-on-genomic-data-sharing-policy-draft/ (accessed on 29 December 2013).
5. Gymrek, M.; McGuire, A.L.; Golan, D.; Halperin, E.; Erlich, Y. Identifying personal genomes by surname inference. *Science* **2013**, *339*, 321–324.
6. Homer, N.; Szelinger, S.; Redman, M.; Duggan, D.; Tembe, W.; Muehling, J.; Pearson, J.V.; Stephan, D.A.; Nelson, S.F.; Craig, D.W. Resolving individuals contributing trace amounts of DNA to highly complex mixtures using high-density SNP genotyping microarrays. *PLoS Genet.* **2008**, *4*, E1000167.
7. Sweeney, L.A.A.; Winn, J. *Identifying Participants in the Personal Genome Project by Name*; Data Privacy Lab, Harvard University, Cambridge, MA, USA, 2013.
8. Lunshof, J.E.; Chadwick, R.; Vorhaus, D.B.; Church, G.M. From genetic privacy to open consent. *Nat. Rev.Genet.* **2008**, *9*, 406–411.
9. Bambauer, J.Y. Tragedy of the data commons. *Harv. J. Law Technol.* **2011**, *25*, doi:10.2139/ssrn.1789749.
10. Geller, L.N.; Alper, J.S.; Billings, P.R.; Barash, C.I.; Beckwith, J.; Natowicz, M.R. Individual, family, and societal dimensions of genetic discrimination: A case study analysis. *Sci. Eng. Ethics* **1996**, *2*, 71–88.
11. McGuire, A.L.; Caulfield, T.; Cho, M.K. Research ethics and the challenge of whole-genome sequencing. *Nat. Rev. Genet.* **2008**, *9*, 152–156.
12. Kaye, J. The tension between data sharing and the protection of privacy in genomics research. *Annu. Rev. Genomics Hum. Genet.* **2012**, *13*, 415–431.
13. *Privacy and Progress in Whole Genome Sequencing*; President's Commission for the Study of Bioethical Issues: Washington, DC, USA, 2012.
14. Knoppers, B.M.; Dove, E.S.; Litton, J.E.; Nietfeld, J.J. Questioning the limits of genomic privacy. *Am. J. Hum. Genet.* **2012**, *91*, 577–578.
15. Rice, T.W. The historical, ethical, and legal background of human-subjects research. *Respir. Care* **2008**, *53*, 1325–1329.
16. Menikoff, J. Research involving human subjects: Ethical and regulatory issues. *Handb. Clin. Neurol.* **2013**, *118*, 289–299.
17. Code of Federal Regulations Title 45, Section 46. In *Federal Policy for the Protection of Human Subjects*; U.S. Department of Health & Human Services: Washington, DC, USA, 2008.
18. The National Commission for the Protection of Human Subjects of Biomedical and Behavioral

Research. *The Belmont Report: Ethical Principles and Guidelines for the Protection of Human Subjects of Research*; Department of Health, Education, and Welfare: Washington, DC, USA, 1979.
19. Meslin, E.M.; Cho, M.K. Research ethics in the era of personalized medicine: Updating science's contract with society. *Public Health Genomics* **2010**, *13*, 378–384.
20. Bunnik, E.M.; de Jong, A.; Nijsingh, N.; de Wert, G.M. The new genetics and informed consent: Differentiating choice to preserve autonomy. *Bioethics* **2013**, *27*, 348–355.
21. Mello, M.M.; Wolf, L.E. The havasupai Indian tribe case—Lessons for research involving stored biologic samples. *NEJM* **2010**, *363*, 204–207.
22. McGuire, A.L.; Oliver, J.M.; Slashinski, M.J.; Graves, J.L.; Wang, T.; Kelly, P.A.; Fisher, W.; Lau, C.C.; Goss, J.; Okcu, M.; *et al.* To share or not to share: A randomized trial of consent for data sharing in genome research. *Genet. Med.* **2011**, *13*, 948–955.
23. Powers, M. Justice and Genetics: Privacy protection and the Moral Basis of Public Policy. In *Genetic Secrets: Protecting Privacy and Confidentiality in the Genetic Era*; Rothstein, M.A., Ed.; Yale University Press: New Haven, CT, USA, 1997; pp. 355–368.
24. *Sharing Clinical Research Data: Workshop Summary*; The National Academy of Sciences: Washington, DC, USA, 2013.
25. Terry, S.F.; Shelton, R.; Biggers, G.; Baker, D.; Edwards, K. The haystack is made of needles. *Genet. Test. Mol. Biomark.* **2013**, *17*, 175–177.
26. Terry, S.F.; Terry, P.F. Power to the people: Participant ownership of clinical trial data. *Sci. Transl. Med.* **2011**, doi:10.1126/scitranslmed.3001857.
27. Ludman, E.J.; Fullerton, S.M.; Spangler, L.; Trinidad, S.B.; Fujii, M.M.; Jarvik, G.P.; Larson, E.B.; Burke, W. Glad you asked: Participants' opinions of re-consent for dbGaP data submission. *JERHRE* **2010**, *5*, 9–16.
28. Lemke, A.A.; Wolf, W.A.; Hebert-Beirne, J.; Smith, M.E. Public and biobank participant attitudes toward genetic research participation and data sharing. *Public Health Genomics* **2010**, *13*, 368–377.
29. Oliver, J.M.; Slashinski, M.J.; Wang, T.; Kelly, P.A.; Hilsenbeck, S.G.; McGuire, A.L. Balancing the risks and benefits of genomic data sharing: Genome research participants' perspectives. *Public Health Genomics* **2012**, *15*, 106–114.
30. Hobbs, A.; Starkbaum, J.; Gottweis, U.; Wichmann, H.E.; Gottweis, H. The privacy-reciprocity connection in biobanking: Comparing german with UK Strategies. *Public Health Genomics* **2012**, *15*, 272–284.
31. Personal Genome Project. Available online: http://www.personalgenomes.org/ (accessed on 17 January 2014).
32. Angrist, M. Eyes wide open: The personal genome project, citizen science and veracity in informed consent. *Per. Med.* **2009**, *6*, 691–699.
33. Reg4All. Available online: https://reg4all.org/ (accessed on 17 January 2014).
34. Eysenbach, G. Medicine 2.0: Social networking, collaboration, participation, apomediation, and openness. *J. Med. Int. Res.* **2008**, *10*, E22.
35. O'Connor, D. The apomediated world: Regulating research when social media has changed research. *J. Law Med. Ethics* **2013**, *41*, 470–483.

36. Wicks, P.; Fountain, N.B. Patient assessment of physician performance of epilepsy quality-of-care measures. *Neurol. Clin. Prac.* **2012**, *2*, 335–342.
37. Wicks, P.; Vaughan, T.E.; Massagli, M.P. The multiple sclerosis rating scale, revised (Msrs-R): Development, refinement, and psychometric validation using an online community. *Health Qual. Life Outcomes* **2012**, doi:10.1186/1477-7525-10-70.
38. Little, M.; Wicks, P.; Vaughan, T.; Pentland, A. Quantifying short-term dynamics of parkinson's disease using self-reported symptom data from an internet social network. *J. Med. Int. Res.* **2013**, *15*, E20.
39. Swan, M. Crowdsourced health research studies: An important emerging complement to clinical trials in the public health research ecosystem. *J. Med. Int. Res.* **2012**, *14*, E46.
40. Inspire. Available online: http://www.inspire.com/ (accessed on 31 October 2013).
41. Trialx. Available online: http://trialx.com/ (accessed on 5 November 2013).
42. Army of women. Available online: http://www.armyofwomen.org/ (accessed on 5 November 2013).
43. Vayena, E.; Tasioulas, J. Adapting standards: Ethical oversight of participant-led health research. *PLoS Med.* **2013**, *10*, E1001402.
44. Vayena, E.; Mastroianni, A.; Kahn, J. Caught in the web: Informed consent for online health research. *Sci. Transl. Med.* **2013**, *5*, 173fs176.
45. Foddy, B. The Ethics of Gamification: Little Rewards for Everything. Available online: http://blog.practicalethics.ox.ac.uk/2011/02/the-ethics-of-gamification-little-rewards-for-everything/ (accessed on 29 December 2013).
46. Gutmann, A.; Wagner, J.W. Found your DNA on the web: Reconciling privacy and progress. *Hastings Center Rep.* **2013**, *43*, 15–18.
47. Roche, P.; Glantz, L.H.; Annas, G.J. The genetic privacy act: A proposal for national legislation. *Am. Bar Assoc.* **1996**, *37*, 1–11.
48. Rothstein, M.A. Is deidentification sufficient to protect health privacy in research? *Am. J. Bioeth.* **2010**, *10*, 3–11.

4

Phenotype-Based Genetic Association Studies (PGAS)—Towards Understanding the Contribution of Common Genetic Variants to Schizophrenia Subphenotypes

Hannelore Ehrenreich [1,2,*] and Klaus-Armin Nave [1,2,*]

[1] Max Planck Institute of Experimental Medicine, Hermann-Rein-Str.3, 37075 Göttingen, Germany
[2] DFG Center for Nanoscale Microscopy and Molecular Physiology of the Brain (CNMPB), Hermann-Rein-Str.3, 37075 Göttingen, Germany

* Authors to whom correspondence should be addressed; E-Mails: ehrenreich@em.mpg.de (H.E.); nave@em.mpg.de (K.-A.N.).

Abstract: Neuropsychiatric diseases ranging from schizophrenia to affective disorders and autism are heritable, highly complex and heterogeneous conditions, diagnosed purely clinically, with no supporting biomarkers or neuroimaging criteria. Relying on these "*umbrella diagnoses*", genetic analyses, including genome-wide association studies (GWAS), were undertaken but failed to provide insight into the biological basis of these disorders. "Risk genotypes" of unknown significance with low odds ratios of mostly <1.2 were extracted and confirmed by including ever increasing numbers of individuals in large multicenter efforts. Facing these results, we have to hypothesize that thousands of genetic constellations in highly variable combinations with environmental co-factors can cause the individual disorder in the sense of a final common pathway. This would explain why the prevalence of mental diseases is so high and why mutations, including copy number variations, with a higher effect size than SNPs, constitute only a small part of variance. Elucidating the contribution of normal genetic variation to (disease) phenotypes, and so re-defining disease entities, will be extremely labor-intense but crucial. We have termed this approach PGAS ("phenotype-based genetic association studies"). Ultimate goal is the definition of biological subgroups of mental diseases. For that purpose, the GRAS (Göttingen Research Association for Schizophrenia) data collection was initiated in 2005. With >3000 phenotypical data points per patient, it comprises the world-wide largest currently available schizophrenia database (N > 1200), combining genome-wide SNP coverage and deep phenotyping under highly standardized conditions. First PGAS results on normal genetic variants, relevant for e.g., cognition or catatonia, demonstrated *proof-of-concept*. Presently, an autistic subphenotype of schizophrenia is being defined where an unfortunate accumulation of normal genotypes, so-called pro-autistic variants of synaptic genes, explains part of the phenotypical variance. Deep phenotyping and comprehensive clinical data sets, however, are expensive and it may take years before PGAS will complement conventional GWAS approaches in psychiatric genetics.

Keywords: mental disease; diagnosis; classification systems; GWAS; monozygotic twins; deep phenotyping; Manhattan plot; genotype; SNP; GRAS data collection

---

1. Schizophrenia Is a Heterogeneous Group of Diseases Diagnosed Purely Clinically

The diagnosis of schizophrenia (as of most mental diseases) is to this day an exclusively clinical one which, based on the leading classification systems, DSM and ICD, demands the simultaneous presence of a number of symptoms that are labeled "positive" or "negative". In addition, persistence of these symptoms for at least 6 months is requested. As such, schizophrenia is the unifying term of a highly complex and heterogeneous group of multigenetic disorders, with an array of environmental hazards having influenced onset and course. Common to this group of disorders are merely some gross phenotypical traits. Biomarkers in unequivocal support of the diagnosis are missing and also modern imaging technologies, despite revealing brain matter loss or various functional alterations, have not yet assisted in better understanding disease etiology or pathogenesis, or in defining objective diagnostic criteria. This conglomerate of uncertainties, however, is the groundwork on which genetic studies have been and are being built on.

2. Despite High Heritability of Schizophrenia No "Disease Genes" Have Been Uncovered

There is no doubt that schizophrenia has a major genetic root. Family studies yielded heritability estimates of up to 80% for schizophrenia [1,2]. Monozygotic twins manifest concordance rates of ~50%, thus—on top of a clear genetic origin—pointing to a considerable influence of non-genetic causes, *i.e.*, environmental and/or epigenetic factors [3–6]. Searching for the genetic part of schizophrenia etiology, a considerable number of explorations have been undertaken, ranging from segregation or linkage analyses to association studies based on candidate genes [7,8]. Most of the so identified "disease genes", e.g., *DISC1* [9], turned out not to respect disease borders. In fact, the *DISC1* translocation was found associated as much with mental health as with affective diseases, schizophrenia, autism, or personality disorders [10]. *DISC1* and others may thus at best deserve the label of a global risk gene for mental disorders.

So far, no definite "schizophrenia genes" of general significance could be identified. Recently reported genome-wide association studies (GWAS) of schizophrenia identified a number of genetic risk markers significantly associated with the disease, but unfortunately could not extract any universally convincing "disease genes". The initially limited reproducibility over studies and ethnicities has improved with increasing numbers to ~60,000 individuals (PGC, Psychiatric Genetic Consortium) [11]. Associations, however, remain hampered by very low odds ratios (OR distinctly <1.2) [12,13]. Altogether, it appears highly unlikely that even larger GWAS, based on the umbrella term "schizophrenia", will succeed in unraveling the genetic basis of these conditions or in identifying relevant disease genes.

3. "GWAS Hits" Do Not Predict Disease Severity or Other Relevant Schizophrenia Phenotypes

Which information do we then gain from the GWAS-identified risk genotypes? In fact, even an accumulation of the GWAS-identified "top 10" risk genotypes [14] does not lead to a more severe disease phenotype. In other words: If an individual possessed all "top 10" risk markers at the same time ("accumulated risk"), his disease will not be more grave than if he carried none of the risk genotypes.

Another attempt to make use of the GWAS-derived information has been the definition of polygenic schizophrenia risk scores (PSS). Besides the genome-wide significant risk loci, a substantial proportion of schizophrenia risk has been hypothesized to lie in markers not achieving genome-wide significance. Thus, quantitative PSS were calculated based on nominal associations of each SNP from the PGC GWAS. These PSS explained up to 6% of variance regarding the diagnosis schizophrenia in an independent sample [11]. Subsequently, PSS effects on various disease-relevant phenotypes, e.g., brain matter dimensions, were explored with variable success [15,16]. Employing the GRAS data collection of >1200 well characterized schizophrenic subjects (see below), we could not uncover the slightest association of PSS with any lead features of schizophrenia. In contrast, we found in the same schizophrenic population dramatic effects of accumulated environmental risk on age at onset of the disease [17].

4. Definition of Biological Subgroups of Schizophrenia: The Essential Next Step

In summary, we note that all genetic "schizophrenia markers" discovered by GWAS up to date do not contribute more to the entirety of this disease than a minimally heightened probability (OR <1.2) of receiving the "umbrella diagnosis schizophrenia". This whole picture could, however, look quite different when considering biological subgroups of schizophrenia as illustrated in Figure 1. GWAS on well-defined disease-relevant subphenotypes may ultimately not only lead to the identification of entirely new risk genotypes but, importantly, also to a re-distribution of some of the identified GWAS risk markers to certain subgroups, resulting then in much higher odds ratios.

In order to define biological subgroups of schizophrenia, the GRAS (*Göttingen Research Association for Schizophrenia*) data collection was initiated in 2004 [18,19]. The lead hypothesis of GRAS states that schizophrenia is caused by thousands of possible combinations of genetic markers interacting with a large array of different environmental risk factors. There may be rare cases where a 'per se' critical genetic load exists and the disease onset is independent of additional external factors. In the vast majority of cases, however, only the interaction between certain genetic predispositions and environmental factors will lead to disease onset. The overlap of genetic risk factors between individual schizophrenic patients/families probably is fairly low. This in turn might explain why it is obviously impossible to obtain common risk genes of schizophrenia with convincing odds ratios. According to the primary GRAS hypothesis, schizophrenia is the result of a combination of 'unfortunate' normal genetic variants exposed to unfavorable environmental influence. Apart from its scientific significance, undoubtedly, this hypothesis strongly supports any anti-stigma campaign.

Figure 1. Schematic presentation of three hypothetical subphenotypes of schizophrenia embedded in a sample Manhattan plot. The non-overlapping red, blue or yellow dots comprise genetic constellations ("assemblies") suggested to account for disease subforms with respective subphenotypes. Note that typical genome-wide association studies (GWAS) "top hits", as defined by the highest significance levels for the clinical endpoint diagnosis ("schizophrenia"), most likely fall into different assemblies. This can explain their low odd ratios in a large and diverse patient group and the apparent lack of interactions. The depicted schema is purely hypothetical and shall illustrate the difficulties to define disease genes by conventional GWAS approaches based on endpoint diagnoses.

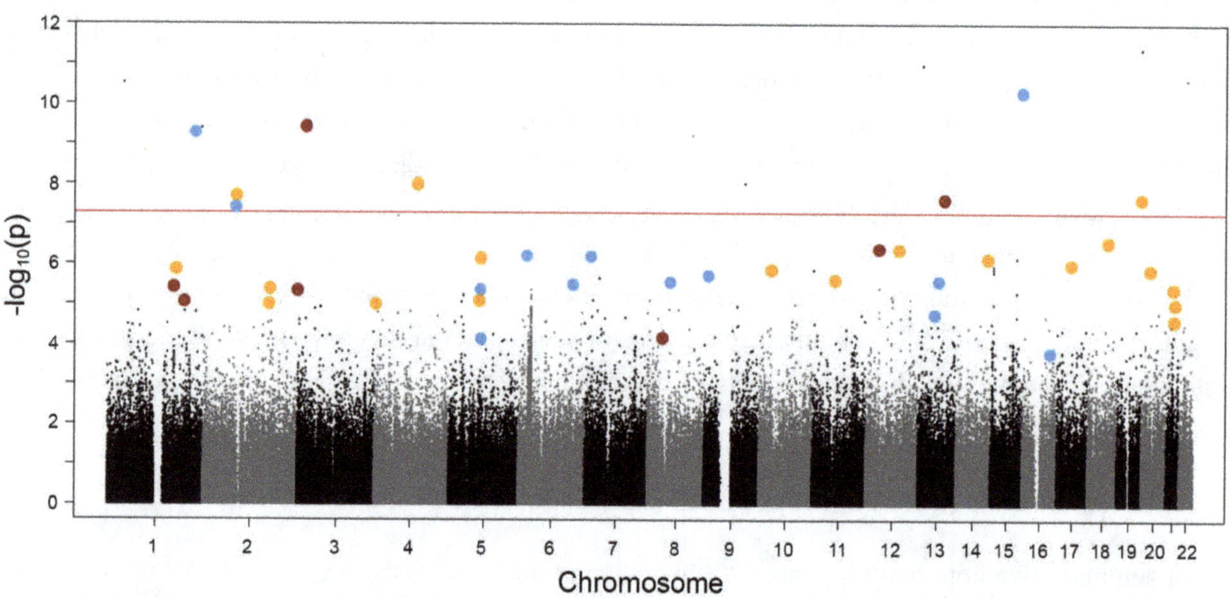

We hypothesize that a considerable proportion of the population across cultures (likely ~50%) may harbor a principal genetic make-up for developing mental diseases whereas the remaining 50% of individuals could never acquire them due to absence of respective genetic prerequisites. Only in a small fraction of risk carriers the disease will break out (~0.5%–1% of the population across cultures) [20], co-induced by a multitude of potential external inputs, while the bulk of carriers will stay healthy and forward their predisposition to their offspring.

5. Deep Phenotyping Is a Prerequisite for PGAS: The GRAS Data Collection

If all human genetic approaches to schizophrenia as a "classic" genetic disease apparently failed, how will it be possible to learn more about the contribution of genes/genotypes to relevant disease subphenotypes? Motivated by this central question we started an alternative and at the same time complementary approach to conventional GWAS. We call this approach PGAS, which stands for *phenotype-based genetic association study*. PGAS enables us to explore the contribution of genes/genetic markers to schizophrenic subphenotypes. To start PGAS, we had to establish a comprehensive standardized phenotypical characterization of schizophrenic patients, the above introduced GRAS data collection. Within a few years we compiled the presently world-wide largest phenotypical database of schizophrenic patients comprising presently ~1200 subjects, with slow steady-state recruitment ongoing. Although much larger data bases of schizophrenic subjects exist, unfortunately, they do not even come close to the deep phenotyping information of GRAS. A total of 23 psychiatric hospitals all over Germany were involved as hosting centers in this effort [18,19]. Inclusion criteria for the patients were confirmed diagnosis of schizophrenia or schizoaffective disorder, and the ability to cooperate at least to a minimal degree. The GRAS population is representative for those schizophrenic patients in central Europe/Germany who are in contact with the healthcare system. Unrivalled in the GRAS data collection is the fact that all cross-sectional examinations of all patients were performed by one and the same travelling team. This fact contributes in a highly significant manner to the homogeneity of the resulting database, containing at present 82% schizophrenic and 18% schizoaffective subjects [18,19].

The GRAS data collection does not only encompass sociodemographic and basic clinical parameters but also, for example, a very comprehensive cognitive test battery, an extensive investigation of neurological signs and symptoms, a thorough medication history, several tests of extrapyramidal side effects of antipsychotic medication, complete information on drug abuse/addiction or other comorbidities, to name just a few of the covered data modules. Apart from the comprehensive cross-sectional analyses, we collected nearly all psychiatric medical reports and discharge letters of all GRAS patients, allowing also for fairly solid longitudinal data analysis, e.g., information on disease course and outcome. Altogether, we have about 3000 data points per patient, which allow us to perform PGAS [18,19]. Importantly, most GRAS patients agreed to be re-contacted for follow-up studies, thereby enabling targeted analyses in the future, e.g., detailed morphometrical or functional magnetic resonance imaging studies, or inducible pluripotent stem cell approaches to study cellular consequences of complex genotypes.

6. Phenotype-Based Genetic Association Studies (PGAS): Proof-of-Principle

First *proof-of-principle* studies for the PGAS approach were performed over the last years. We could show that genetic variants within the *complexin2* gene (6 single nucleotide polymorphisms) influence the cognitive capability of schizophrenic patients [18]. This gene encodes a synaptic protein that is crucial for the regulation of neurotransmitter release. Similarly, the calcium-activated potassium channel SK3 is involved in the modulation of neuroplasticity and influences cognition of schizophrenic individuals genotype-dependently (as a function of the length of a CAG repeat in the *N*-terminal region) [21]. Moreover, a distinct combination of erythropoietin (EPO) and EPO receptor (EPOR) genotypes leads to a remarkable cognitive benefit in schizophrenic subjects as compared to all other possible EPO/EPOR genotype combinations [22]. Of note, the association of certain genetic variants with cognition in schizophrenia may likewise hold true for healthy individuals and other disease populations [22]. This clearly emphasizes the importance of investigating the genotype contribution to phenotypes in general, rather than immediately focusing on risk genes in complex diseases.

Among the discussed candidate risk genes of schizophrenia is *NRG1* [23–25]. Analyzing its disease-relevance in the GRAS population yielded for the first time an association with age at disease onset and severity of positive symptoms [26]. A myelin-associated gene, *CNP*, turned out to co-determine the occurrence of a catatonia-depression syndrome upon aging [27]. Importantly, in most of these PGAS approaches, either replication in an independent human cohort and/or behavioral phenotypes in corresponding mouse mutants were demonstrated. In fact, the translation to mouse models, and even to cultured cells, plays an important part in the search for mechanistic comprehension of the observed phenomena.

7. PGAS Will Be Instrumental for Definition of Biological Schizophrenia Subgroups

Even though the above mentioned first *proof-of-principle* studies were important to set the stage for PGAS, it is now mandatory to work on the "bigger picture". The availability of genome-wide genetic data on the GRAS sample enables us to better understand the genetic basis of neuropsychiatric phenotypes. Using the Axiom^{TM} Caucasian European array as a backbone, we have enriched it by including additional markers of putative functional importance [28]. At present, we are in the process of defining biological subphenotypes of schizophrenia which are less complex as compared to the 'umbrella diagnosis schizophrenia' and can be described and well quantified by respective phenotype scores. Prerequisite for applying these scores to genetic studies is a high internal consistency of the selected score items, reflected by a demanded Cronbach's alpha of >80%. As a first example, we have operationalized an autistic subphenotype of schizophrenia. This subphenotype, which comprises lead features of high-functioning autism (deficits in social interaction and communication, as well as repetitive behaviors/stereotypies), shows a normal distribution in the GRAS population, thus allowing for extreme group comparisons with respect to the associated genotypical information. Our primary target genes for the autistic subphenotype are synaptic genes where we find an unfortunate accumulation of normal genotypes, that we consider as "pro-autistic" variants of these genes, to explain an appreciable part of the phenotypical variance. A sample of Asperger autists is presently recruited with the aim to replicate the genotype-phenotype associations found in the autistic subgroup of schizophrenic individuals. Similarly, a psychomotor and a schizoaffective subphenotype of schizophrenia will be phenotypically as well as—subsequently—genetically characterized, targeting myelin- and ion channel-associated genes, respectively [29].

8. Conclusions

To conclude, the lack of objective diagnoses and the non-existence of classical disease genes have forced us to re-consider the genetic approach to mental diseases. Deep clinical phenotyping as prerequisite for PGAS, combined with genome-wide SNP coverage, makes it possible to improve our understanding of the molecular-genetic architecture of schizophrenia and likely other mental diseases via systematic phenotype-based approaches. These in turn necessitate highly labor-intense groundwork since they depend on the availability of comprehensive phenotypical databases comparable to the GRAS data collection. Some efforts in a similar direction are launched in other disciplines, for instance in a first trial to integrate available electronic medical record data and GWAS information [30]. Regarding the depth of phenotyping, however, this approach certainly needs to be substantially improved. Even if "deep phenotyping" is much more tedious than collecting thousands of subjects with merely a certain diagnostic label versus healthy individuals, it will be the only way to reach the goal, *i.e.*, novel definitions of biologically sound disease subgroups. It can only be hoped that big consortia and sponsors will follow this rationale such that the number of deeply phenotyped individuals will grow. At the end of the road, PGAS will permit better insights into the complex genotype-phenotype interactions of schizophrenia (and other neuropsychiatric diseases) and in this way open up more targeted therapeutic strategies for biological subphenotypes of the disease.

Acknowledgments

This work has received continuous support by the Max Planck Society, the Max Planck Förderstiftung, and the DFG (CNMPB).

Author Contributions

Both authors discussed the contents, designed the paper outline, and jointly wrote the manuscript.

References

1. Cardno, A.G.; Gottesman, I.I. Twin studies of schizophrenia: From bow-and-arrow concordances to star wars mx and functional genomics. *Am. J. Med. Genet.* **2000**, *97*, 12–17.
2. Lichtenstein, P.; Yip, B.H.; Bjork, C.; Pawitan, Y.; Cannon, T.D.; Sullivan, P.F.; Hultman, C.M. Common genetic determinants of schizophrenia and bipolar disorder in swedish families: A population-based study. *Lancet* **2009**, *373*, 234–239.
3. Cardno, A.G.; Marshall, E.J.; Coid, B.; Macdonald, A.M.; Ribchester, T.R.; Davies, N.J.; Venturi, P.; Jones, L.A.; Lewis, S.W.; Sham, P.C.; *et al.* Heritability estimates for psychotic disorders: The maudsley twin psychosis series. *Arch. Gen. Psychiatry* **1999**, *56*, 162–168.
4. Franzek, E.; Beckmann, H. Different genetic background of schizophrenia spectrum psychoses: A twin study. *Am. J. Psychiatry* **1998**, *155*, 76–83.
5. Svrakic, D.M.; Zorumski, C.F.; Svrakic, N.M.; Zwir, I.; Cloninger, C.R. Risk architecture of schizophrenia: The role of epigenetics. *Curr. Opin. Psychiatry* **2013**, *26*, 188–195.
6. Van Os, J.; Kenis, G.; Rutten, B.P. The environment and schizophrenia. *Nature* **2010**, *468*, 203–212.
7. Giusti-Rodriguez, P.; Sullivan, P.F. The genomics of schizophrenia: Update and implications. *J. Clin. Invest.* **2013**, *123*, 4557–4563.
8. Sullivan, P.F. The genetics of schizophrenia. *PLoS Med.* **2005**, *2*, e212.
9. St. Clair, D.; Blackwood, D.; Muir, W.; Carothers, A.; Walker, M.; Spowart, G.; Gosden, C.; Evans, H.J. Association within a family of a balanced autosomal translocation with major mental illness. *Lancet* **1990**, *336*, 13–16.
10. Brandon, N.J.; Millar, J.K.; Korth, C.; Sive, H.; Singh, K.K.; Sawa, A. Understanding the role of disc1 in psychiatric disease and during normal development. *J. Neurosci.* **2009**, *29*, 12768–12775.
11. Ripke, S.; O'Dushlaine, C.; Chambert, K.; Moran, J.L.; Kahler, A.K.; Akterin, S.; Bergen, S.E.; Collins, A.L.; Crowley, J.J.; Fromer, M.; *et al.* Genome-wide association analysis identifies 13 new risk loci for schizophrenia. *Nat. Genet.* **2013**, *45*, 1150–1159.
12. Consortium, I.S.; Purcell, S.M.; Wray, N.R.; Stone, J.L.; Visscher, P.M.; O'Donovan, M.C.;

Sullivan, P.F.; Sklar, P. Common polygenic variation contributes to risk of schizophrenia and bipolar disorder. *Nature* **2009**, *460*, 748–752.

13. O'Donovan, M.C.; Craddock, N.; Norton, N.; Williams, H.; Peirce, T.; Moskvina, V.; Nikolov, I.; Hamshere, M.; Carroll, L.; Georgieva, L.; *et al.* Identification of loci associated with schizophrenia by genome-wide association and follow-up. *Nat. Genet.* **2008**, *40*, 1053–1055.
14. Papiol, S.; Malzahn, D.; Kästner, A.; Sperling, S.; Begemann, M.; Stefansson, H.; Bickeboller, H.; Nave, K.A.; Ehrenreich, H. Dissociation of accumulated genetic risk and disease severity in patients with schizophrenia. *Transl. Psychiatry* **2011**, *1*, e45.
15. Papiol, S.; Mitjans, M.; Assogna, F.; Piras, F.; Hammer, C.; Caltagirone, C.; Arias, B.; Ehrenreich, H.; Spalletta, G. Polygenic determinants of white matter volume derived from gwas lack reproducibility in a replicate sample. *Transl. Psychiatry* **2014**, doi:10.1038/tp.2013.126.
16. Terwisscha van Scheltinga, A.F.; Bakker, S.C.; van Haren, N.E.; Derks, E.M.; Buizer-Voskamp, J.E.; Boos, H.B.; Cahn, W.; Hulshoff Pol, H.E.; Ripke, S.; Ophoff, R.A.; *et al.* Genetic schizophrenia risk variants jointly modulate total brain and white matter volume. *Biol. Psychiatry* **2013**, *73*, 525–531.
17. Stepniak, B.; Papiol, S.; Hammer, C.; Ramin, A.; Everts, S.; Hennig, L.; Begemann, M.; Ehrenreich, H. Accumulated environmental risk determining the onset of schizophrenia. Submitted for publication, 2014.
18. Begemann, M.; Grube, S.; Papiol, S.; Malzahn, D.; Krampe, H.; Ribbe, K.; Friedrichs, H.; Radyushkin, K.A.; El-Kordi, A.; Benseler, F.; *et al.* Modification of cognitive performance in schizophrenia by complexin 2 gene polymorphisms. *Arch. Gen. Psychiatry* **2010**, *67*, 879–888.
19. Ribbe, K.; Friedrichs, H.; Begemann, M.; Grube, S.; Papiol, S.; Kästner, A.; Gerchen, M.F.; Ackermann, V.; Tarami, A.; Treitz, A.; *et al.* The cross-sectional gras sample: A comprehensive phenotypical data collection of schizophrenic patients. *BMC Psychiatry* **2010**, *10*, 91.
20. Van Os, J.; Kapur, S. Schizophrenia. *Lancet* **2009**, *374*, 635–645.
21. Grube, S.; Gerchen, M.F.; Adamcio, B.; Pardo, L.A.; Martin, S.; Malzahn, D.; Papiol, S.; Begemann, M.; Ribbe, K.; Friedrichs, H.; *et al.* A cag repeat polymorphism of kcnn3 predicts sk3 channel function and cognitive performance in schizophrenia. *EMBO Mol. Med.* **2011**, *3*, 309–319.
22. Kästner, A.; Grube, S.; El-Kordi, A.; Stepniak, B.; Friedrichs, H.; Sargin, D.; Schwitulla, J.; Begemann, M.; Giegling, I.; Miskowiak, K.W.; *et al.* Common variants of the genes encoding erythropoietin and its receptor modulate cognitive performance in schizophrenia. *Mol. Med.* **2012**, *18*, 1029–1040.
23. Gong, Y.G.; Wu, C.N.; Xing, Q.H.; Zhao, X.Z.; Zhu, J.; He, L. A two-method meta-analysis of neuregulin 1(nrg1) association and heterogeneity in schizophrenia. *Schizophr. Res.* **2009**, *111*, 109–114.
24. Li, D.; Collier, D.A.; He, L. Meta-analysis shows strong positive association of the neuregulin 1 (nrg1) gene with schizophrenia. *Hum. Mol. Genet.* **2006**, *15*, 1995–2002.
25. Munafo, M.R.; Thiselton, D.L.; Clark, T.G.; Flint, J. Association of the nrg1 gene and schizophrenia: A meta-analysis. *Mol. Psychiatry* **2006**, *11*, 539–546.
26. Papiol, S.; Begemann, M.; Rosenberger, A.; Friedrichs, H.; Ribbe, K.; Grube, S.; Schwab, M.H.;

Jahn, H.; Gunkel, S.; Benseler, F.; *et al.* A phenotype-based genetic association study reveals the contribution of neuregulin1 gene variants to age of onset and positive symptom severity in schizophrenia. *Am. J. Med. Genet. Part B Neuropsychiatry Genet.* **2011**, *156*, 340–345.

27. Hagemeyer, N.; Goebbels, S.; Papiol, S.; Kästner, A.; Hofer, S.; Begemann, M.; Gerwig, U.C.; Boretius, S.; Wieser, G.L.; Ronnenberg, A.; *et al.* A myelin gene causative of a catatonia-depression syndrome upon aging. *EMBO Mol. Med.* **2012**, *4*, 528–539.
28. Hammer, C.; Stepniak, B.; Schneider, A.; Papiol, S.; Tantra, M.; Begemann, M.; Siren, A.L.; Pardo, L.A.; Sperling, S.; Mohd Jofrry, S.; *et al.* Neuropsychiatric disease relevance of circulating anti-nmda receptor autoantibodies depends on blood-brain barrier integrity. *Mol. Psychiatry* **2013**, doi:10.1038/mp.2013.110.
29. Ehrenreich, H.; Nave, K.-A. Max Planck Institute of Experimental Medicine, and DFG Center for Nanoscale Microscopy and Molecular Physiology of the Brain (CNMPB), Hermann-Rein-Str.3, 37075 Göttingen, Germany. Unpublished work, 2014.
30. Denny, J.C.; Bastarache, L.; Ritchie, M.D.; Carroll, R.J.; Zink, R.; Mosley, J.D.; Field, J.R.; Pulley, J.M.; Ramirez, A.H.; Bowton, E.; *et al.* Systematic comparison of phenome-wide association study of electronic medical record data and genome-wide association study data. *Nat. Biotechnol.* **2013**, *31*, 1102–1111.

5

Association Claims in the Sequencing Era

Sara L. Pulit [1], Maarten Leusink [1,2,3], Androniki Menelaou [1] and Paul I. W. de Bakker [1,3,*]

[1] Department of Medical Genetics, Institute for Molecular Medicine, University Medical Center Utrecht, Universiteitsweg 100, 3584 CG, Utrecht, The Netherlands; E-Mails: s.l.pulit@umcutrecht.nl (S.L.P.); m.leusink@uu.nl (M.L.); a.menelaou@umcutrecht.nl (A.M.)

[2] Division of Pharmacoepidemiology & Clinical Pharmacology, Utrecht Institute for Pharmaceutical Sciences, Utrecht University, Universiteitsweg 99, 3584 CG, Utrecht, The Netherlands

[3] Julius Center for Health Sciences and Primary Care, University Medical Center Utrecht, Universiteitsweg 100, 3584 CG, Utrecht, The Netherlands

* Author to whom correspondence should be addressed; E-Mail: pdebakker@umcutrecht.nl;

Abstract: Since the completion of the Human Genome Project, the field of human genetics has been in great flux, largely due to technological advances in studying DNA sequence variation. Although community-wide adoption of statistical standards was key to the success of genome-wide association studies, similar standards have not yet been globally applied to the processing and interpretation of sequencing data. It has proven particularly challenging to pinpoint unequivocally disease variants in sequencing studies of polygenic traits. Here, we comment on a number of factors that may contribute to irreproducible claims of association in scientific literature and discuss possible steps that we can take towards cultural change.

Keywords: association; GWAS; next-generation sequencing; significance; bias; complex traits

---

1. The Evolution of Genetic Association Studies

The Human Genome Project [1] remains the largest scientific endeavor in the biological sciences, spanning thirteen years, requiring hundreds of researchers around the globe, and costing $3 billion. Called the "most important, most wondrous map ever produced by human kind" [2] by U.S. President

Clinton upon its completion, the map of the genome catapulted the investigation of human disease into a new era. The study of complex traits and disease had previously been limited to genetic linkage studies, typically laborious efforts limited to constructing linkage maps in families and powered for discovering highly penetrant variants. Now, geneticists could identify single-base changes in the genome (single nucleotide polymorphisms, or SNPs) and, by measuring the frequencies of these changes in large groups of cases and controls, test SNPs for association with susceptibility to any number of diseases.

Upon the completion of the Human Genome Project, the field of genetics was faced with determining how best to hunt for such associations. The first widely used method was candidate gene studies, in which genes with suspected biological or functional relevance to the disease in question were selected for testing. These studies were applied to a variety of traits [3], but in the absence of community-wide standards for analysis, they were plagued with problems. Reviews of candidate gene studies found that small sample sizes, population stratification issues, weak effects, and a lack of statistical evidence for the claimed associations were common [3–6]. All of these problems contributed to the irreproducibility of initial findings; one review found that of 166 associations with more than two follow-up studies, only six (3.6%) replicated [3].

Despite limited success in candidate gene studies, the number of replicated associations indicated that common variation (frequency >5%) indeed plays a role in genetic susceptibility to common disease (involving tens of genes and environmental factors) [7]. In 2005, the approach in genetics began shifting towards genome-wide association studies (GWAS) [8], which require no prior hypothesis about which genes are likely to influence disease. Instead, geneticists could test millions of common variants across the genome for association with the trait of interest. Systematic cataloguing of these variants [9] allowed for the design of genome-wide SNP arrays, ultimately allowing for low-cost capture of tens of thousands of variants in large samples.

The beginnings of GWAS were slow; initial studies produced few if any associated loci, and it quickly became clear that larger samples would be necessary for sufficient power to detect susceptibility variants [10]. With the formation of international consortia, collection of large samples, and assembly of imputation panels that allowed for testing of variants not present on SNP arrays came an explosion in discovered loci (Figure 1). Along with the rapid increase in the number of GWAS being performed came a large-scale effort to standardize the method. The community adopted the genome-wide significance p-value threshold of 5×10^{-8}, a p-value that reflects a Bonferroni correction for the approximately one million independent tests performed in a GWAS [11,12]. Methods for handling population stratification were developed [13], as were approaches for finding and removing poorly captured genotypes [14]. Replication of discovered loci also became a criterion for declaring a SNP to be associated with a disease [15,16]. With best practices in place, GWAS has become an efficient and robust method for discovering the contribution of common variation to susceptibility in common disease. The total number of SNPs associated with complex traits (at genome-wide significance) was only seven by the end of 2006, but by 2008, an additional 637 associations had been discovered [17]. Today, more than 6,000 disease-SNP associations have been reported.

Figure 1. Disease-susceptibility loci discovered to date in various complex traits, as reported in the National Human Genome Research Institute (NHGRI) genome-wide association studies (GWAS) catalog [17]. Early genome-wide association studies interrogated small samples and uncovered few, if any, loci associated with the trait of interest. However, collaborative efforts to assemble large-scale samples improved power and implicated tens and even hundreds of susceptibility loci, revealing a (roughly) linear relationship between sample size and associated loci [18].

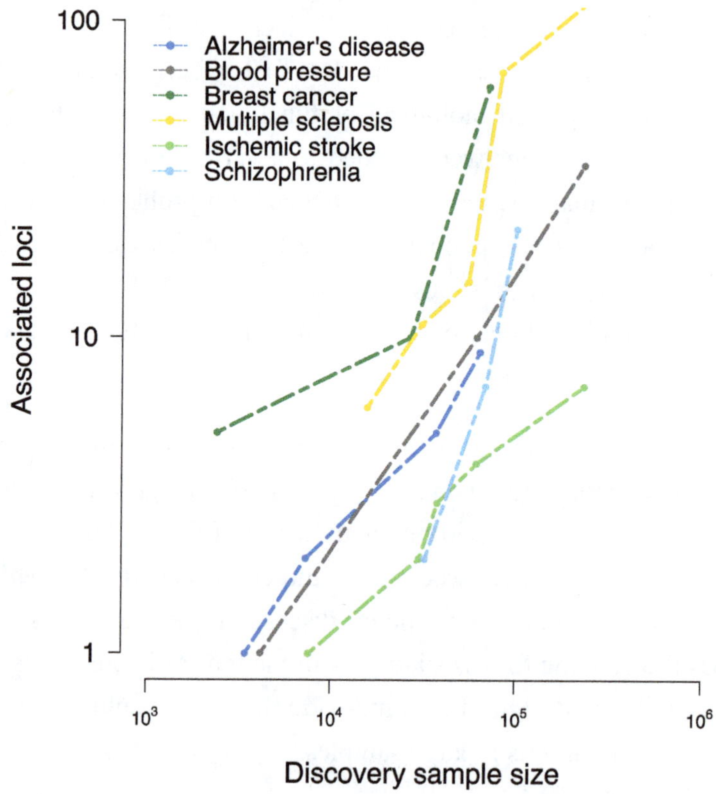

Because of established best practices and statistical thresholds, most GWAS findings are robust and reproducible. The method is, of course, not immune to error; for example, one GWAS looking for variants associated with longevity failed to correct for technical bias introduced by the use of two different SNP arrays, inducing spurious association [19]. The mistake went unnoticed until the results were published and several geneticists caught the error. The false report received attention from the genetics community and the media alike [20] and was later retracted [21].

More recently, scientists and the media have cast a critical eye towards the aspects of scientific culture that also give rise to false-positive findings in published work. The *Economist* recently published an article suggesting that science is not as self-correcting as many assume [22,23], and journals such as *Nature* and *Science* are publishing columns on scientific misconduct, peer-review, and other issues that contribute to the reporting of inaccurate results [24–28]. The last four years have also seen several published studies on the prevalence of false discoveries and misconduct, indicating that as many as "1% of published papers are fraudulent" (about 20,000 papers each year) [25]. Given the increased attention from the public and scientists alike, and because genetics is in a transition phase as it moves from performing GWAS data to studying next-generation sequencing data, now is an ideal time to address some of these shortcomings.

2. Sources of Error and Bias in Genetic Research

A number of factors contribute to false-positive findings in published human genetic research. Technical artifact, such as mishandling of population stratification, poorly genotyped SNPs, and batch effects introduced by different SNP arrays or genotyping runs can cause spurious results, though a number of methods have been designed for detecting them [13,29–32]. Study design is also crucial in avoiding false positives. SNPs implicated in disease susceptibility typically have modest effect sizes (odds ratios ranging from 1.1–1.5) [18]; insufficient sample size to detect such effects can substantially reduce a study's power and increase the likelihood of discovering an artifact rather than a true association.

Although technical error can lead to false positives, a number of other forces in research culture also contribute to the number of published erroneous findings. In the span of just seven years, GWAS moved from small-scale efforts to studies of thousands of samples, leading to a heavy reliance on statistical methods to study large datasets. Yet, researchers' understanding of statistics has not kept pace with data generation, making them more likely to apply inappropriate statistical tests or perform tests they do not fully understand. The same holds true for the many programs written to analyze genome-wide datasets. Not all researchers using this software will fully understand the underlying methods, increasing the chance of false positives going unnoticed.

The peer-review process is also partially to blame for the introduction of false findings into scientific literature. Though researchers would seem ideal candidates for catching the mistakes of their peers, studies suggest that they often fail to catch errors (even when instructed that there are errors to find) [22]. Further, peer-reviewers are not provided all data underlying a paper and therefore cannot reproduce analyses to verify findings.

Studies also indicate that misconduct (which includes plagiarism, fraud, and duplicate publications) has been on the rise in recent years [33] and accounts for the majority of retracted papers in the life sciences [34]. A number of aspects of research culture likely contribute to such misconduct. Scientific research has become a field in which the number of articles a researcher produces is a primary measure of success. Career opportunities in the sciences also often hinge on the number of publications a scientist has produced. Conflating financial interests (whether in the form of employment or grants) and pressure to publish is a factor that potentially gives rise to false associations reported in scientific literature [35]. Anyone pressed to produce such a high volume of results, preferably at great speed, is more likely to miss mistakes in her own work.

Publication and funding bias are also problematic [36]. Although replicating a result is the backbone of establishing the veracity of a scientific claim, replication studies are less likely to be funded than discovery-focused experiments, and journals are less likely to publish them. Journals also typically do not publish negative studies, preferring to focus instead on novel results. The establishment of the impact factor system has further intensified journals' bias towards novelty. While selecting manuscripts for publication, journal editors consider not only the content of the manuscript, but also the potential that the paper will improve the journal's impact factor. The impact factor ranking system also heightens publishing competition for researchers. Researchers seek publication in only a very small set of highly selective journals in pursuit of many citations, widespread attention from the scientific community and popular media, and potential career advancement. In molecular biology and genetics, just six journals accounted for 85 of the 100 most-cited articles between 1998 and 2008 [37].

Increasing the prevalence of false positives are those journals that seek to turn a profit by preying off a research culture that keeps publishing at its center. The number of journals accepting manuscripts is large and rapidly growing. The Directory of Open Access Journals, which tracks open access journals and their credibility, has tracked the swift global expansion of the number of open access journals, including an addition of over 1000 journals in 2012 alone [38] (Figure 2). A "sting" conducted on open-access journals revealed that many of them are uninterested in scientific veracity; a paper concocted wholesale and containing glaring errors was accepted by more than 52% of the targeted journals [39]. The "sting" in part implicated a flawed peer-review system, as 40% of the submissions were reviewed and then accepted. The other 60% of submissions, however, were accepted without any indication of peer review, suggesting that many of the journals the "sting" targeted are focused more on profit rather than on scientific rigor, encouraging a culture that values a published finding over a robust result.

Figure 2. The growth of open access journals over time and around the world. (**a**) The number of open-access journals, as tracked by the Directory of Open Access Journals, over the last 140 years. (**b**) Countries with over 100 open-access journals accepting manuscripts.

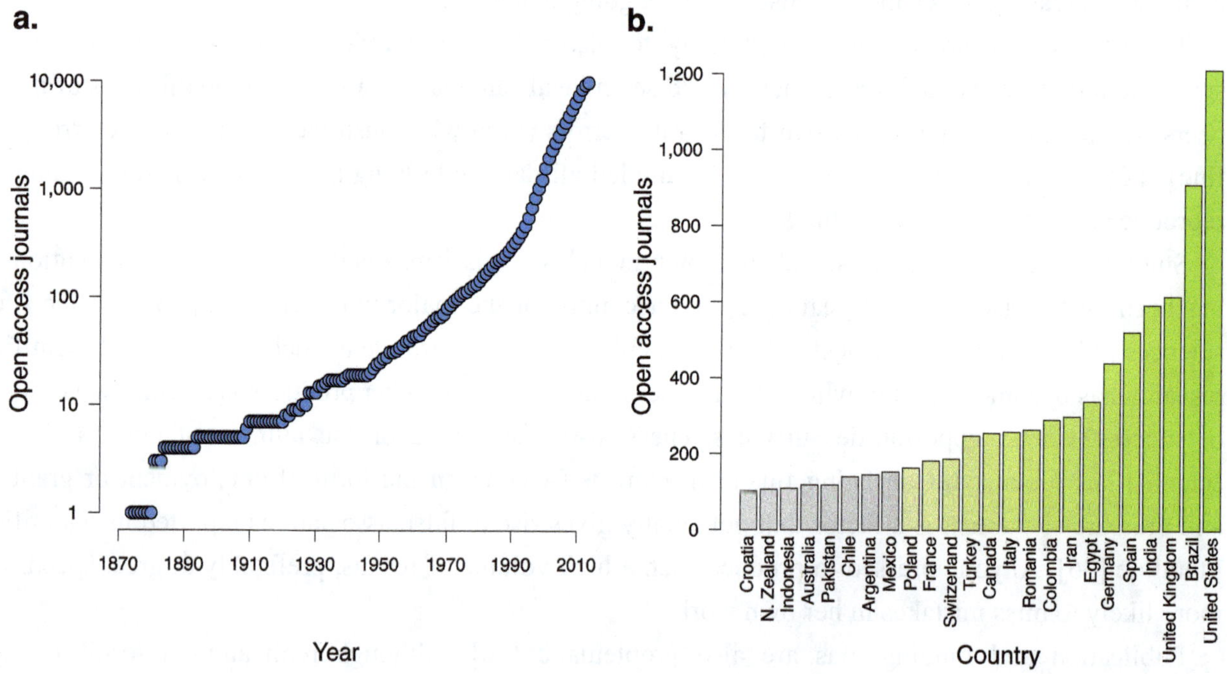

3. Next-Generation Sequencing: New Technology, New Challenges

The advent of next-generation sequencing (NGS) technology has ushered in a new wave of studies in human genetic research. Given that many complex traits involve tens and sometimes hundreds of loci [40,41], lower-frequency variants may also contribute to the architecture of human disease. However, these variants are weakly tagged by common SNPs and have therefore gone untested by GWAS efforts. Researchers using NGS data can test (nearly) the entire set of variants in a single genome, helping to complete the picture of the role of genetic variation in common disease. Yet, this new technology brings with it many challenges, giving rise to additional forces that may lead to false positives in scientific literature.

A number of technical errors can give rise to a false-positive association in a NGS disease study. Determining genotypes from sequencing reads is more challenging than determining genotypes from SNP array data [42,43]. Rather than measuring probe intensities, as is done with SNP arrays to determine genotypes, extracting genotypes from sequencing data involves multiple steps, including mapping sequencing reads to a reference, detecting bases that do not match the reference, and determining the genotypes of each individual at each base; errors can occur at any of these steps and are particularly likely in regions that are difficult to capture, such as those rich in GC content or that are highly repetitive. Further, determining inclusion and exclusion criteria for variants based on a host of sequencing metrics can be difficult, sometimes requiring manual review of each of the metrics to determine appropriate filters [44]. Even the most conservative variant calling and quality control (QC) cannot guarantee that "variants" that are actually artifacts will be removed from the dataset.

Study design flaws can also prompt false-positive results. Although methods for detecting population stratification have been developed and widely used in GWAS, our understanding of population stratification in rare variants is limited and may confound association results [45]. To reduce costs, an investigator may choose to sequence only cases and use external (previously sequenced) controls; this approach may introduce stratification because the controls may be sequenced using a different platform [46] or genotype-called using outdated software. Alternatively, an investigator may want to sequence only cases and then genotype the discovered variants in controls or have cases substantially outnumber the controls, but these approaches also inflate type I error [47].

So-called loss-of-function mutations are of particular interest in NGS studies since they truncate proteins and are thus good candidates for likely pathogenic mutations. However, determining the deleteriousness of a loss-of-function mutation can be challenging, and this class of variants is enriched for artifacts [48]. It may be tempting to relax statistical thresholds for loss-of-function mutations or produce functional results for them before assembling appropriate statistical evidence from the genetic data, but doing so can lead to error.

Though there are many challenges in performing an NGS study, the single largest problem plaguing sequencing disease studies to date is low statistical power to detect an association. The hypothesis that rare variants influencing susceptibility to common disease would have larger effect sizes than those discovered by GWAS has gone largely unsupported, and analyses indicate that NGS studies will require tens and even hundreds of thousands of samples to be well powered [46,49] (Figure 3). Consequently, assembling adequate sample sizes for NGS studies will take time, but researchers remain under pressure to publish. As a result, they will likely be anxious to push forward results that are not fully understood, lack statistical evidence, and may not even be real.

Finally, publishing bias will prove even more problematic in the sequencing age. The optimal choices for designing and performing a sequencing study—from phenotype ascertainment to selecting algorithms for calling and filtering variants and deciding which association tests to apply—remain difficult to discern. Sequencing studies that lack "positive" findings are highly informative but publishing bias will prevent such information from being widely disseminated. It will be more difficult to establish a standardized methodology for NGS studies, as was done for GWAS, because it is likely that "failed" experiments will not be given the same attention as studies reporting new and exciting results.

Figure 3. (**a**) The power to detect a genetic association is a function of sample size (N), effect size (γ), the frequency of the associated allele (p), and linkage disequilibrium (LD) between the tested and causal variants (R^2), assuming an additive model [50]. (**b**) For sequencing studies, many thousands of samples will be necessary to detect single, low-frequency variants associated with disease risk at genome-wide significance (black line). RAF, risk allele frequency.

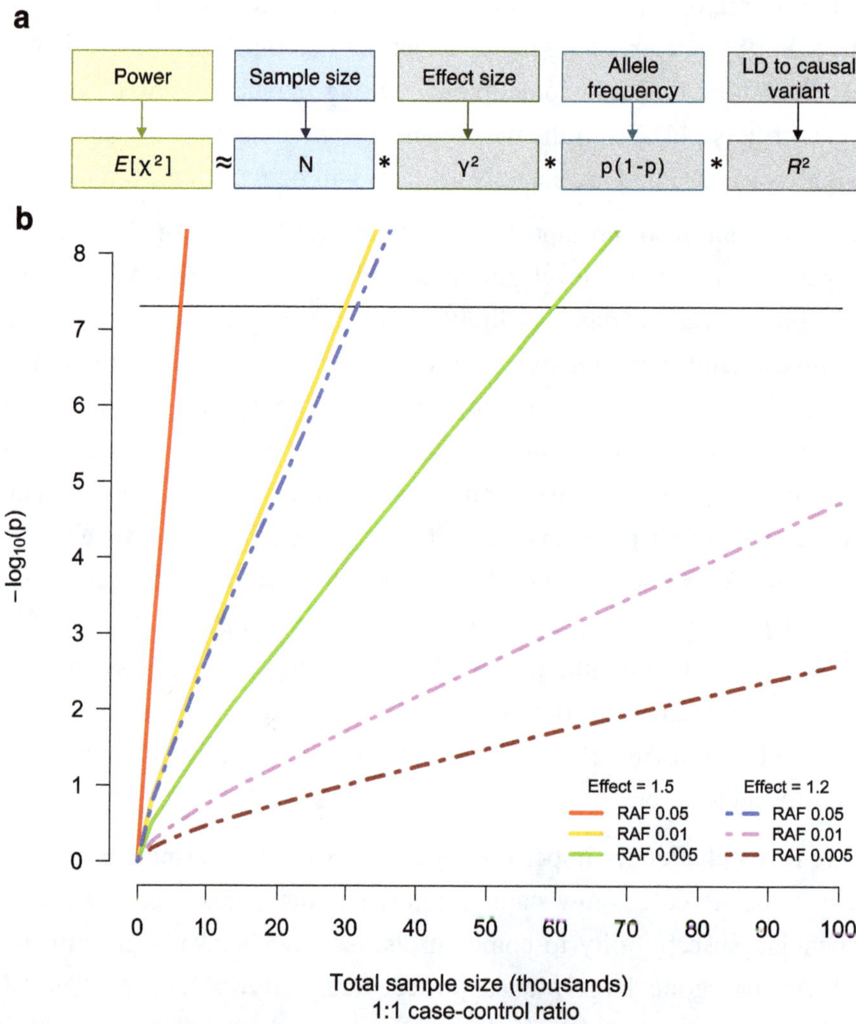

4. Lacking Evidence: Examples from NGS Studies

To publish disease-associated loci discovered through genome-wide association studies, it has become standard practice to meet basic criteria for discovery: appropriate sample and genotype cleaning, a SNP at genome-wide significance with a reasonable effect size and frequency, and replication in independent samples. Such standards do not yet exist for sequencing studies. Without criteria for claiming an association combined with publication bias and the pervasive pressure to "publish or perish," some NGS studies in complex traits have been published despite a paucity of statistical evidence.

A targeted sequencing project in anorexia nervosa (AN) patients claims an association between AN and the epoxide hydrolase 2 (*EPHX2)* gene, though the burden test *p*-value of the gene failed to meet exome-wide significance (discovery, $p = 4 \times 10^{-4}$; replication, $p = 6.2 \times 10^{-3}$) [51]. Because *EPHX2* has been linked to lipid traits and hypercholesterolemia is common in AN patients, the authors performed a variety of interaction tests between variants in *EPHX2* and cholesterol and body mass index. They suggest that the results of the interaction tests of seven SNPs (p = 0.004–0.045) are additional evidence for the role of *EPHX2* in AN, but show no correction for multiple testing. A small sample size (1205 cases and 1719 controls), a large case-control ratio (~3:1) in discovery, and mismatched ancestries between cases and controls (indicated by a principal component plot) may have also confounded the results. Although the authors acknowledge the need for additional replication, declaring the association between *EPHX2* and AN "statistically compelling" seems premature.

Another study examined whole-exome sequencing (WES) from four samples with multiple sclerosis (MS) selected from a family with more than 15 affected individuals [52]. The group found one novel missense mutation in the tyrosine kinase 2 (*TYK2*) gene, an MS-susceptibility locus established through GWAS [53,54]. The authors performed genotyping of the variant in all remaining family members and report the percentage of affected and unaffected individuals carrying the variant (10/14 (71.4%) and 28/60 (46.7%), respectively); the difference is not statistically significant (p = 0.17; not reported). Follow-up genotyping in an additional 2,104 cases and 1,543 controls revealed that the variant had a frequency of 0.8% in cases and 0.6% in controls, also not a statistically significant difference (which the authors state themselves). In fact, these frequencies are consistent with observed frequencies in several sets of healthy individuals [55]. Nonetheless, the authors conclude the variant has a modest effect on MS risk. Even though GWAS has established *TYK2* as an MS-associated locus, the particular variant implicated by this study is severely lacking in statistical evidence and seems highly unlikely to confer disease susceptibility.

Some sequencing studies rely on functional follow-up of a variant or gene in the absence of statistically compelling genetic evidence. One recent paper examining a family with 22 members with early-onset myocardial infarction (EOMI) implicated a frameshift insertion in the guanyl cyclase 1, soluble alpha 3 (*GUCY1A3*) gene and a nonsynonymous single nucleotide variant (SNV) in the chaperonin containing TCP1 (*CCT7*) gene in susceptibility to disease [56]. The frequency of the variants in affected *versus* unaffected family members was not statistically significant, as the sample size was small. A search for other susceptibility variants in these two genes in 252 EOMI cases and 800 controls yielded counts that were also not statistically significant (Fisher's $p = 0.023$ and $p = 0.12$ for *GUCY1A3* and *CCT7*, respectively). Functional work on both variants is provided as additional evidence, indicating that mutations in both *GUCY1A3* and *CCT7* in mice induce a protein deficiency that can accelerate the formation of clots that potentially cause infarction. Guanyl cyclase is a key gene in signal transduction involved in vasodilation, and it is possible that these two loci influence susceptibility to EOMI. However, the statistical evidence from the genetic data alone is not strong enough to claim association between the mutations and the trait. Moreover, it is not immediately obvious whether the conclusions of functional work in mouse models will be pertinent to humans or whether findings from such an extreme family will be extendable to EOMI in the general population [57].

A WES study in sporadic amyotrophic lateral sclerosis (ALS) trios also relied on functional work to bolster the finding in the absence of statistically compelling results [58]. The group identified *de novo* mutations in 47 ALS patients and discovered 25 *de novo* variants in 25 different genes, a distribution consistent with the (null) distribution in healthy individuals [59]. A pathway analysis indicated enrichment for chromatin regulator genes, such as the synovial sarcoma translocation (*SS18L1* or *CREST*) gene, which contained a single *de novo* event in one sample. The authors found that variants in *CREST* inhibited neurite growth in animal models and claimed that *de novo* mutations in the gene confer ALS risk. However, a single *de novo* event from one individual, even if the variant is functional, is insufficient evidence to claim a role in disease susceptibility [60], and extensive replication efforts will be necessary to determine whether the ALS-*CREST* association is real.

A similar WES effort in Hirschsprung's disease sequenced two affected (related) samples and also used pathway analysis to investigate the role of the neuregulin 3 (*NRG3*) gene in disease susceptibility [61]. Three variants found in *NRG3* in the initial samples were followed up in 96 cases and 110 controls; the variants were carried by a few cases and no controls (Fisher's $p = 0.021$, $p = 0.22$, $p = 0.021$, for the three variants, respectively). The authors discuss the biological plausibility of the *NRG3* association based on the gene's role in the nervous system, but due to the small sample size, the genetic evidence for the association is lacking.

In each of these studies, the statistical evidence based on the provided data was insufficient. Although it is certainly possible that replication will show these findings to be real, their publication before gathering more persuasive statistical evidence from genetic data seems a symptom of a much larger problem. Researchers eager to publish novel results, peer reviewers not questioning the dearth of evidence, and journals enthusiastic to publish exciting stories that will garner both attention and citations all combine to allow findings that have not yet been demonstrated as statistically robust to enter scientific literature as such.

5. Conclusions: Moving Towards Permanent Change

A number of changes, addressing both technical challenges and cultural characteristics of research, can be implemented in order to improve the veracity of claims in published research. Though association testing approaches may differ between sequencing studies (such as selecting single-variant testing or gene-based testing), the universal application of particular thresholds will help to ensure the robustness of claimed associations. The statistical stringency of genome-wide significance at 5×10^{-8} has served the genetics community well and has ensured the robustness of the majority of GWAS findings. Single-variant testing across whole-genome data should maintain this threshold, if not establish a more conservative one given the increased number of variants tested. Similarly, burden tests of genes should be held to an exome-wide significance level, estimated to be approximately 5×10^{-7} [44]. Relaxing these thresholds, particularly tempting in the case of loss-of-function mutations or studies that analyze only a small set of genes, is unwise. A number of mechanisms make interpreting the true deleteriousness of a variant difficult; functional annotation alone does not guarantee a true loss-of-function mutation [44,48]. Findings from other association testing approaches, such as pathway analyses or polygenic modeling, should be evaluated based on appropriate Bonferroni correction for independent tests. Some may argue that it is overly conservative to hold a small set of genes or variants to an exome- or genome-wide threshold, but these smaller analyses are simply subsets of what will likely become exhaustive genome-wide searches. Contingent on the assembly of large datasets, testing every gene or variant is inevitable. False-positive signals can occur anywhere in the search space, and assuming these false positives will not occur in smaller, earlier searches is faulty logic [62].

Designing and performing a study correctly is also crucial to avoiding spurious associations. Large samples are of the utmost priority and will lead to additional discoveries, as demonstrated by a recent WES project in Alzheimer's that discovered rare variants in the phospholipase D family (Member 3) (*PLD3*) gene by whole-exome sequencing of 14 families and replicated the finding in a large-scale cohort (11,000 European-ancestry cases and controls, gene burden odds ratio = 2.75, $p = 1.44 \times 10^{-11}$; 302 African-ancestry cases and controls, gene burden odds ratio = 5.48, $p = 1.4 \times 10^{-3}$) [63]. Leveraging population isolates [64] and assembling non-European samples [46] may also help in improving power, as was true in the GWAS era [65]. Because of varying study designs and technologies, standardizing NGS data processing and analysis is difficult and will likely take time as methods continue to develop. However, some best practices already exist [42,43,55] and should be followed. Further, the methods sections of papers should be as clear and explicit as possible, reading as "how-to" guides to allow peer reviewers to catch technical mistakes and aid external replication efforts. Ideally, data should be made publicly available whenever possible. Each of these steps will improve the quality of NGS studies as they are increasingly used in the future to study complex traits (Figure 4a,b).

Designing a sequencing study not only involves consideration of the technical aspects of the study but also of the analysis team. Diversifying the team to include scientists with an array of expertise will improve the study's execution and increase the likelihood that the findings prove replicable. In addition to geneticists, computer scientists, statisticians, biologists, engineers, and clinicians can all contribute to different aspects of a sequencing study, from assembling the raw sequencing data to interpreting results. Methods for sequencing analysis are evolving rapidly, and multifaceted teams representing many scientific fields are optimal for keeping pace with this changing landscape.

Future scientific publications will also be improved by an open peer review system. Some journals, such as the journal of the European Molecular Biology Organization (EMBO), have already begun making peer review a public process to great success [66]. The exchange that occurs between scientists during peer review can be valuable not only for the authors, but also for other researchers. If reviews, responses and revisions are made public the scientific community will benefit as a whole by being able to design better studies and avoid errors that other researchers have made. Further, peer review is crucial to the scientific process, yet is rarely ever taught. Making peer review reports accessible to all can serve as a teaching tool, particularly for younger scientists who may be unfamiliar with the process [66].

Peer review is not the only aspect of publishing in scientific journals that should be open. Though some open-access journals were discredited in *Science*'s recent "sting", several open-access journals have become highly reputable in the scientific community, and other journals are giving authors the option to make their article open-access. This trend towards open-access science should continue. Rather than hiding important discoveries behind exorbitantly priced subscriptions, journals should be making these findings easily accessible so that they may be discussed and retested by the scientific community and available to the public.

While changes such as statistical stringency, public peer review, and open-access articles are readily implementable, other changes in research culture will enhance the robustness of published findings, but will require a larger community effort. Journals should require researchers to increase transparency and meet certain criteria before submitting a manuscript for consideration. The Nature Publishing Group recently updated its editorial policies [67], requiring authors to fill out a checklist that accompanies their submitted manuscript. The checklist addresses areas of manuscripts, such as study design and analysis, which the editors have noticed are often not reported completely. The checklist also lends particular emphasis to justifying statistical analyses; the Nature Publishing Group will now consult with statisticians if there seem to be glaring analytic issues or if a referee suggests outside consultation. Other journals should adopt a similar strategy.

Figure 4. Trending topics on PubMed, 2000–2013. The cumulative number of times certain phrases appear per one million abstracts in PubMed. (**a**) Earlier genetic association studies (search terms: "candidate gene"; "linkage analysis" or "linkage study"; "genome-wide association study"). (**b**) Next-generation sequencing studies (search terms: "exome sequencing"; "whole-genome sequencing"). (**c**) Retraction notices (search terms: "retraction notice" or "notice of retraction"). (**d**) Claims of novelty and statistical significance (search terms: "novel"; "not statistically significant" or "not significant"; "statistically significant" removing "not statistically significant").

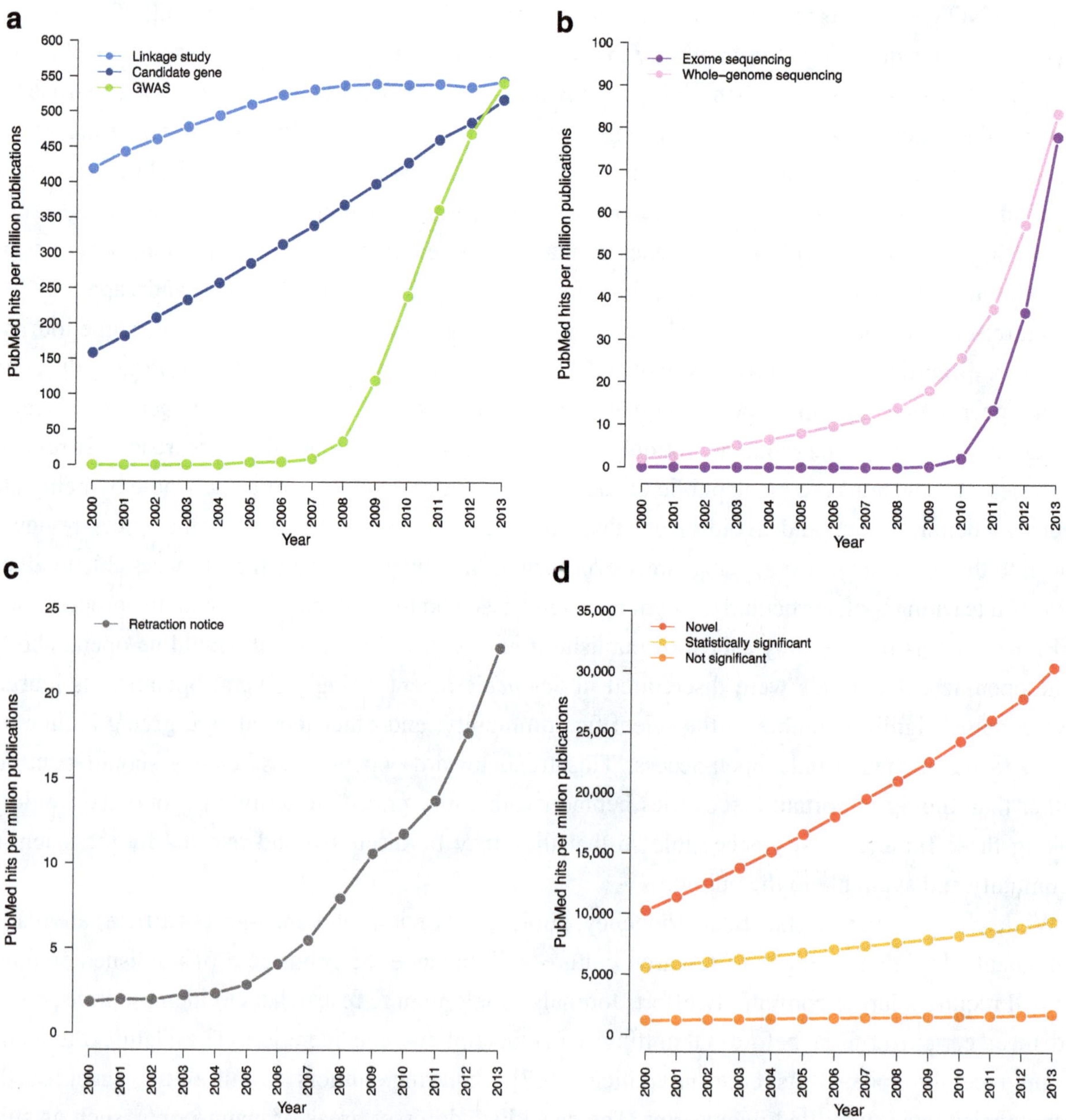

Retractions of genome-wide association and next-generation sequencing studies have been limited and should remain as such. To further increase the responsibility of journals in helping prevent retractions, the genetics community could establish a retraction index (similar to impact factor) for each journal. Such an index would track retraction rate and thus encourage journal editors to rigorously review findings and pursue a thorough peer-review process. Of course, researchers must take on greater responsibility for retractions, as well. Currently, retraction notices can be limited in information or completely ambiguous [34]. Instead, authors should be required to provide retraction notices that detail the steps that led to the retraction. The notice should be published as a brief article for the journal's readership to see and appended to the original manuscript. Requiring a detailed notice will encourage authors to be particularly critical of their own work and may help deter future retractions (Figure 4c).

Journals should also work to rectify publication bias towards novel findings (Figure 4d). A study on *de novo* variation in Autism Spectrum Disorder (which did not discover any disease-susceptibility loci) extensively investigated population stratification in rare variants and the effects of different meta-analysis approaches on power [68]. A recent WES paper on Type 2 Diabetes also did not discover disease-associated genes, but performed a host of analyses to determine likely etiological architecture [69]. Such findings should be made accessible to the entire community, as they can clarify our understanding of the genetic architecture of complex traits and improve future studies.

Journals can also be used to educate the scientific community in areas where it is lacking, such as applications of statistics and data interpretation. Three papers published together in the *Journal of the American Medical Association* explain how to interpret genome-wide association studies [70–72]. This fall, *Nature Methods* began publishing the column *Points of Significance* [73–76], which addresses important statistical concepts such as *p*-values, significance, and the relationship between sample size and power. *Nature* also recently ran a column that discussed the role of bias, the inexact nature of measurements, and the important distinction between correlation and causation [77]. Such articles do a great community service and will remain invaluable teaching tools to the many genetics researchers without formal training in statistics, allowing them to evaluate claims of novelty, both in their work and the work of others.

Of course, in addition to all of these larger changes, the field will hugely benefit from changes on a smaller scale. Senior scientists should encourage younger members of the field to conduct rigorous experiments, remain vigilant to prevent error, and seek additional help when in doubt about an analysis or result. Mistakes or "negative" findings should be met with discussion about how to improve a scientific question or refine an experiment. Randy Schekman, who won the Nobel Prize for physiology or medicine in 2013, has recently sparked public debate by announcing that he and all members of his lab will no longer be publishing in *Nature*, *Science*, and *Cell* [78]. Schekman's announcement, while controversial because of its potentially detrimental effects on the careers of his younger lab members, has an important motivation: to publicly address the adverse effects that publication bias and lack of open-access have had on the field. Making younger scientists aware of the aspects of scientific culture that are damaging to the quality of published and public science and prompting them to push for change can only alter the field for the better.

Fortunately, science in general and genetics in particular has proven to be a highly adaptive field. Peer-review is a relatively new process, becoming standard practice for most scientific journals in the second half of the 20th century [79]. In less than a decade, the foundation of the Public Library of

Science (PLoS) has helped to revolutionize publishing in science, providing the scientific community with a family of open-access journals and encouraging other journals to follow suit in its pursuit of freely available scientific literature [80]. In genetics specifically, the community acknowledged that candidate gene studies were poorly performing and established standards for more robust genome-wide association studies. Again, the tide has begun to shift as communities to discuss pre-prints continue to grow [81,82], researchers increasingly use social media to discuss many of the issues addressed here, and journals begin altering their editorial and publishing processes. With a concerted effort to improve published work, the field of human genetics is capable of permanent change.

Acknowledgements

The authors would like to thank Ron Do and Jan Veldink for helpful conversations during the writing of this manuscript.

Author Contributions

Conceived and wrote the paper: Sara L. Pulit, Maarten Leusink, Androniki Menelaou and Paul I. W. de Bakker.

References

1. Lander, E.S.; Linton, L.M.; Birren, B.; Nusbaum, C.; Zody, M.C.; Baldwin, J.; Devon, K.; Dewar, K.; Doyle, M.; FitzHugh, W.; *et al.* Initial sequencing and analysis of the human genome. *Nature* **2001**, *409*, 860–921.
2. Clinton, W.J. Remarks made by the President on the Completion of the First Survey of the Entire Human Genome Project. The White House Office of the Press Secretary: Washington, DC, USA, 2014; pp. 1–7.
3. Hirschhorn, J.N.; Lohmueller, K.; Byrne, E.; Hirschhorn, K. A comprehensive review of genetic association studies. *Genet. Med.* **2002**, *4*, 45–61.
4. Hirschhorn, J.N.; Altshuler, D. Once and again-issues surrounding replication in genetic association studies. *J. Clin. Endocrinol. Metab.* **2002**, *87*, 4438–4441.
5. Ioannidis, J.P.; Ntzani, E.E.; Trikalinos, T.A.; Contopoulos-Ioannidis, D.G. Replication validity of genetic association studies. *Nat. Genet.* **2001**, *29*, 306–309.
6. Kathiresan, S.; Newton-Cheh, C.; Gerszten, R.E. On the interpretation of genetic association studies. *Eur. Heart J.* **2004**, *25*, 1378–1381.
7. Lohmueller, K.E.; Pearce, C.L.; Pike, M.; Lander, E.S.; Hirschhorn, J.N. Meta-analysis of genetic association studies supports a contribution of common variants to susceptibility to common disease. *Nat. Genet.* **2003**, *33*, 177–182.
8. Hirschhorn, J.N.; Daly, M.J. Genome-wide association studies for common diseases and complex traits. *Nat. Rev. Genet.* **2005**, *6*, 95–108.
9. The International HapMap Consortium. A haplotype map of the human genome. *Nature* **2005**, *437*, 1299–1320.
10. Wang, W.Y.S.; Barratt, B.J.; Clayton, D.G.; Todd, J.A. Genome-wide association studies:

Theoretical and practical concerns. *Nat. Rev. Genet.* **2005**, *6*, 109–118.
11. Dudbridge, F.; Gusnanto, A. Estimation of significance thresholds for genomewide association scans. *Genet. Epidemiol.* **2008**, *32*, 227–234.
12. Pe'er, I.; Yelensky, R.; Altshuler, D.; Daly, M.J. Estimation of the multiple testing burden for genomewide association studies of nearly all common variants. *Genet. Epidemiol.* **2008**, *32*, 381–385.
13. Price, A.L.; Patterson, N.J.; Plenge, R.M.; Weinblatt, M.E.; Shadick, N.A.; Reich, D. Principal components analysis corrects for stratification in genome-wide association studies. *Nat. Genet.* **2006**, *38*, 904–909.
14. Ioannidis, J.P.A.; Boffetta, P.; Little, J.; O'Brien, T.R.; Uitterlinden, A.G.; Vineis, P.; Balding, D.J.; Chokkalingam, A.; Dolan, S.M.; Flanders, W.D.; *et al.* Assessment of cumulative evidence on genetic associations: interim guidelines. *Int. J. Epidemiol.* **2008**, *37*, 120–132.
15. Kraft, P.; Zeggini, E.; Ioannidis, J.P.A. Replication in genome-wide association studies. *Stat. Sci.* **2009**, *24*, 561–573.
16. Chanock, S.J.; Manolio, T.; Boehnke, M.; Boerwinkle, E.; Hunter, D.J.; Thomas, G.; Hirschhorn, J.N.; Abecasis, G.; Altshuler, D.; Bailey-Wilson, J.E.; *et al.* Replicating genotype-phenotype associations. *Nature* **2007**, *447*, 655–660.
17. Hindorff, L.A.; MacArthur, J.; Morales, J.; Junkins, H.A.; Hall, P.N.; Klemm, A.K.; Manolio, T.A. A Catalog of Published Genome-Wide Association Studies. Available online: www.genome.gov/gwastudies/ (accessed on 2 December 2013).
18. Visscher, P.M.; Brown, M.A.; McCarthy, M.I.; Yang, J. Five years of GWAS discovery. *Am. J. Hum. Genet.* **2012**, *90*, 7–24.
19. Sebastiani, P.; Solovieff, N.; Puca, A.; Hartley, S.W.; Melista, E.; Andersen, S.; Dworkis, D.A.; Wilk, J.B.; Myers, R.H.; Steinberg, M.H.; *et al.* Genetic signatures of exceptional longevity in humans. *Science* **2010**, doi:10.1126/science.1190532.
20. Carmichael, M. The little flaw in the longevity-gene study that could be a big problem. Available online: http://www.newsweek.com/little-flaw-longevity-gene-study-could-be-big-problem-74703/ (accessed on 10 December 2013).
21. Sebastiani, P.; Solovieff, N.; Puca, A.; Hartley, S.W.; Melista, E.; Andersen, S.; Dworkis, D.A.; Wilk, J.B.; Myers, R.H.; Steinberg, M.H.; *et al.* Letters: Retraction. *Science* **2011**, *333*, 404.
22. Unreliable research: Trouble at the Lab. *Economist* **2013**, *409*, 27.
23. Problems with scientific reasearch: How science goes wrong. *Economist* **2013**, *409*, 12.
24. Neaves, W. The roots of research misconduct. *Nature* **2012**, *488*, 121–122.
25. Macilwain, C. The time is right to confront misconduct. *Nature* **2012**, *488*, 7.
26. Corbyn, Z. Misconduct is the main cause of life-sciences retractions. *Nature* **2012**, *490*, 21.
27. Macilwain, C. Scientific misconduct: More cops, more robbers? *Cell* **2012**, *149*, 1417–1419.
28. Yong, E.; Ledford, H.; van Noorden, R. Research ethics: 3 ways to blow the whistle. *Nature* **2013**, *503*, 454–457.
29. The Wellcome Trust Case Control Consortium. Genome-wide association study of 14,000 cases of seven common diseases and 3000 shared controls. *Nature* **2007**, *447*, 661–678.
30. Zheng, G.; Freidlin, B.; Gastwirth, J.L. Robust genomic control for association studies. *Am. J. Hum. Genet.* **2006**, *78*, 350–356.

31. Clayton, D.G.; Walker, N.M.; Smyth, D.J.; Pask, R.; Cooper, J.D.; Maier, L.M.; Smink, L.J.; Lam, A.C.; Ovington, N.R.; Stevens, H.E.; *et al.* Population structure, differential bias and genomic control in a large-scale, case-control association study. *Nat. Genet.* **2005**, *37*, 1243–1246.
32. Plagnol, V.; Cooper, J.D.; Todd, J.A.; Clayton, D.G. A method to address differential bias in genotyping in large-scale association studies. *PLoS Genet.* **2007**, *3*, e74.
33. Steen, R.G. Retractions in the scientific literature: Is the incidence of research fraud increasing? *J. Med. Ethics* **2011**, *37*, 249–253.
34. Fang, F.C.; Steen, R.G.; Casadevall, A. Misconduct accounts for the majority of retracted scientific publications. *Proc. Natl. Acad. Sci. USA* **2012**, *109*, 17028–17033.
35. Ioannidis, J.P.A. Why most published research findings are false. *PLoS Med.* **2005**, *2*, e124.
36. Young, N.S.; Ioannidis, J.P.A.; Al-Ubaydli, O. Why current publication practices may distort science. *PLoS Med.* **2008**, *5*, e201.
37. Ioannidis, J.P.A. Concentration of the most-cited papers in the scientific literature: Analysis of journal ecosystems. *PLoS One* **2006**, *1*, e5.
38. Bjornshauge, L.; Brage, S.; Mitchell, D.; Zeylon, R. Directory of Open Access Journals. Available online: www.doaj.org/ (accessed on 23 December 2013).
39. Bohannon, J. Who's afraid of peer review? *Science* **2013**, *342*, 60–65.
40. Stahl, E.A.; Wegmann, D.; Trynka, G.; Gutierrez-Achury, J.; Do, R.; Voight, B.F.; Kraft, P.; Chen, R.; Kallberg, H.J.; Kurreeman, F.A.S.; *et al.* Bayesian inference analyses of the polygenic architecture of rheumatoid arthritis. *Nat. Genet.* **2012**, *44*, 483–489.
41. Hemani, G.; Yang, J.; Vinkhuyzen, A.; Powell, J.E.; Willemsen, G.; Hottenga, J.-J.; Abdellaoui, A.; Mangino, M.; Valdes, A.M.; Medland, S.E.; *et al.* Inference of the genetic architecture underlying BMI and height with the use of 20,240 sibling pairs. *Am. J. Hum. Genet.* **2013**, *93*, 865–875.
42. DePristo, M.A.; Banks, E.; Poplin, R.; Garimella, K.V.; Maguire, J.R.; Hartl, C.; Philippakis, A.A.; del Angel, G.; Rivas, M.A.; Hanna, M.; *et al.* A framework for variation discovery and genotyping using next-generation DNA sequencing data. *Nat. Genet.* **2011**, *43*, 491–498.
43. McKenna, A.; Hanna, M.; Banks, E.; Sivachenko, A.; Cibulskis, K.; Kernytsky, A.; Garimella, K.; Altshuler, D.; Gabriel, S.; Daly, M.; *et al.* The Genome Analysis Toolkit: A MapReduce framework for analyzing next-generation DNA sequencing data. *Genome Res.* **2010**, *20*, 1297–1303.
44. Do, R.; Kathiresan, S.; Abecasis, G.R. Exome sequencing and complex disease: Practical aspects of rare variant association studies. *Hum. Mol. Genet.* **2012**, *21*, R1–R9.
45. Mathieson, I.; McVean, G. Differential confounding of rare and common variants in spatially structured populations. *Nat. Genet.* **2012**, *44*, 243–246.
46. Kiezun, A.; Garimella, K.; Do, R.; Stitziel, N.O.; Neale, B.M.; McLaren, P.J.; Gupta, N.; Sklar, P.; Sullivan, P.F.; Moran, J.L.; *et al.* Exome sequencing and the genetic basis of complex traits. *Nat. Genet.* **2012**, *44*, 623–630.
47. Li, B.; Leal, S.M. Discovery of rare variants via sequencing: Implications for the design of complex trait association studies. *PLoS Genet.* **2009**, *5*, e1000481.
48. MacArthur, D.G.; Balasubramanian, S.; Frankish, A.; Huang, N.; Morris, J.; Walter, K.; Jostins, L.; Habegger, L.; Pickrell, J.K.; Montgomery, S.B.; *et al.* A systematic survey of loss-of-function variants in human protein-coding genes. *Science* **2012**, *335*, 823–828.

49. Kryukov, G.V.; Shpunt, A.; Stamatoyannopoulos, J.A.; Sunyaev, S.R. Power of deep, all-exon resequencing for discovery of human trait genes. *Proc. Natl. Acad. Sci. USA* **2009**, *106*, 3871–3876.
50. Chapman, J.M.; Cooper, J.D.; Todd, J.A.; Clayton, D.G. Detecting disease associations due to linkage disequilibrium using haplotype tags: A class of tests and the determinants of statistical power. *Hum. Hered.* **2003**, *56*, 18–31.
51. Scott-Van Zeeland, A.A.; Bloss, C.S.; Tewhey, R.; Bansal, V.; Torkamani, A.; Libiger, O.; Duvvuri, V.; Wineinger, N.; Galvez, L.; Darst, B.F.; *et al.* Evidence for the role of EPHX2 gene variants in anorexia nervosa. *Mol. Psychiatry* **2013**, doi:10.1038/mp.2013.91.
52. Dyment, D.A.; Cader, M.Z.; Chao, M.J.; Lincoln, M.R.; Morrison, K.M.; Disanto, G.; Morahan, J.M.; de Luca, G.C.; Sadovnick, A.D.; Lepage, P.; *et al.* Exome sequencing identifies a novel multiple sclerosis susceptibility variant in the TYK2 gene. *Neurology* **2012**, *79*, 406–411.
53. Ban, M.; Goris, A.; Lorentzen, A.R.; Baker, A.; Mihalova, T.; Ingram, G.; Booth, D.R.; Heard, R.N.; Stewart, G.J.; Bogaert, E.; *et al.* Replication analysis identifies TYK2 as a multiple sclerosis susceptibility factor. *Eur. J. Hum. Genet.* **2009**, *17*, 1309–1313.
54. Australia and New Zealand Multiple Sclerosis Genetics Consortium. Genome-wide association study identifies new multiple sclerosis susceptibility loci on chromosomes 12 and 20. *Nat. Genet.* **2009**, *41*, 824–828.
55. Abecasis, G.R.; Auton, A.; Brooks, L.D.; DePristo, M.A.; Durbin, R.M.; Handsaker, R.E.; Kang, H.M.; Marth, G.T.; McVean, G.A. An integrated map of genetic variation from 1092 human genomes. *Nature* **2012**, *491*, 56–65.
56. Erdmann, J.; Stark, K.; Esslinger, U.B.; Rumpf, P.M.; Koesling, D.; de Wit, C.; Kaiser, F.J.; Braunholz, D.; Medack, A.; Fischer, M.; *et al.* Dysfunctional nitric oxide signalling increases risk of myocardial infarction. *Nature* **2013**, *504*, 432–436.
57. Flannick, J.; Beer, N.L.; Bick, A.G.; Agarwala, V.; Molnes, J.; Gupta, N.; Burtt, N.P.; Florez, J.C.; Meigs, J.B.; Taylor, H.; *et al.* Assessing the phenotypic effects in the general population of rare variants in genes for a dominant Mendelian form of diabetes. *Nat. Genet.* **2013**, *45*, 1380–1385.
58. Chesi, A.; Staahl, B.T.; Jovičić, A.; Couthouis, J.; Fasolino, M.; Raphael, A.R.; Yamazaki, T.; Elias, L.; Polak, M.; Kelly, C.; *et al.* Exome sequencing to identify *de novo* mutations in sporadic ALS trios. *Nat. Neurosci.* **2013**, *16*, 851–855.
59. Neale, B.M.; Kou, Y.; Liu, L.; Ma'ayan, A.; Samocha, K.E.; Sabo, A.; Lin, C.-F.; Stevens, C.; Wang, L.-S.; Makarov, V.; *et al.* Patterns and rates of exonic *de novo* mutations in autism spectrum disorders. *Nature* **2012**, *485*, 242–245.
60. Gratten, J.; Visscher, P.M.; Mowry, B.J.; Wray, N.R. Interpreting the role of de novo protein-coding mutations in neuropsychiatric disease. *Nat. Genet.* **2013**, *45*, 234–238.
61. Yang, J.; Duan, S.; Zhong, R.; Yin, J.; Pu, J.; Ke, J.; Lu, X.; Zou, L.; Zhang, H.; Zhu, Z.; *et al.* Exome sequencing identified NRG3 as a novel susceptible gene of Hirschsprung's disease in a Chinese population. *Mol. Neurobiol.* **2013**, *47*, 957–966.
62. Lander, E.; Kruglyak, L. Genetic dissection of complex traits: Guidelines for interpreting and reporting linkage results. *Nat. Genet.* **1995**, *11*, 241–247.
63. Cruchaga, C.; Karch, C.M.; Jin, S.C.; Benitez, B.A.; Cai, Y.; Guerreiro, R.; Harari, O.; Norton, J.; Budde, J.; Bertelsen, S.; *et al.* Rare coding variants in the phospholipase D3 gene confer risk for

Alzheimer's disease. *Nature* **2013**, *505*, 550–554.

64. Panoutsopoulou, K.; Tachmazidou, I.; Zeggini, E. In search of low-frequency and rare variants affecting complex traits. *Hum. Mol. Genet.* **2013**, *22*, R16–R21.
65. Pulit, S.L.; Voight, B.F.; de Bakker, P.I.W. Multiethnic genetic association studies improve power for locus discovery. *PLoS One* **2010**, *5*, e12600.
66. Pulverer, B. Transparency showcases strength of peer review. *Nature* **2010**, *468*, 29–31.
67. Nature Neuroscience Editors. Raising standards. *Nat. Neurosci.* **2013**, *16*, 517.
68. Liu, L.; Sabo, A.; Neale, B.M.; Nagaswamy, U.; Stevens, C.; Lim, E.; Bodea, C.A.; Muzny, D.; Reid, J.G.; Banks, E.; *et al.* Analysis of rare, exonic variation amongst subjects with autism spectrum disorders and population controls. *PLoS Genet.* **2013**, *9*, e1003443.
69. Lohmueller, K.E.; Sparsø, T.; Li, Q.; Andersson, E.; Korneliussen, T.; Albrechtsen, A.; Banasik, K.; Grarup, N.; Hallgrimsdottir, I.; Kiil, K.; *et al.* Whole-exome sequencing of 2000 danish individuals and the role of rare coding variants in type 2 diabetes. *Am. J. Hum. Genet.* **2013**, *93*, 1072–1086.
70. Attia, J.; Ioannidis, J.P.A.; Thakkinstian, A.; McEvoy, M.; Scott, R.J.; Minelli, C.; Thompson, J.; Infante-Rivard, C.; Guyatt, G. How to use an article about genetic association: A: Background concepts. *JAMA* **2009**, *301*, 74–81.
71. Attia, J.; Ioannidis, J.P.A.; Thakkinstian, A.; McEvoy, M.; Scott, R.J.; Minelli, C.; Thompson, J.; Infante-Rivard, C.; Guyatt, G. How to use an article about genetic association: B: Are the results of the study valid? *JAMA* **2009**, *301*, 191–197.
72. Attia, J.; Ioannidis, J.P.A.; Thakkinstian, A.; McEvoy, M.; Scott, R.J.; Minelli, C.; Thompson, J.; Infante-Rivard, C.; Guyatt, G. How to use an article about genetic association: C: What are the results and will they help me in caring for my patients? *JAMA* **2009**, *301*, 304–308.
73. Krzywinski, M.; Altman, N. Points of significance: Importance of being uncertain. *Nat. Methods* **2013**, *10*, 809–810.
74. Krzywinski, M.; Altman, N. Points of significance: error bars. *Nat. Methods* **2013**, *10*, 921–922.
75. Krzywinski, M.; Altman, N. Points of significance: Significance, P values and t-tests. *Nat. Methods* **2013**, *10*, 1041–1042.
76. Krzywinski, M.; Altman, N. Points of significance: Power and sample size. *Nat. Methods* **2013**, *10*, 1139–1140.
77. Sutherland, W.J.; Spiegelhalter, D.; Burgman, M.A. Policy: Twenty tips for interpreting scientific claims. *Nature* **2013**, *503*, 335–337.
78. Schekman, R. "How journals like Nature, Cell and Science are damaging science." Available online: http://www.theguardian.com/commentisfree/2013/dec/09/how-journals-nature-science-cell-damage-science/ (accessed on 20 December 2013).
79. Chapelle, F.H. The history and practice of peer review. *Ground Water* **2014**, *52*, 1.
80. Van Noorden, R. Open access: The true cost of science publishing. *Nature* **2013**, *495*, 426–429.
81. Coop, G.; Howie, B.; Pickrell, J. Haldane's Sieve. Available online: haldanessieve.org/ (accessed on 20 January 2014).
82. Cornell University Library. arXiv. Available online: www.arxiv.org/ (accessed on 20 January 2014).

6

Changing Histopathological Diagnostics by Genome-Based Tumor Classification

Michael Kloth [1,2,]* and Reinhard Buettner [1,2]

[1] Institute of Pathology, University Hospital Cologne, Kerpener Str. 62, Cologne D-50937, Germany; E-Mail: reinhard.buettner@uk-koeln.de

[2] Center for Integrated Oncology Cologne-Bonn, Cologne D-50937, Germany

* Author to whom correspondence should be addressed; E-Mail: michael.kloth@uk-koeln.de;

Abstract: Traditionally, tumors are classified by histopathological criteria, *i.e.*, based on their specific morphological appearances. Consequently, current therapeutic decisions in oncology are strongly influenced by histology rather than underlying molecular or genomic aberrations. The increase of information on molecular changes however, enabled by the Human Genome Project and the International Cancer Genome Consortium as well as the manifold advances in molecular biology and high-throughput sequencing techniques, inaugurated the integration of genomic information into disease classification. Furthermore, in some cases it became evident that former classifications needed major revision and adaption. Such adaptations are often required by understanding the pathogenesis of a disease from a specific molecular alteration, using this molecular driver for targeted and highly effective therapies. Altogether, reclassifications should lead to higher information content of the underlying diagnoses, reflecting their molecular pathogenesis and resulting in optimized and individual therapeutic decisions. The objective of this article is to summarize some particularly important examples of genome-based classification approaches and associated therapeutic concepts. In addition to reviewing disease specific markers, we focus on potentially therapeutic or predictive markers and the relevance of molecular diagnostics in disease monitoring.

Keywords: human genome project; cancer genomics; tumor classification; targeted therapies

1. Approaches to a Genome-Based Tumor Classification

1.1. The 2008 WHO Classification of Hematological Malignancies

The 2008 WHO classification of chronic myeloid malignancies is at present the most evolved approach to a taxonomy considering defined molecular aberrations [1]. The malignancies that are included are now classified into five categories:

(1) Acute myeloid leukemia (AML) and related precursor neoplasms;
(2) Myelodysplastic syndromes (MDS);
(3) Myeloproliferative neoplasms (MPN);
(4) Myelodysplastic/Myeloproliferative neoplasms (MDS/MPN);
(5) Myeloid and lymphoid neoplasms with eosinophilia and abnormalities of *PDGFRA*, *PDGFRB*, or *FGFR1*.

The integration of histology and genetics is particularly visible in the category of myeloproliferative neoplasms (MPN). Classification of specific entities into MPN is dependent on presence or absence of *BCR-ABL1*, the disease-causing translocation in CML [2]. The first description of the associated karyotype t(9;22)(q34;q11), according to an abnormally short chromosome 22, was published as early as 1960 and is widely known as the Philadelphia (Ph) chromosome [3]. Due to the high specificity of *BCR-ABL1,* its detection is mandatory for diagnosis of CML and is further underscored by the influence on therapy with the small molecule tyrosine kinase inhibitor (TKI) Imatinib [4]. Nevertheless, the remarkable journey, from the genomic aberration to the specific therapy, required approximately forty years and started long before the elucidation of the human genome (Figure 1).

Figure 1. Timeline of the elucidation of genomic alterations in myeloproliferative neoplasms. Major breakthroughs in the understanding of the Ph+ neoplasm CML are depicted below the line, those in the understanding of Ph- neoplasms above. Note the significant impact of the human genome project on the elucidation on Ph- specific genomic alterations.

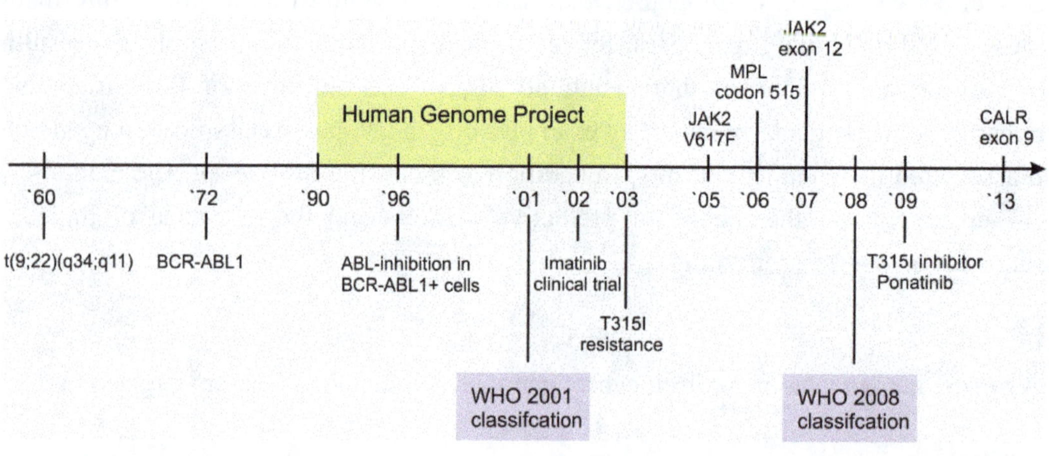

Further myeloproliferative neoplasms are characterized by the absence of *BCR-ABL1*, including the three Ph-negative classic MPNs—polycythemia vera (PV), essential thrombocythemia (ET) and progressive myelofibrosis (PMF). Despite the absence of *BCR-ABL1*, it has consecutively been shown that these MPNs are themselves also characterized by additional recurrent aberrations [5,6] (Figure 1).

The *JAK2 V617F* mutation is detectable in approximately 95% of all PV patients and 50% of both ET and PMF patients [7–9]. It is further detected in patients with refractory anemia with ring sideroblasts and thrombocytosis (RARS-T), but in less than 5% of patients with acute myeloid leukemia (AML) or myeloid dysplastic syndrome (MDS) and not in solid tumors [10,11]. Although, the diagnostic process is initially dominated by peripheral blood cell count and serum erythropoetin (EPO) levels, *JAK2 V617F* or less common *JAK2 Exon 12* mutations [12] confirm the diagnosis of a suspected PV without the need of a bone marrow biopsy [13]. Beside the two most important genetic aberrations in the diagnostic algorithm of MPNs, namely *BCR-ABL1* and *JAK2 V617F*, several other potentially helpful recurrent aberrations are known. These include the presence of recurrent mutations in *CALR exon 9* [14] and *MPL codon 515* [15], essentially occurring in *JAK2 V617F* negative cases (Figure 2).

Figure 2. Diagnostic algorithm of classic myeloproliferative neoplasms using specific molecular aberrations. Detection of the molecular aberrations depicted above is highly suggestive for the suspected myeloproliferative disorder. Nevertheless, at least in the case of absence of these specific aberrations, a bone marrow biopsy should be performed.

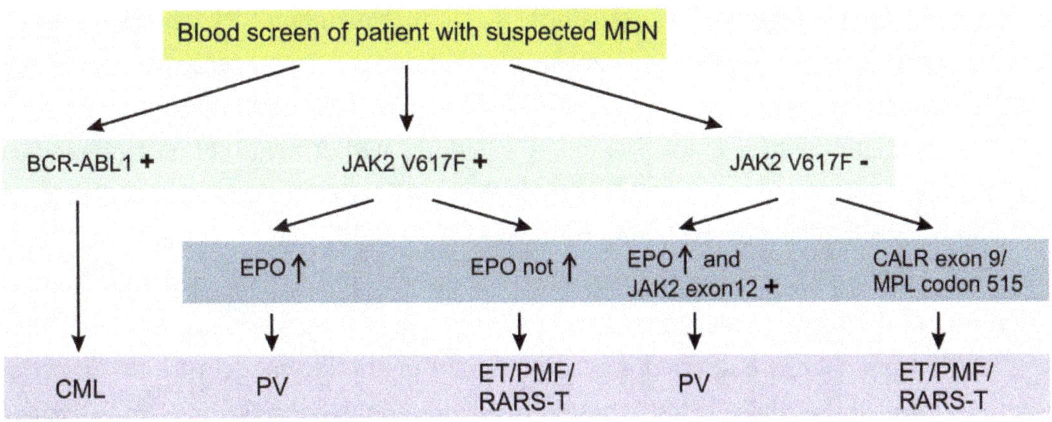

Another important example of the integration of molecular aberrations in the 2008 WHO classification of hematological malignancies is the newly introduced group of myeloid and lymphoid neoplasms with eosinophilia and abnormalities of *PDGFRA*, *PDGFRB*, or *FGFR1*. This reclassification highlights the consideration of the targetable alterations *FIP1L1-PDGFRA* or *PDGFRB*-rearrangements and those harboring *FGFR1*-rearrangements, indicating response or resistance to Imatinib [2,16].

1.2. Lung Cancer as a Paradigm: Advances in the Molecular Characterization of Solid Malignancies

The ongoing comprehensive characterizations of solid tumors, such as those conducted by The Cancer Genome Atlas (TCGA) will significantly impact upcoming tumor classifications. This also includes lung cancer, the leading cause of cancer death worldwide [17] and an example for substantial advances by genome-based therapy approaches [18]. In general, NSCLC is subclassified into adenocarcinoma, squamous cell carcinoma and large cell carcinoma. Future classifications need to characterize clinically relevant subtypes, instead of the traditional distinction of non-small cell lung cancers (NSCLC) and small-cell lung cancer (SSLC). Concepts for a reclassification of lung adenocarcinoma were suggested, particularly with respect to reclassify large cell carcinoma based on genomic aberrations into adenocarcinoma, squamous cell carcinoma and large cell neuroendocrine carcinoma, respectively. Recommendations for (immuno)histological diagnostic work-up, and also for determining specific molecular aberrations in lung adenocarcinoma subtypes have been published [19].

Major rationales for changes in adenocarcinoma histologic variants are that the invasive mucinous type shows frequent mutations in *KRAS* and hardly any in *EGFR*, whereas non-mucinous adenocarcinoma of predominant lepidic subtype is characterized by frequent mutations in *EGFR* and fewer in *KRAS* [20]. In this context, it is worth noting and described in more detail below, that clinical studies with the TKIs Gefitinib and Erlotinib showed significantly improved survival in patients suffering from lung adenocarcinoma harboring mutations in the kinase domain of *EGFR* [21,22]. Besides the mutations in *EGFR* and *KRAS*, the targetable translocation *EML4-ALK* recurrently occurs in lung adenocarcinoma [23] and is most frequent in tumours with mucinous signet-ring appearance. Clinical trials using the tyrosine kinase inhibitors Crizotinib and Ceritinib have now shown improved progression-free survival in those patients [24,25].

In contrast, squamous cell carcinoma of the lung is commonly associated with mutations in the *NFE2L2/KEAP*-axis [26] and often harbors targetable alterations of *FGFR1* [27] and in fewer cases mutations in *DDR2*. Interestingly, DDR2-transformed cell lines maintain SRC phosphorylation and are sensitive to Dasatinib [28], proved by the response in squamous cell lung cancer patients [29].

Beyond the combined loss of *RB1 and TP53* in neuroendocrine pulmonary tumors [30], it was shown that low- and intermediate-grade pulmonary cacinoids harbor recurrent mutations in chromatin remodeling complexes [31], whereas the high-grade neuroendocrine tumor, small cell lung cancer, is associated with sequential changes, including the deletion of *PTEN* [32].

Additionally, we assessed cancer genome alterations linked to histomorphological and immunohistochemical features, considering high therapeutic relevance and improved patient outcome [33] (Figure 3). By this approach, we devised a genomic-based prediction model of lung cancer subtypes. This model shows that the majority of large cell cancers could be reassigned to adenocarcinoma, squamous cell carcinoma or small-cell lung cancer. By the combined analysis of immunohistochemical, genomic and clinical features it becomes further evident that personalized approaches significantly improve the outcome of patients with advanced lung cancer and other solid cancers.

Figure 3. Frequencies of significant genomic alterations in histological subgroups of lung cancer. Colors of histological subtypes are encoded as follows: green—squamous cell lung cancer (SQ), purple—carcinoid tumor, light blue—large cell lung cancer (LC), red—adenocarcinoma of the lung (AD), dark blue—small cell lung cancer (SCLC). Data adapted from [33].

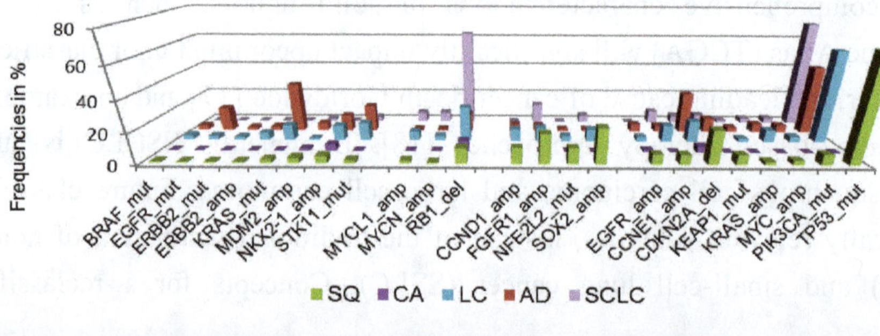

Beyond the principally well-characterized situation in lung cancer, we already know of highly recurrent genomic alterations in many other solid malignancies. Prostate cancer is the most common cancer in men and is characterized by fewer mutations in typical cancer genes, when compared to other solid cancers [34–36]. However, genomic alterations in androgen signaling, the rearrangement of *ETS* transcription factors, especially the fusion of *TMPRSS2-ERG* [37], as well as the deletion of *PTEN* [38] are known to be highly recurrent in primary cancers [39], early-onset cancers [40] and castration-resistant prostate cancers [41]. Beside the outstanding therapeutic role of androgen deprivation, recent efforts investigate a mechanistic rationale of PARP inhibition in *ETS*-rearranged prostate cancers [42].

1.3. EWS and the Importance of Translocations in the Diagnostic Workup of Mesenchymal Malignancies

Tumors ascribed to the family of Ewing's sarcomas or primitive neuroectodermal tumors (PNET) are the second most common bone tumors in children, but can also arise from any other tissue. The recent molecular understanding of these aggressive tumors has greatly advanced new therapeutic approaches. However, whereas chemotherapy improved the survival rate from 10% to 70%–80% in localized disease, the survival of patients with distant metastases is still poor [43]. The WHO classification of Ewing sarcoma (ES) as a single entity is underlined by its cytogenetic signature t(11;12)(q24;q12), according to the translocation *EWSR1-FLI1* in approximately 85% of patients. Although, *EWSR1* translocations can be found in many other mesenchymal tumors, almost all of the remaining ES cases are characterized by further ES-specific *EWSR1* translocations [44,45]. This also includes the second most common cytogenetic aberration t(21;22)(q22;q12), corresponding to *EWSR1-ERG* [46] (Table 1).

Intriguingly, the well-characterized molecular situation of ES is in contrast to the fact that the cell of origin of the small round cell appearing tumor is not known. A further complication of the histological diagnosis is caused by atypical ES, including large/epithelioid/clear/spindle cell ES, vascular-like ES, adamantinoma-like ES, ES with neuroectodermal features, synovial sarcoma-like PNET and sclerosing PNET [47]. Beyond this difficult histopathological classification, sarcomas may be genetically classified from a near-diploid karyotpe to a more complex genomic instability. The first is characterized by highly recurrent translocations, the latter by numerical and structural abnormalities affecting multiple chromosomes [48]. In the case of small round cell appearing sarcoma, recently, new subtypes were defined by recurrent translocations of BCOR-CCNB3 [49] and CIC-DUX4 [50].

Furthermore, we already know several diagnostically relevant genomic rearrangements in soft tissue malignancies [51]. The recurrently occurring reciprocal translocation t(X;18) is characteristic of synovial sarcoma, and leads to the potentially therapeutic relevant SSX-S18 protein [52]. The translocation *FUS-CHOP* is detected in myxoid liposarcoma [53] and recurrent amplifications of the E3-Ubiqutitin ligase *MDM2*, as well as *CDK4*, in well differentiated and dedifferentiated liposarcoma [54,55]. Interestingly, ongoing efforts investigate the reactivation of p53 by MDM2-p53 interaction inhibitors [56].

Table 1. Fusion partners of *EWSR1*-rearrangements in different soft tissue tumors. Overlapping rearrangements in different histopathological entities are in bold. Adapted from [44,45].

Histological type	Translocation	EWS-rearrangements
Ewing's Sarcoma	t(11;22)(q24;q12)	EWSR1-FLI1
	t(21;22)(q22;q12)	EWSR1-ERG
	t(7;22)(q22;q12)	EWSR1-ETV1
	t(17;22)(q21;q12)	EWSR1-ETV4
	t(2;22)(q36;q12)	EWSR1-FEV
	inv(22)(q12q12)	EWSR1-PATZ1
	t(2;22)(q31;q12)	EWSR1-SP3
	t(20;22)(q13;q12)	EWSR1-NFATC2
	t(4;22)(q31;12)	EWSR1-SMARCA5
	t(17;22)(q12;q12)	EWSR1-E1AF
	inv(22)(q21;12)	EWSR1-ZSG
Angiomatoid fibrous histiocytoma	t(12;22)(q13;q12)	**EWSR1-ATF1**
	t(2;22)(q33;q12)	**EWSR1-CREB1**
Clear cell sarcoma	t(12;22)(q13;q12)	**EWSR1-ATF1**
	t(2;22)(q33;q12)	**EWSR1-CREB1**
Malignant gastrointestinal neuroectodermal tumor	t(12;22)(q13;q12)	**EWSR1-ATF1**
	t(2;22)(q33;q12)	**EWSR1-CREB1**
Myoepithelial tumor of soft tissue and bone	t(1;22)(q23;q12)	EWSR1-PBX1
	t(19;22)(q13;q12)	EWSR1-ZNF444
	t(6;22)(p21;q12)	EWSR1-POU5F1
Extraskeletal myxoid chondrosarcoma	t(9;22)(q22;q12)	EWSR1-NR4A3
Myxoid liposarcoma	t(12;22)(q13;q12)	EWSR1-DD1T3

1.4. Cancer of Unknown Primary Origin (CUP)

Carcinoma of an unknown primary origin (CUP) is descriptive of a metastatic cancer without an identifiable primary tumor site. CUP accounts for 3%–5% of all cancer diagnoses and is usually characterized by an aggressive metastatic growth and a challenging clinical presentation. In theory, CUP could be considered as a unique biological entity, or in the opposite view, as a group of different entities. The classification of CUP is essentially based on the prognostic outcome, thereby distinguishing favorable and unfavorable carcinomas [57–59] (Table 2).

Table 2. Classification of cancers of unknown primary origin.

Clinically favorable CUP	Clinically unfavorable CUP
Extragonadal germ-cell cancer	Metastatic adenocarcinoma
Peritoneal papillary adenocarcinoma	Non papillary malignant ascites
Adenocarcinoma in axillary lymph nodes	Multiple cerebral metastases
Cervical squamous-cell carcinoma	Squamous-cell carcinoma of the abdominopelvic cavity
Neuroendocrine carcinoma	Lytic bone metastases
Blastic bone metastases and PSA elevation	

This classification predominantly depends on the morphological and immunohistochemical appearance. At present, the specimen is investigated by the use of specific antibodies (investigated epitopes in parentheses):

(1) Identification of the cancer type:
Carcinoma (CK AE1/3), mesothelioma (Calretinin, BerEP4), sarcoma (Vimentin), lymphoma (LCA), melanoma (HMB-45, MITF, S100);

(2) Identification of the subtype:
Adenocarcinoma (CK7, CK20), squamous cell carcinoma (CK5/6, p40, p63), hepatocellular carcinoma (Heparl), renal cell carcinoma (RCC, PAX8, CA9), urothelial carcinoma (GATA3, S100P, Uroplakin), thyroid cancer (hTG, TTF1), neuroendocrine cancer (CD56, Synaptophysin, ChromoA), germ-cell tumor (PLAP);

(3) Identification of the origin:
Lung (TTF1, NapsinA), colorectal cancer (CDX2, CK20), breast (ER, PR), pancreas (CDX2, CK7, CK20), ovary (Ca125, ER, WT1), prostate (PSA, PSAP, AR).

However, only 30% of all cases studied can be assigned to a primary tumor and even fewer benefit from a change in therapy regimen. Over the last decade, the molecular characterization by gene-expression profiling in various tumor types has led to the development of several gene signature assays with identification rates of a putative tissue of origin in up to 90% of cases. Furthermore, it was shown that CUP can be treated along actionable genomic alterations and recent whole exome sequencing revealed the existence of known recurrent mutations [60]. These include therapeutically relevant mutations in *PIK3CA*, *MET*, *FGFR3*, *IDH1* as well as several others and it has been already demonstrated that targeted therapies can significantly influence a more favorable outcome in CUP patients [61]. As another variation on this theme, a drug-sensitizing genome alteration in one tumor type may not confer drug susceptibility in another histology, as has been observed in the case of *BRAF* mutations that confer *MEK* and *BRAF* dependency in melanomas [62,63] but not in colorectal carcinomas resulting from *EGFR* activation [64]. These interactions highlight the need for classifications integrating cancer genome alterations, but also histomorphological and immunohistochemical features.

2. Monitoring of Malignancies

Molecular monitoring of CML is the most advanced routinely used surveillance strategy and reflects the need for standardization and quality controls of diagnostic tests. A main objective is the identification of patients, which have a worse response or resistance to therapy with tyrosine kinase inhibitors. The quantification of *BCR-ABL1* transcripts in peripheral blood is thereby of outstanding value in the early phase after initial drug administration, corresponding to the molecular response rate (MMR) upon TKI therapy [65–67]. The MMR is calculated in relation to a control gene (e.g., *BCR* or *ABL1*) and was standardized between laboratories using the international scale (IS) [68]. Essentially, the WHO has undertaken extensive efforts to simplify and standardize the assay by providing reference reagents. The importance of such efforts is reflected by a comparable poor overall survival of patients with an early MMR >10%, in turn leading to universally valid changes in the associated NCCN and ELN guidelines [66].

Despite significant progress in therapy of lung adenocarcinoma, all patients with *EGFR* mutations and *ALK* or *ROS1* translocations receiving specific tyrosine kinase inhibitors will ultimately experience relapse. Recent work highlights the potential of noninvasive detection and monitoring of resistance mutations in free circulating plasma DNA of lung cancer patients. The most prominent example is the known resistance mechanisms mediated by T790M in *EGFR* [69]. Beside the potential of directly targeting cancers harboring T790M by new tyrosine kinase inhibitors [70], it is important to note that T790M cells proliferate more slowly, thereby enabling resensitizing of tumors to primarily used TKIs after temporary withdrawal of the drug. Since it is difficult in clinical practice to undertake repeat biopsies at multiple metastatic sites, noninvasive targeted sequencing techniques may further enable additional therapeutic strategies [71]. It thereby becomes evident that new diagnostic techniques will not only lead to major influences in treatment and monitoring guidelines, but will influence therapeutic guidelines and also the classification of tumors. Especially, the opportunity of non-invasive testing seems to be an attractive and emerging field in diagnostic and therapeutic concepts, as further delineated by testing for *TMPRSS2-ERG* in the case of suspicion of prostate cancer [72].

3. Clinical Success of Targeted Therapeutic Approaches Based on Molecular Biomarkers

3.1. BCR-ABL1

As depicted above, *BCR-ABL1* is the driving lesion in CML, leading to the development of the first small molecule TKI, Imatinib [4,73]. Several multicenter studies confirmed that the overall survival of advanced-stage cancer patients in a chronic disease phase climbed from approximately 50% before 2002 up to approximately 90% after introduction of Imatinib in 2002 [74]. The success of the targeted approach in CML patients is further underscored by ongoing studies investigating the discontinuation of Imatinib [75] and the groundbreaking question of a potential cure [76]. Despite these promising developments, several further therapeutic possibilities for second line therapies are underway. These include the use of second generation TKIs in case of therapeutic failure or intolerance, e.g., Dasatinib, Nilotinib, Bosutinib and Ponatinib, the latter particularly used in case of a secondary resistance by T315I mutation [75] (Figure 1).

3.2. BRAF

The most common melanoma mutation in *BRAF exon 15*, the activating mutation V600E, leads to response rates of more than 50% of all patients treated with the specific TKI Vemurafenib. Further, therapy with Vemurafenib is associated with a relative reduction in death of 63% when compared to standard therapeutic regime with Dacarbazine [62]. In contrast to the activating biology of *BRAF V600E*, ongoing clinical trials investigate the potency of Dasatinib in tumors harboring inactivating *BRAF exon 11* mutations (NCT01514864 clinicaltrials.gov).

3.3. EGFR-Family and KRAS Mutational Status

Genetic aberrations in members of the *EGFR*-family are well known for targeted therapies, including HER2- and EGFR-targeted inhibition of downstream signaling cascades. *HER2/ERBB2* is primary known as being amplified and activated in breast cancer causing high recurrence rates and increased mortality in approximately 15% of all patients [77]. Patients with amplification of *HER2/ERBB2* treated with the monoclonal antibody Trastuzumab in combination with chemotherapy showed improved outcome in several studies [78]. Beyond Trastuzumab several ongoing studies are investigating further drug therapies targeting the HER2-axis. These include the combined inhibition of HER2/HER3-heterodimerization and activation by Trastuzumab/Pertuzumab [79] and the use of the covalent immunoconjugate Trastuzumab-Emtansine (T-DM1) [80] (Figure 4). Moreover, several trials evaluate the therapeutic significance of small molecule inhibition in *HER2*-positive breast cancer, e.g., Lapatinib, Afatinib, Pazopanib and Neratinib [81].

Figure 4. Predictive Biomarkers for Targeted and Selective Therapies. Signaling of EGFR-family receptors is characterized by homo-/heterodimerization and subsequent activation of the targetable downstream signaling pathways RAS/RAF and PI3K/AKT. Present therapeutic approaches focus on the inhibition of ligand-dependent activation, dimerization and receptor tyrosine kinases. Immunoconjugates, e.g., T-DM1, specifically deliver chemotherapeutic agents by the process of receptor internalization. As described in the text in more detail, ongoing efforts investigate the effectiveness of combined or dual approaches.

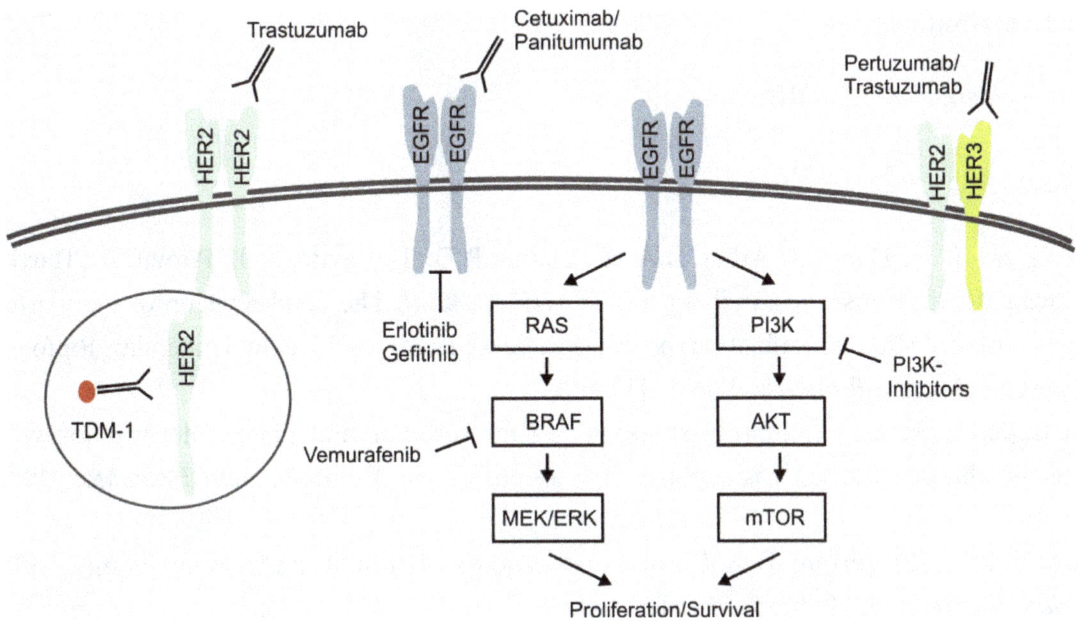

Comparable to the situation in *HER2*-amplified breast cancer, substantial progress has been made by the introduction of EGFR-targeted therapies in the treatment of lung cancer and colorectal cancer. These efforts become evident by comparing the median overall survival of lung adenocarcinoma patients under standard therapeutic regimes of approximately 12 months with approximately 2 years under EGFR-targeted therapy with Erlotinib or Gefitinib in *EGFR*-mutated cancers [82]. As depicted above, further efforts are made to overcome primary and secondary therapeutic resistance by next generation TKIs [70]. Similar positive achievements were made for treatment of metastatic colorectal cancer (mCRC) by an improvement of survival from 12 months with fluoruracil monotherapy up to approximately 2 years with EGFR/VEGFR-targeted therapy combined with chemotherapy [83].

Beyond that, it recently became evident that we need to predict therapeutic response to cetuximab/panitumumab in mCRC not only by *KRAS* mutational status, but also by *NRAS* mutational status [84], highlighting the increasing importance of mutations in downstream or interacting pathways. As depicted above, it becomes also clear that combined approaches, like the inhibition of the EGFR/BRAF-axis in *BRAF V600E* mutated colorectal cancers [64], could be used to overcome primary resistance in histological subtypes.

4. Conclusions

The purpose of integrating pathogenetic and molecular information into disease classification systems, exemplified by the 2008 WHO classification of hematological malignancies, reflects the high clinical relevance for predicting therapy outcome and prognosis. The human genome project and emerging technologies in the last decade have led to fundamental pathogenetic breakthroughs, which substantially improved the translation into clinical practice and individual therapeutic possibilities. Altogether, this data underlines the significant influence of cancer genomics and the substantial increase in genomic information on the process of defining tumor entities and effective and selective treatment approaches.

Author Contributions

Wrote the paper: MK and RB.

References

1. Vardiman, J.W.; Thiele, J.; Arber, D.A.; Brunning, R.D.; Borowitz, M.J.; Porwit, A.; Harris, N.L.; le Beau, M.M.; Hellstrom-Lindberg, E.; Tefferi, A.; *et al.* The 2008 revision of the world health organization (who) classification of myeloid neoplasms and acute leukemia: Rationale and important changes. *Blood* **2009**, *114*, 937–951.
2. Cross, N.C.; Reiter, A. Fibroblast growth factor receptor and platelet-derived growth factor receptor abnormalities in eosinophilic myeloproliferative disorders. *Acta Haematol.* **2008**, *119*, 199–206.
3. Nowell, P.C.; Hungerford, D.A. Chromosome studies on normal and leukemic human leukocytes. *J. Natl. Cancer Inst.* **1960**, *25*, 85–109.
4. Kantarjian, H.; Sawyers, C.; Hochhaus, A.; Guilhot, F.; Schiffer, C.; Gambacorti-Passerini, C.; Niederwieser, D.; Resta, D.; Capdeville, R.; Zoellner, U.; *et al.* Hematologic and cytogenetic responses to imatinib mesylate in chronic myelogenous leukemia. *N. Engl. J. Med.* **2002**, *346*, 645–652.
5. Tefferi, A.; Skoda, R.; Vardiman, J.W. Myeloproliferative neoplasms: Contemporary diagnosis using histology and genetics. *Nat. Rev. Clin. Oncol.* **2009**, *6*, 627–637.
6. Vannucchi, A.M.; Guglielmelli, P.; Tefferi, A. Advances in understanding and management of myeloproliferative neoplasms. *CA Cancer J. Clin.* **2009**, *59*, 171–191.
7. Levine, R.L.; Wadleigh, M.; Cools, J.; Ebert, B.L.; Wernig, G.; Huntly, B.J.; Boggon, T.J.;

Wlodarska, I.; Clark, J.J.; Moore, S.; *et al.* Activating mutation in the tyrosine kinase jak2 in polycythemia vera, essential thrombocythemia, and myeloid metaplasia with myelofibrosis. *Cancer Cell* **2005**, *7*, 387–397.

8. Kralovics, R.; Passamonti, F.; Buser, A.S.; Teo, S.S.; Tiedt, R.; Passweg, J.R.; Tichelli, A.; Cazzola, M.; Skoda, R.C. A gain-of-function mutation of jak2 in myeloproliferative disorders. *N. Engl. J. Med.* **2005**, *352*, 1779–1790.
9. James, C.; Ugo, V.; Le Couedic, J.P.; Staerk, J.; Delhommeau, F.; Lacout, C.; Garcon, L.; Raslova, H.; Berger, R.; Bennaceur-Griscelli, A.; *et al.* A unique clonal jak2 mutation leading to constitutive signalling causes polycythaemia vera. *Nature* **2005**, *434*, 1144–1148.
10. Szpurka, H.; Tiu, R.; Murugesan, G.; Aboudola, S.; Hsi, E.D.; Theil, K.S.; Sekeres, M.A.; Maciejewski, J.P. Refractory anemia with ringed sideroblasts associated with marked thrombocytosis (rars-t), another myeloproliferative condition characterized by jak2 v617f mutation. *Blood* **2006**, *108*, 2173–2181.
11. Nishii, K.; Nanbu, R.; Lorenzo, V.F.; Monma, F.; Kato, K.; Ryuu, H.; Katayama, N. Expression of the jak2 v617f mutation is not found in de novo aml and mds but is detected in mds-derived leukemia of megakaryoblastic nature. *Leukemia* **2007**, *21*, 1337–1338.
12. Scott, L.M.; Tong, W.; Levine, R.L.; Scott, M.A.; Beer, P.A.; Stratton, M.R.; Futreal, P.A.; Erber, W.N.; McMullin, M.F.; Harrison, C.N.; *et al.* Jak2 exon 12 mutations in polycythemia vera and idiopathic erythrocytosis. *N. Engl. J. Med.* **2007**, *356*, 459–468.
13. Tefferi, A.; Vainchenker, W. Myeloproliferative neoplasms: Molecular pathophysiology, essential clinical understanding, and treatment strategies. *J. Clin. Oncol.* **2011**, *29*, 573–582.
14. Klampfl, T.; Gisslinger, H.; Harutyunyan, A.S.; Nivarthi, H.; Rumi, E.; Milosevic, J.D.; Them, N.C.; Berg, T.; Gisslinger, B.; Pietra, D.; *et al.* Somatic mutations of calreticulin in myeloproliferative neoplasms. *N. Engl. J. Med.* **2013**, *369*, 2379–2390.
15. Pardanani, A.D.; Levine, R.L.; Lasho, T.; Pikman, Y.; Mesa, R.A.; Wadleigh, M.; Steensma, D.P.; Elliott, M.A.; Wolanskyj, A.P.; Hogan, W.J.; *et al.* Mpl515 mutations in myeloproliferative and other myeloid disorders: A study of 1182 patients. *Blood* **2006**, *108*, 3472–3476.
16. Gotlib, J. World health organization-defined eosinophilic disorders: 2011 update on diagnosis, risk stratification, and management. *Am. J. Hematol.* **2011**, *86*, 677–688.
17. Siegel, R.; Naishadham, D.; Jemal, A. Cancer statistics, 2013. *CA Cancer J. Clin.* **2013**, *63*, 11–30.
18. Buettner, R.; Wolf, J.; Thomas, R.K. Lessons learned from lung cancer genomics: The emerging concept of individualized diagnostics and treatment. *J. Clin. Oncol.* **2013**, *31*, 1858–1865.
19. Travis, W.D.; Brambilla, E.; Noguchi, M.; Nicholson, A.G.; Geisinger, K.R.; Yatabe, Y.; Beer, D.G.; Powell, C.A.; Riely, G.J.; van Schil, P.E.; *et al.* International association for the study of lung cancer/american thoracic society/european respiratory society international multidisciplinary classification of lung adenocarcinoma. *J. Thorac. Oncol.* **2011**, *6*, 244–285.
20. Hata, A.; Katakami, N.; Fujita, S.; Kaji, R.; Imai, Y.; Takahashi, Y.; Nishimura, T.; Tomii, K.; Ishihara, K. Frequency of egfr and kras mutations in japanese patients with lung adenocarcinoma with features of the mucinous subtype of bronchioloalveolar carcinoma. *J. Thorac. Oncol.* **2010**, *5*, 1197–1200.
21. Roberts, P.J.; Stinchcombe, T.E. Kras mutation: Should we test for it, and does it matter? *J. Clin. Oncol.* **2013**, *31*, 1112–1121.

22. Rosell, R.; Carcereny, E.; Gervais, R.; Vergnenegre, A.; Massuti, B.; Felip, E.; Palmero, R.; Garcia-Gomez, R.; Pallares, C.; Sanchez, J.M.; *et al.* Erlotinib *versus* standard chemotherapy as first-line treatment for european patients with advanced egfr mutation-positive non-small-cell lung cancer (eurtac): A multicentre, open-label, randomised phase 3 trial. *Lancet Oncol.* **2012**, *13*, 239–246.
23. Kwak, E.L.; Bang, Y.J.; Camidge, D.R.; Shaw, A.T.; Solomon, B.; Maki, R.G.; Ou, S.H.; Dezube, B.J.; Janne, P.A.; Costa, D.B.; *et al.* Anaplastic lymphoma kinase inhibition in non-small-cell lung cancer. *N. Engl. J. Med.* **2010**, *363*, 1693–1703.
24. Doebele, R.C. Targeted therapies: Time to shift the burden of proof for oncogene-positive cancer? *Nat. Rev. Clin. Oncol.* **2013**, *10*, 492–493.
25. Shaw, A.T.; Kim, D.W.; Mehra, R.; Tan, D.S.; Felip, E.; Chow, L.Q.; Camidge, D.R.; Vansteenkiste, J.; Sharma, S.; de Pas, T.; *et al.* Ceritinib in alk-rearranged non-small-cell lung cancer. *N. Engl. J. Med.* **2014**, *370*, 1189–1197.
26. Comprehensive genomic characterization of squamous cell lung cancers. *Nature* **2012**, *489*, 519–525.
27. Weiss, J.; Sos, M.L.; Seidel, D.; Peifer, M.; Zander, T.; Heuckmann, J.M.; Ullrich, R.T.; Menon, R.; Maier, S.; Soltermann, A.; *et al.* Frequent and focal fgfr1 amplification associates with therapeutically tractable fgfr1 dependency in squamous cell lung cancer. *Sci. Transl. Med.* **2010**, *2*, 62ra93.
28. Hammerman, P.S.; Sos, M.L.; Ramos, A.H.; Xu, C.; Dutt, A.; Zhou, W.; Brace, L.E.; Woods, B.A.; Lin, W.; Zhang, J.; *et al.* Mutations in the ddr2 kinase gene identify a novel therapeutic target in squamous cell lung cancer. *Cancer Discov.* **2011**, *1*, 78–89.
29. Pitini, V.; Arrigo, C.; di Mirto, C.; Mondello, P.; Altavilla, G. Response to dasatinib in a patient with sqcc of the lung harboring a discoid-receptor-2 and synchronous chronic myelogenous leukemia. *Lung Cancer* **2013**, *82*, 171–172.
30. Peifer, M.; Fernandez-Cuesta, L.; Sos, M.L.; George, J.; Seidel, D.; Kasper, L.H.; Plenker, D.; Leenders, F.; Sun, R.; Zander, T.; *et al.* Integrative genome analyses identify key somatic driver mutations of small-cell lung cancer. *Nat. Genet.* **2012**, *44*, 1104–1110.
31. Fernandez-Cuesta, L.; Peifer, M.; Lu, X.; Sun, R.; Ozretic, L.; Seidel, D.; Zander, T.; Leenders, F.; George, J.; Muller, C.; *et al.* Frequent mutations in chromatin-remodelling genes in pulmonary carcinoids. *Nat. Commun.* **2014**, *5*, doi:10.1038/ncomms4518.
32. McFadden, D.G.; Papagiannakopoulos, T.; Taylor-Weiner, A.; Stewart, C.; Carter, S.L.; Cibulskis, K.; Bhutkar, A.; McKenna, A.; Dooley, A.; Vernon, A.; *et al.* Genetic and clonal dissection of murine small cell lung carcinoma progression by genome sequencing. *Cell* **2014**, *156*, 1298–1311.
33. The Clinical Lung Cancer Genome Project (CLCGP); Network Genomic Medicine (NGM). A genomics-based classification of human lung tumors. *Sci. Transl. Med.* **2013**, *5*, 209ra153.
34. Taylor, B.S.; Schultz, N.; Hieronymus, H.; Gopalan, A.; Xiao, Y.; Carver, B.S.; Arora, V.K.; Kaushik, P.; Cerami, E.; Reva, B.; *et al.* Integrative genomic profiling of human prostate cancer. *Cancer Cell* **2010**, *18*, 11–22.
35. Barbieri, C.E.; Baca, S.C.; Lawrence, M.S.; Demichelis, F.; Blattner, M.; Theurillat, J.P.; White, T.A.;

Stojanov, P.; van Allen, E.; Stransky, N.; *et al.* Exome sequencing identifies recurrent spop, foxa1 and med12 mutations in prostate cancer. *Nat. Genet.* **2012**, *44*, 685–689.

36. Baca, S.C.; Prandi, D.; Lawrence, M.S.; Mosquera, J.M.; Romanel, A.; Drier, Y.; Park, K.; Kitabayashi, N.; MacDonald, T.Y.; Ghandi, M.; *et al.* Punctuated evolution of prostate cancer genomes. *Cell* **2013**, *153*, 666–677.
37. Tomlins, S.A.; Rhodes, D.R.; Perner, S.; Dhanasekaran, S.M.; Mehra, R.; Sun, X.W.; Varambally, S.; Cao, X.; Tchinda, J.; Kuefer, R.; *et al.* Recurrent fusion of tmprss2 and ets transcription factor genes in prostate cancer. *Science* **2005**, *310*, 644–648.
38. Li, J.; Yen, C.; Liaw, D.; Podsypanina, K.; Bose, S.; Wang, S.I.; Puc, J.; Miliaresis, C.; Rodgers, L.; McCombie, R.; *et al.* Pten, a putative protein tyrosine phosphatase gene mutated in human brain, breast, and prostate cancer. *Science* **1997**, *275*, 1943–1947.
39. Berger, M.F.; Lawrence, M.S.; Demichelis, F.; Drier, Y.; Cibulskis, K.; Sivachenko, A.Y.; Sboner, A.; Esgueva, R.; Pflueger, D.; Sougnez, C.; *et al.* The genomic complexity of primary human prostate cancer. *Nature* **2011**, *470*, 214–220.
40. Weischenfeldt, J.; Simon, R.; Feuerbach, L.; Schlangen, K.; Weichenhan, D.; Minner, S.; Wuttig, D.; Warnatz, H.J.; Stehr, H.; Rausch, T.; *et al.* Integrative genomic analyses reveal an androgen-driven somatic alteration landscape in early-onset prostate cancer. *Cancer Cell* **2013**, *23*, 159–170.
41. Grasso, C.S.; Wu, Y.M.; Robinson, D.R.; Cao, X.; Dhanasekaran, S.M.; Khan, A.P.; Quist, M.J.; Jing, X.; Lonigro, R.J.; Brenner, J.C.; *et al.* The mutational landscape of lethal castration-resistant prostate cancer. *Nature* **2012**, *487*, 239–243.
42. Brenner, J.C.; Ateeq, B.; Li, Y.; Yocum, A.K.; Cao, Q.; Asangani, I.A.; Patel, S.; Wang, X.; Liang, H.; Yu, J.; *et al.* Mechanistic rationale for inhibition of poly(adp-ribose) polymerase in ets gene fusion-positive prostate cancer. *Cancer Cell* **2011**, *19*, 664–678.
43. Ross, K.A.; Smyth, N.A.; Murawski, C.D.; Kennedy, J.G. The biology of ewing sarcoma. *ISRN Oncol.* **2013**, *2013*, doi:10.1155/2013/759725.
44. Riggi, N.; Cironi, L.; Suva, M.L.; Stamenkovic, I. Sarcomas: Genetics, signalling, and cellular origins. Part 1: The fellowship of tet. *J. Pathol.* **2007**, *213*, 4–20.
45. Demicco, E.G. Sarcoma diagnosis in the age of molecular pathology. *Adv. Anat. Pathol.* **2013**, *20*, 264–274.
46. Wang, W.L.; Patel, N.R.; Caragea, M.; Hogendoorn, P.C.; Lopez-Terrada, D.; Hornick, J.L.; Lazar, A.J. Expression of erg, an ets family transcription factor, identifies erg-rearranged ewing sarcoma. *Mod. Pathol.* **2012**, *25*, 1378–1383.
47. Machado, I.; Noguera, R.; Mateos, E.A.; Calabuig-Farinas, S.; Lopez, F.I.; Martinez, A.; Navarro, S.; Llombart-Bosch, A. The many faces of atypical ewing's sarcoma. A true entity mimicking sarcomas, carcinomas and lymphomas. *Virchows Arch.* **2011**, *458*, 281–290.
48. Taylor, B.S.; Barretina, J.; Maki, R.G.; Antonescu, C.R.; Singer, S.; Ladanyi, M. Advances in sarcoma genomics and new therapeutic targets. *Nat. Rev. Cancer* **2011**, *11*, 541–557.
49. Pierron, G.; Tirode, F.; Lucchesi, C.; Reynaud, S.; Ballet, S.; Cohen-Gogo, S.; Perrin, V.; Coindre, J.M.; Delattre, O. A new subtype of bone sarcoma defined by bcor-ccnb3 gene fusion. *Nat. Genet.* **2012**, *44*, 461–466.
50. Choi, E.Y.; Thomas, D.G.; McHugh, J.B.; Patel, R.M.; Roulston, D.; Schuetze, S.M.; Chugh, R.; Biermann, J.S.; Lucas, D.R. Undifferentiated small round cell sarcoma with t(4;19)(q35;q13.1)

cic-dux4 fusion: A novel highly aggressive soft tissue tumor with distinctive histopathology. *Am. J. Surg. Pathol.* **2013**, *37*, 1379–1386.

51. Tanas, M.R.; Goldblum, J.R. Fluorescence *in situ* hybridization in the diagnosis of soft tissue neoplasms: A review. *Adv. Anat. Pathol.* **2009**, *16*, 383–391.
52. Trautmann, M.; Sievers, E.; Aretz, S.; Kindler, D.; Michels, S.; Friedrichs, N.; Renner, M.; Kirfel, J.; Steiner, S.; Huss, S.; *et al.* Ss18-ssx fusion protein-induced wnt/beta-catenin signaling is a therapeutic target in synovial sarcoma. *Oncogene* **2013**, doi:10.1038/onc.2013.443.
53. Shing, D.C.; McMullan, D.J.; Roberts, P.; Smith, K.; Chin, S.F.; Nicholson, J.; Tillman, R.M.; Ramani, P.; Cullinane, C.; Coleman, N. Fus/erg gene fusions in ewing's tumors. *Cancer Res.* **2003**, *63*, 4568–4576.
54. Leach, F.S.; Tokino, T.; Meltzer, P.; Burrell, M.; Oliner, J.D.; Smith, S.; Hill, D.E.; Sidransky, D.; Kinzler, K.W.; Vogelstein, B. P53 mutation and mdm2 amplification in human soft tissue sarcomas. *Cancer Res.* **1993**, *53*, 2231–2234.
55. Pilotti, S.; Della Torre, G.; Lavarino, C.; Sozzi, G.; Minoletti, F.; Vergani, B.; Azzarelli, A.; Rilke, F.; Pierotti, M.A. Molecular abnormalities in liposarcoma: Role of mdm2 and cdk4-containing amplicons at 12q13–22. *J. Pathol.* **1998**, *185*, 188–190.
56. Shangary, S.; Wang, S. Small-molecule inhibitors of the mdm2-p53 protein-protein interaction to reactivate p53 function: A novel approach for cancer therapy. *Annu. Rev. Pharmacol. Toxicol.* **2009**, *49*, 223–241.
57. Pavlidis, N.; Pentheroudakis, G. Cancer of unknown primary site. *Lancet* **2012**, *379*, 1428–1435.
58. Massard, C.; Loriot, Y.; Fizazi, K. Carcinomas of an unknown primary origin—Diagnosis and treatment. *Nat. Rev. Clin. Oncol.* **2011**, *8*, 701–710.
59. Stella, G.M.; Senetta, R.; Cassenti, A.; Ronco, M.; Cassoni, P. Cancers of unknown primary origin: Current perspectives and future therapeutic strategies. *J. Transl. Med.* **2012**, *10*, 12.
60. Tothill, R.W.; Li, J.; Mileshkin, L.; Doig, K.; Siganakis, T.; Cowin, P.; Fellowes, A.; Semple, T.; Fox, S.; Byron, K.; *et al.* Massively-parallel sequencing assists the diagnosis and guided treatment of cancers of unknown primary. *J. Pathol.* **2013**, *231*, 413–423.
61. Tan, D.S.; Montoya, J.; Ng, Q.S.; Chan, K.S.; Lynette, O.; Sakktee Krisna, S.; Takano, A.; Lim, W.T.; Tan, E.H.; Lim, K.H. Molecular profiling for druggable genetic abnormalities in carcinoma of unknown primary. *J. Clin. Oncol.* **2013**, *31*, e237–e239.
62. Chapman, P.B.; Hauschild, A.; Robert, C.; Haanen, J.B.; Ascierto, P.; Larkin, J.; Dummer, R.; Garbe, C.; Testori, A.; Maio, M.; *et al.* Improved survival with vemurafenib in melanoma with braf v600e mutation. *N. Engl. J. Med.* **2011**, *364*, 2507–2516.
63. Flaherty, K.T.; Puzanov, I.; Kim, K.B.; Ribas, A.; McArthur, G.A.; Sosman, J.A.; O'Dwyer, P.J.; Lee, R.J.; Grippo, J.F.; Nolop, K.; *et al.* Inhibition of mutated, activated braf in metastatic melanoma. *N. Engl. J. Med.* **2010**, *363*, 809–819.
64. Prahallad, A.; Sun, C.; Huang, S.; di Nicolantonio, F.; Salazar, R.; Zecchin, D.; Beijersbergen, R.L.; Bardelli, A.; Bernards, R. Unresponsiveness of colon cancer to braf(v600e) inhibition through feedback activation of egfr. *Nature* **2012**, *483*, 100–103.
65. Hanfstein, B.; Muller, M.C.; Hehlmann, R.; Erben, P.; Lauseker, M.; Fabarius, A.; Schnittger, S.; Haferlach, C.; Gohring, G.; Proetel, U.; *et al.* Early molecular and cytogenetic response is predictive for long-term progression-free and overall survival in chronic myeloid leukemia (cml).

Leukemia **2012**, *26*, 2096–2102.
66. Oehler, V.G. Update on current monitoring recommendations in chronic myeloid leukemia: Practical points for clinical practice. *Hematol. Am. Soc. Hematol. Educ. Program* **2013**, *2013*, 176–183.
67. Merx, K.; Muller, M.C.; Kreil, S.; Lahaye, T.; Paschka, P.; Schoch, C.; Weisser, A.; Kuhn, C.; Berger, U.; Gschaidmeier, H.; *et al.* Early reduction of bcr-abl mrna transcript levels predicts cytogenetic response in chronic phase cml patients treated with imatinib after failure of interferon alpha. *Leukemia* **2002**, *16*, 1579–1583.
68. Cross, N.C.; White, H.E.; Muller, M.C.; Saglio, G.; Hochhaus, A. Standardized definitions of molecular response in chronic myeloid leukemia. *Leukemia* **2012**, *26*, 2172–2175.
69. Maheswaran, S.; Sequist, L.V.; Nagrath, S.; Ulkus, L.; Brannigan, B.; Collura, C.V.; Inserra, E.; Diederichs, S.; Iafrate, A.J.; Bell, D.W.; *et al.* Detection of mutations in egfr in circulating lung-cancer cells. *N. Engl. J. Med.* **2008**, *359*, 366–377.
70. Walter, A.O.; Sjin, R.T.; Haringsma, H.J.; Ohashi, K.; Sun, J.; Lee, K.; Dubrovskiy, A.; Labenski, M.; Zhu, Z.; Wang, Z.; *et al.* Discovery of a mutant-selective covalent inhibitor of egfr that overcomes t790m-mediated resistance in nsclc. *Cancer Discov.* **2013**, *3*, 1404–1415.
71. Forshew, T.; Murtaza, M.; Parkinson, C.; Gale, D.; Tsui, D.W.; Kaper, F.; Dawson, S.J.; Piskorz, A.M.; Jimenez-Linan, M.; Bentley, D.; *et al.* Noninvasive identification and monitoring of cancer mutations by targeted deep sequencing of plasma DNA. *Sci. Transl. Med.* **2012**, *4*, 136ra168.
72. Tomlins, S.A.; Aubin, S.M.; Siddiqui, J.; Lonigro, R.J.; Sefton-Miller, L.; Miick, S.; Williamsen, S.; Hodge, P.; Meinke, J.; Blase, A.; *et al.* Urine tmprss2:Erg fusion transcript stratifies prostate cancer risk in men with elevated serum psa. *Sci. Transl. Med.* **2011**, *3*, 94ra72.
73. Druker, B.J.; Tamura, S.; Buchdunger, E.; Ohno, S.; Segal, G.M.; Fanning, S.; Zimmermann, J.; Lydon, N.B. Effects of a selective inhibitor of the abl tyrosine kinase on the growth of bcr-abl positive cells. *Nat. Med.* **1996**, *2*, 561–566.
74. Kantarjian, H.; O'Brien, S.; Jabbour, E.; Garcia-Manero, G.; Quintas-Cardama, A.; Shan, J.; Rios, M.B.; Ravandi, F.; Faderl, S.; Kadia, T.; *et al.* Improved survival in chronic myeloid leukemia since the introduction of imatinib therapy: A single-institution historical experience. *Blood* **2012**, *119*, 1981–1987.
75. Baccarani, M.; Deininger, M.W.; Rosti, G.; Hochhaus, A.; Soverini, S.; Apperley, J.F.; Cervantes, F.; Clark, R.E.; Cortes, J.E.; Guilhot, F.; *et al.* European leukemianet recommendations for the management of chronic myeloid leukemia: 2013. *Blood* **2013**, *122*, 872–884.
76. Mahon, F.X. Is going for cure in chronic myeloid leukemia possible and justifiable? *Hematol. Am. Soc. Hematol. Educ. Program* **2012**, *2012*, 122–128.
77. Ross, J.S.; Slodkowska, E.A.; Symmans, W.F.; Pusztai, L.; Ravdin, P.M.; Hortobagyi, G.N. The her-2 receptor and breast cancer: Ten years of targeted anti-her-2 therapy and personalized medicine. *Oncologist* **2009**, *14*, 320–368.
78. Arteaga, C.L.; Sliwkowski, M.X.; Osborne, C.K.; Perez, E.A.; Puglisi, F.; Gianni, L. Treatment of her2-positive breast cancer: Current status and future perspectives. *Nat. Rev. Clin. Oncol.* **2012**, *9*, 16–32.
79. Swain, S.M.; Kim, S.B.; Cortes, J.; Ro, J.; Semiglazov, V.; Campone, M.; Ciruelos, E.; Ferrero, J.M.; Schneeweiss, A.; Knott, A.; *et al.* Pertuzumab, trastuzumab, and docetaxel for her2-positive

metastatic breast cancer (cleopatra study): Overall survival results from a randomised, double-blind, placebo-controlled, phase 3 study. *Lancet Oncol.* **2013**, *14*, 461–471.

80. Hurvitz, S.A.; Dirix, L.; Kocsis, J.; Bianchi, G.V.; Lu, J.; Vinholes, J.; Guardino, E.; Song, C.; Tong, B.; Ng, V.; *et al.* Phase ii randomized study of trastuzumab emtansine *versus* trastuzumab plus docetaxel in patients with human epidermal growth factor receptor 2-positive metastatic breast cancer. *J. Clin. Oncol.* **2013**, *31*, 1157–1163.
81. Incorvati, J.A.; Shah, S.; Mu, Y.; Lu, J. Targeted therapy for her2 positive breast cancer. *J. Hematol. Oncol.* **2013**, *6*, 38.
82. Pao, W.; Chmielecki, J. Rational, biologically based treatment of egfr-mutant non-small-cell lung cancer. *Nat. Rev. Cancer* **2010**, *10*, 760–774.
83. Van Cutsem, E.; Kohne, C.H.; Hitre, E.; Zaluski, J.; Chang Chien, C.R.; Makhson, A.; D'Haens, G.; Pinter, T.; Lim, R.; Bodoky, G.; *et al.* Cetuximab and chemotherapy as initial treatment for metastatic colorectal cancer. *N. Engl. J. Med.* **2009**, *360*, 1408–1417.
84. Douillard, J.Y.; Oliner, K.S.; Siena, S.; Tabernero, J.; Burkes, R.; Barugel, M.; Humblet, Y.; Bodoky, G.; Cunningham, D.; Jassem, J.; *et al.* Panitumumab-folfox4 treatment and ras mutations in colorectal cancer. *N. Engl. J. Med.* **2013**, *369*, 1023–1034.

7

Illuminating the Transcriptome through the Genome

David J. Elliott

Institute of Genetic Medicine, Newcastle University, Newcastle, NE1 3BZ, UK;
E-Mail: David.Elliott@ncl.ac.uk.

Abstract: Sequencing the human genome was a huge milestone in genetic research that revealed almost the total DNA sequence required to create a human being. However, in order to function, the DNA genome needs to be expressed as an RNA transcriptome. This article reviews how knowledge of genome sequence information has led to fundamental discoveries in how the transcriptome is processed, with a focus on new system-wide insights into how pre-mRNAs that are encoded by split genes in the genome are rearranged by splicing into functional mRNAs. These advances have been made possible by the development of new post-genome technologies to probe splicing patterns. Transcriptome-wide approaches have characterised a "splicing code" that is embedded within and has a significant role in deciphering the genome, and is deciphered by RNA binding proteins. These analyses have also found that most human genes encode multiple mRNA isoforms, and in some cases proteins, leading in turn to a re-assessment of what exactly a gene is. Analysis of the transcriptome has given insights into how the genome is packaged and transcribed, and is helping to explain important aspects of genome evolution.

Keywords: transcriptome; RNA splicing; post-genome

---

1. Introduction

The completion of the human genome sequence [1,2] brought together key scientific and philosophical questions, including exactly what we are as a species and individuals. However, in order to function, the genome has to be expressed. The primary expression product of the genome is RNA, and the complete set of all RNA molecules made through copying the genome into RNA is called the transcriptome (Figure 1). After transcription in the nucleus, mRNAs are translated into protein in the cytoplasm to yield the proteome while other RNAs have noncoding functions [3].

Figure 1. Information flow from the genome to the proteome. The genome represents an archive of information embedded in DNA. This information is transcribed as RNA to give the transcriptome, and then translated into protein to give the proteome.

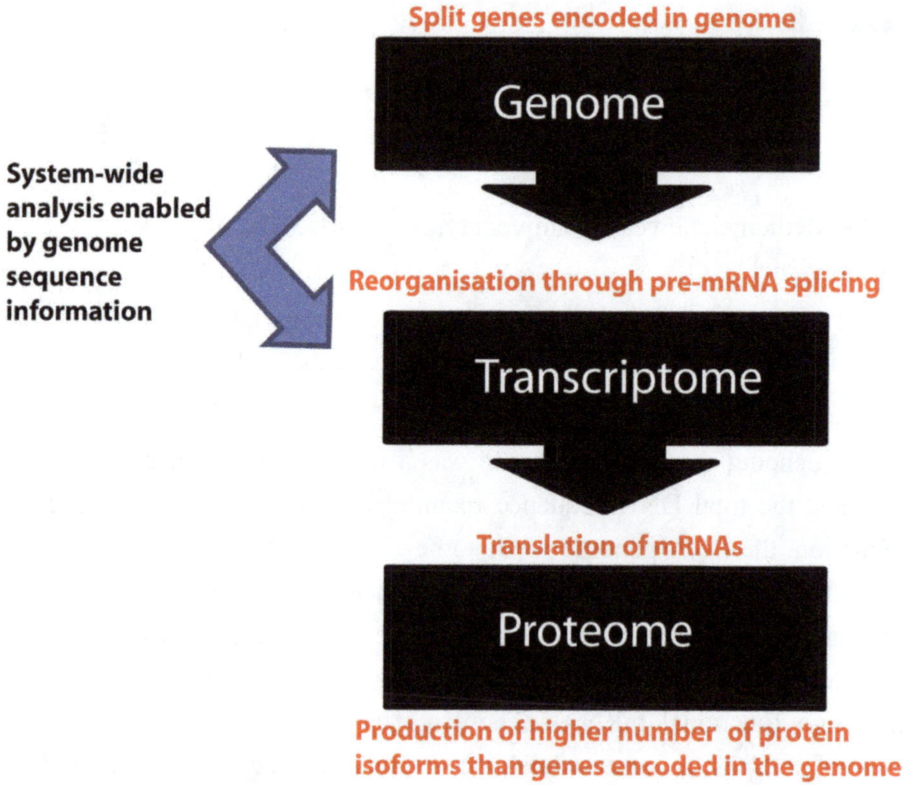

A key feature of human (and other eukaryotic) genes is their split exon-intron structure [4,5]. Figure 2 shows the exon-intron structure of a typical human gene displayed on a genome browser [6]. The exons include the protein coding information of the gene while introns are the intervening sequences between them. The term exon refers to the fact that exon sequences are expressed in the mRNA made from the gene, as opposed to introns which are removed (intron refers to intragenic regions) [7]. The presence of introns within genes and the long intergenic sequences between genes mean that only a small fraction of the human genome is truly protein-encoding. To put some figures on this, human protein-encoding genes contribute ~33.5% of the human genome sequence [1,2] but exons alone comprise 2.94% of the genome [8]. Protein coding exons make up a smaller proportion still (1.2%) of the genome. This is because there are at least partially untranslated exons in every mRNA (some of which can have important regulatory roles), and some exons remain entirely untranslated (see below).

Genes need to be transcribed over their full length (including both introns and exons) to generate precursor mRNAs (pre-mRNAs). Transcription of long genes represents a considerable energy investment by the cell. One of the longest genes in the human genome, *DYSTROPHIN* takes in the order of 16 hours to transcribe, yet produces a final mRNA of just ~14 kb that would have just taken ~7 minutes to transcribe by itself, assuming an elongation rate of 2 kb/minute [9]. It is calculated that ~95% of RNA does not leave the nucleus [10]. Nuclear-retained RNA includes intron sequences and some long ncRNAs that are also spliced but retained in the nucleus.

Figure 2. The intron-exon structure of a typical human gene displayed on the UCSC genome browser [6]. Introns are shown as lines (the "arrowheads" in the lines indicate the direction of transcription). Exons are shown as vertical bars. Coding exons are shown as thicker vertical bars than non-coding exon sequence. This example shows the *NASP* gene. The gene structures shown are "Refseq genes" that represent known human protein-coding and non-protein-coding genes taken from the NCBI RNA reference sequences collection. Notice that this single gene locus contains three distinct Refseq annotations containing different exon structures. Conserved sequences detected by comparative genomic information from 100 vertebrate genome sequences are shown at the bottom as a Phastcons plot—the higher values are most conserved, and often correspond to exons.

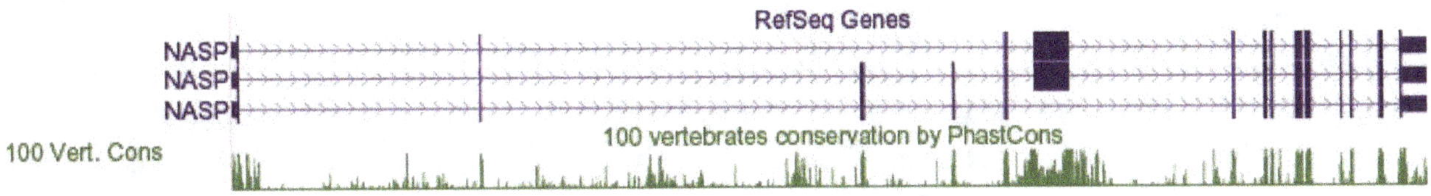

The splicing reaction is catalysed by the spliceosome. While introns are generally discarded after splicing, some introns can yield functional RNAs after splicing (e.g., miRNAs) [11,12]. Spliceosomes themselves are multi-component machines containing five snRNAs and at least 200 proteins, making them one of the most complicated assemblies in the cell [13,14]. Spliceosomes assemble *de novo* around each intron to be removed. A typical gene containing eight exons would require the assembly of eight spliceosomes to create a functional mRNA. Input of energy is required for spliceosomes to properly assemble through multiple ATP-dependent RNA helicases and other energy consuming proteins (including GTPases) [15].

Prior to completion of the human genome sequence, research into splicing typically looked at single genes and exons as models. However, while these detailed studies continue to be very important, the advances in genomics catalysed by genome sequencing projects have spawned parallel advances in transcriptomics, enabling a much broader system-wide dissection of RNA processing pathways. Here I review some of these new insights. While the focus here is on pre-mRNA splicing, transcriptome-wide analyses have also been directed at other aspects of genome expression, including RNA editing, RNA stability, expression of ncRNAs, polyadenylation and translation.

2. The Human Genome Sequence Has Led to New Global Insights into the Control of Splicing

In the 1980s examination of a limited number of genes led to the identification of short conserved sequences called 5' and 3' splice sites at exon-intron junctions [16]. The availability of the human and other genome sequences have enabled these studies to be extended genome-wide [17]. Most human exons are spliced together by a single kind of major spliceosome that recognises most 5' and 3' splice sites. In addition a second minor spliceosome exists in parallel that decodes a smaller subset of intron–exon junctions [18]. This minor spliceosome has a different but overlapping complement of snRNAs to the major spliceosome. Recent transcriptome-wide data show a key snRNA component in the minor spliceosome (called $U6_{ATAC}$) is an important gene expression switch controlling patterns of splicing [19]. In the rest of this review the activities of the major and minor spliceosomes are not separately distinguished.

The conserved 5' and 3' splice site sequences encoded in the genome at exon-intron junctions are quite short. Furthermore, scattered within introns are short sequence elements called pseudoexons. Pseudoexons "look like" exons in that they have 5' and 3' splice sites, but are not selected as exons by the spliceosome. Estimates from model human genes suggest pre-mRNA splicing is remarkably accurate [20]. However, transcriptome-wide analyses of splicing patterns using RNAseq do detect some errors in splicing (at a rate of around 0.7% errors/intron)—these errors have been termed "noisy splicing", and might contribute to gene and protein evolution by enabling new mRNA isoforms to be made even at low frequencies [21].

The spliceosome uses several mechanisms to accurately decode exon/intron structure using information embedded in the transcriptome. Firstly, in humans and most vertebrates, spliceosomes recognise exons from introns through a process called exon definition [22–24]. The advantage of exon definition is that since exons are quite small they should be easier to identify as discrete units compared most (considerably longer) vertebrate introns. In exon definition, early spliceosome components bind to the pre-mRNA and "flag" exons to be spliced together. Following exon definition, pairs of exons are then joined together by splicing which removes the intervening intron sequences. Amongst the early binding components of the spliceosome involved in exon definition are the U1 snRNP RNA-protein complex that recognises the 5' splice site, and a protein dimer which recognises 3' splice sites called U2AF (U2AF65 and U2AF35 are the two proteins in the dimer).

A second mechanism that facilitates accurate decoding of the genome is the presence of a splicing code that helps to differentiate between exons and introns in pre-mRNA. Before the sequencing of the human and other eukaryotic genomes, the important sequences controlling splicing of an exon were usually worked out on a gene by gene basis, using a finite number of model exons. The importance of exon sequences outside the splice junctions for splicing were first identified in pioneering experiments using model exons in the *FIBRONECTIN* and β globin genes [25,26]. It is now known that pre-mRNAs each contain multiple short nucleotide sequences that can enhance or silence splicing of their associated exons [27–29]. Exon sequences can function as exonic splicing enhancers (abbreviated ESEs) that help the spliceosome to recognise exons, or exonic splicing silencers (abbreviated ESSs) that inhibit spliceosome recognition by the spliceosome. Similarly, intron associated sequences can function as intronic splicing enhancers (abbreviated ISE) or Intronic Splicing Silencers (abbreviated ISSs). Splicing enhancers are bound by proteins or complexes of proteins, including the SR proteins that contain domains enriched in serine and arginine residues, and splicing silencers are frequently bound by heterogeneous ribonucleoproteins (abbreviated hnRNPs).

The availability of genome sequences have allowed system-wide approaches to identify splicing enhancers and silencers that control splicing and led to significant insights into the "splicing code" [29]. These approaches have included machine learning approaches to utilize hundreds of features in pre-mRNAs including motifs bound by RNA binding proteins and RNA secondary structure predictions to predict *in vivo* splicing decisions [30]. Computer programmes have also been devised that can computationally predict positions of predicted splicing enhancers and silencers and the target sequences of RNA binding proteins in an input genomic sequence (e.g., Figure 3) [30–33]. The combination of these system-wide experimental and bioinformatic analyses show the splicing code is maintained as nucleotide information in the genome. The splicing code has similar importance to the genetic code that is deciphered to read amino acid sequences from mRNAs. However, the splicing code is much more complex than the genetic code. While the genetic code uses triplet codons to specify amino acids, in the splicing code multiple sequence elements act in combination to decipher the exon/intron structure of pre-mRNAs [30].

Figure 3. Transcriptome-wide data can be used to predict the splicing code in specific genes. In this example sequences within a cassette exon in the mouse NASP gene have been analysed using genome and transcriptome wide datasets to pinpoint splicing control sequences. Firstly a Chasin Z-score plot was used that can identify sequences predictive of exonic splicing enhancers and silencers [32,34]: the four exonic splicing enhancer (ESE) sequences identified are shown as peaks in the plot above the sequence. In this example, these ESE sequences were individually mutated to test function in minigenes (the Chasin profiles of the mutants M1-M4 are shown compared to the wild type sequence: notice the change in the Z-score plot removes predicted ESE activity for each mutant). The positions of these ESEs mapped to binding sites for the splicing factor Tra2β, both individually identified using cross linking immunoprecipitation (CLIP) in the mouse testis and predicted from the *in vivo* binding site generated from transcriptome-wide Tra2β binding data from the mouse testis. This figure is adapted from [32].

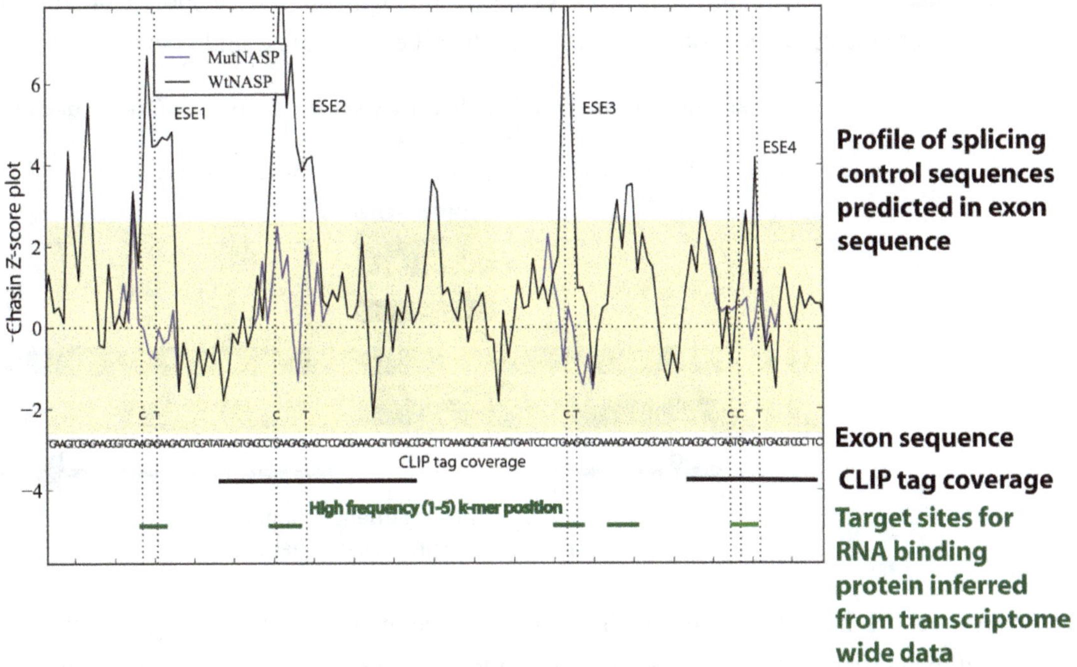

Because they are needed for exons to be spliced into mRNAs, ESE sequences have been maintained in exons as well as the codons that specify amino acid sequences [35]. The intronic sequences that flank exons are often also strongly conserved between species (Figure 4 shows as an example conserved nucleotide sequences flanking an exon in the mouse *Neurexin3* gene downloaded from the UCSC genome browser). Comparison of the human and mouse genomes [1,36] which diverged 75 million years ago show that alternative exons are usually flanked by much longer stretches of conserved intron sequences than constitutive exons, consistent with more elaborate control mechanisms [37,38]. Conservation in these exon flanking intron sequences in some cases is much higher than in promoters, suggesting that one of the main functions of conserved noncoding sequences between mouse and human is the regulation of alternative splicing [37]. Even non-protein coding exons can be highly conserved in the genome. Noncoding exons include highly conserved "poison exons" (for example see Figure 5), that when included insert premature translational termination codons and lead to mRNA decay [32,39]. "Poison exons" are very important for auto-regulation of RNA binding proteins that control splicing [39,40]. Together these studies show the maintenance of splicing control sequences has had a significant impact in constraining genome evolution.

Figure 4. A functional requirement to maintain splicing control sequences constrains evolution of the genome. This screenshot is downloaded from the UCSC genome browser [6] and shows conserved intron sequences flanking the alternatively spliced AS4 exon from the mouse *Nrxn3* gene. The conserved sequences are shaded. At the top, the UCSC gene annotations show that this cassette exon is included in ¾ mRNA isoforms made from this gene. At the bottom the Phastcons plot shows that the flanking intron sequences are also highly conserved as well as the exon sequence (exons might be conserved because of their protein-coding content). Conservation of these intron sequences are likely important to control tissue-specific splicing of this exon by the spliceosome. Known alternative events are annotated on the UCSC track "Alt events", and can be shown alongside the gene structure (here the cassette exon is annotated in the alt events track, and is in purple).

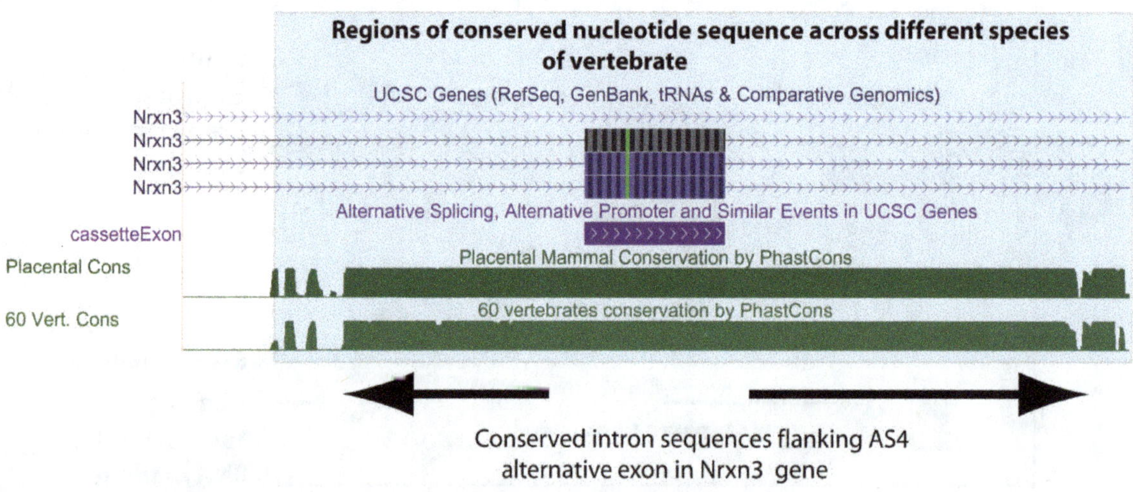

Genome sequences have been used to help develop technologies aimed to globally dissect RNA processing pathways [41]. These technologies can identify the target sites of RNA binding proteins transcriptome-wide. In cross linking immunoprecipitation (abbreviated CLIP) experiments RNA binding proteins are cross-linked *in situ* to their target RNAs within intact cells using ultra violet irradiation, followed by immunoprecipitation of the RNA protein complexes and amplification by PCR [42–44]. Once unique target sites are identified by next generation sequencing (these are called CLIP tags), these can be mapped onto genome sequences to reveal the initial binding sites in the transcriptome. Transcriptome-wide CLIP analyses have enabled maps to be drawn of the target sites of RNA binding proteins relative to regulated exons, and these maps can be used to predict mechanisms of splicing control [45]. For example, CLIP tags of the RNA binding protein Tra2β that is needed for splicing inclusion for a regulated cassette exon in the NASP gene are shown in Figure 3 [32].

Figure 5. Non protein coding exons are conserved in the genome. Some non-coding exons are highly conserved in the human genome and play important roles in controlling the expression levels of splicing regulator proteins. The *TRA2A* gene encodes the splicing regulator protein Tra2α and contains a poison exon that contains multiple stop codons and is only inserted into some mRNAs. Despite not containing coding information, the *TRA2A* poison exon is highly conserved across species (indicated by the Phastcons score). Notice that the TRA2A gene encodes also alternative 5' splice sites and uses alternative promoters. This screenshot was downloaded from the UCSC genome browser [6].

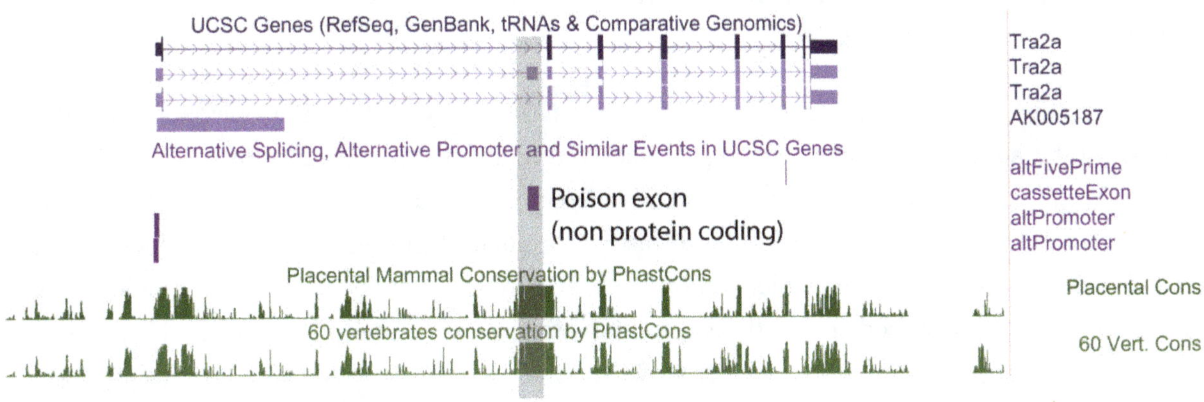

The completion of the human genome sequence enabled the development of comprehensive microarrays to interrogate gene expression. These global techniques include the development of splice-sensitive microarrays. These microarrays either detect specific exons in the transcriptome or the use of specific splice junctions and report splicing patterns in mRNA [46,47]. Transcriptome-wide patterns of alternative splicing can also be detected by RNAseq [48,49]. These technologies have been used to search for exons mis-spliced after depletion of particular RNA binding proteins from cells. By analysing thousands of exons in parallel the splicing events that are regulated by specific RNA binding proteins can be comprehensively identified. These experiments have shown some individual RNA binding proteins bind to and regulate similar mRNAs. For example, the NOVA protein regulates splicing of a functionally coherent set of genes involved in synapse function [50]. Other splicing regulators might similarly have coherent RNA functional targets, including T-STAR and SAM68 which regulate regional alternative splicing of the synaptic neurexin genes in the brain [51,52].

Knowing which splicing events are regulated by what proteins at a global level has been used to derive general rules. For example, binding of the NOVA proteins upstream of exons tends to block splicing, while binding of the same proteins downstream of exons enhance splicing [44,45]. These rules governing the RNA motifs bound by NOVA and their role in activating or repressing associated exon inclusion are conserved between flies and mammals, although the actual target mRNAs are different [53]. Similar rules have also been uncovered for PTB and some other splicing regulator proteins [45,54,55]. Recent developments to understand the splicing code have compared binding maps for different RNA binding proteins, and show that some functionally collaborate with each other to generate tissue specific splicing patterns [56].

Transcriptome-wide insights have have also revealed the involvement of RNA binding proteins in other aspects of genome biology. Alu sequences are retro-transposable elements that frequently insert into introns, and have sequence similarities to exons [57]. There are over 15,000 Alu sequences in the human genome, many of which are inserted into introns. The RNA binding protein hnRNP C has an important role for in protecting the transcriptome from potential mis-splicing caused by the insertion of Alu retrotransposable elements into genomic introns [58]. Depletion of hnRNPC leads to aberrant inclusion of around 1000 Alu-derived exons into the transcriptome. Alu sequences contain polypyrimidine tracts that potentially bind U2AF65, leading to them being aberrantly included into mRNAs, but this splicing is blocked by hnRNPC.

3. Analysis of the Human Transcriptome Led to the Realisation That Most Human Genes Encode Alternatively Spliced mRNAs

Historically genes have been defined in different ways by different people at different times. Following the human genome sequencing project, the fundamental definition of what a gene actually is has been enriched, and to some extent clouded, by comparison of genome and transcriptome sequences [59,60]. Alternative mRNA isoforms can originate from the same genetic loci through use of alternative promoters and polyadenylation sites, and by the process of alternative splicing through which exons can be spliced into different combinations to give multiple mRNAs. Hence a "single gene" can encode multiple products. Alternative splicing fits into four different categories, depending on how variable splice junctions are utilised (Figure 6). In the simplest form of alternative splicing, called exon skipping, whole exons are either spliced into the mRNA or skipped (ignored by the spliceosome). Exon sequences can also be spliced into mRNA using different splice sites (alternative 5' splice site or 3' splice sites can be selected). Whole introns can also be left in the mRNA (this is called intron retention). Transcriptome-wide analyses show that in humans exon skipping is the most frequent form of alternative splicing, and intron retention the least frequent [61].

Figure 6. Types of alternative splicing events detected in the human transcriptome. Exons are shown as boxes, introns as lines, and splicing patterns as broken lines. mRNAs can be made from individual genes can using multiple alternative events, including different types of splicing, to build up complex patterns of alternative spliced mRNAs.

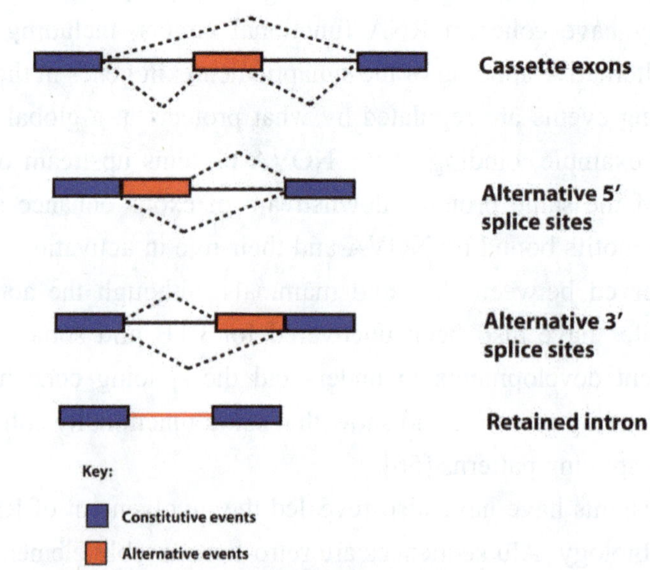

Evidence provided by comparative genome and transcriptome sequences has shown alternative splicing to be extremely frequent. Before the human genome was published, random sequencing of human mRNAs from their 3' ends using oligo dT priming suggested ~40% of human genes encoded alternatively spliced mRNAs [62,63]. During the human genome sequencing project, reconstruction of mRNAs from the gene rich chromosomes 19 and 22 upped this estimate to 59% of genes encode alternative mRNAs, with 2–3 mRNA isoforms made/gene [1]. The use of microarrays to detect global patterns of alternative splicing indicated 73%–74% of human genes express alternatively spliced mRNA isoforms [46,64]. The most recently reported RNAseq analyses of the human transcriptome is consistent with ~95% of multi-exonic genes expressing variant mRNA isoforms, with a plateaux of 10–11 isoforms/gene/cell line [65]. Usually one or two major mRNA isoforms predominate in a given cell line so many cell types will just express one major mRNA isoform [60,65,66]. 86% of genes have a minor isoform frequency of 15% or more, and more than 50% of alternative exons are tissue specific in expression [48,67]. Alternative events are now annotated on genome browsers like the UCSC genome browser (e.g., Figure 4) [6].

4. To What Extent Can Human Complexity Be Ascribed to Alternative Splicing?

Proteins make up large components of human bodies. Prior to the genome sequence estimates of human gene numbers went as high as 100,000. An initial surprise from the human genome sequence was a much lower protein coding gene number, initially counted at around 23,000 [1,2]. The most recent gene counts from the ENCODE consortia suggest only 20,687 human protein coding genes [8]. The number of proteins expressed in a human cell is in contrast estimated to be ~100,000 [68,69]. This protein number represents an amplification factor of 5-fold compared to the counted number of genes. This total gene number in humans does not seem to be exceptionally higher than seemingly less sophisticated organisms. The genome of *Haemophilus influenza* contains 1743 predicted genes, the yeast *Saccharomyces cerevisiae* ~6000 genes, *Drosophila melanogaster* ~13,600 genes, and nematode worms 18,425 genes [70–72].

This apparent failure of gene numbers to correlate with complexity has been called the gene number paradox, and counted as one of the major surprises from the human genome sequence. To what extent might alternative splicing and the resulting expansion in protein coding information help explain developmental and physiological complexity in humans (Figure 7)? Until recently this question has been difficult to address, since the higher number of mRNA and EST sequences available from humans made comparisons of alternative splicing frequency with other species biased. However, a recent modencode project based on RNAseq analysis to look at alternative splicing in *C. elegans* found ~25% alternative splicing in 5000 genes, with around 30% of these being alternatively spliced between different developmental stages [73]. In depth experiments using RNAseq and tiling arrays do show a lower frequency (60.7%) of genes in the fruitfly are alternatively spliced than in humans, often in a developmental or sex-specific fashion [74].

Figure 7. Alternative splicing amplifies genome information. Alternative splicing amplifies information in the human transcriptome relative to the genome.

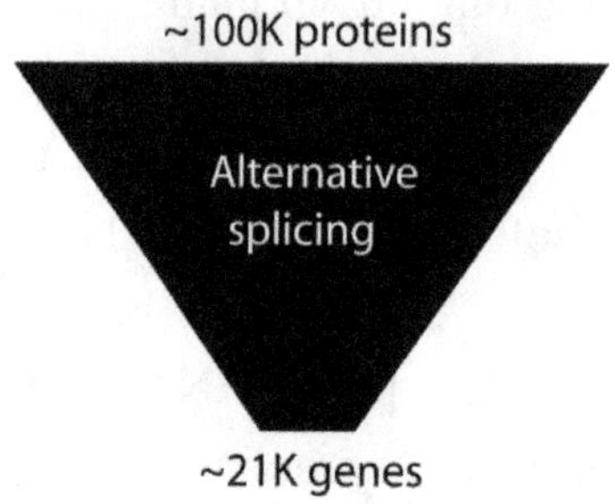

Hence amongst multicellular animals investigated in detail, humans do seem to exhibit higher levels of alternative splicing, which might lend credence to the idea that alternative splicing may be a factor contributing to human sophistication. If phylogenetic differences in alternative splicing frequency correlate with complexity, one might expect a decreased level in single celled organisms compared with multicellular organisms. On the one level less introns are found in the single celled baker's yeast *Saccharomyces cerevisiae*: only 5% of genes contain introns in this yeast (290/6000 genes). However, these lower intron numbers are a bit misleading. The reason for a low overall intron number in this yeast is that many introns have been lost because of reverse transcriptase activity converting mRNAs into cDNAs, which then re-integrate into the genome through high levels of homologous recombination replacing originally intron containing genes [75]. While intron-containing genes generally are rare in yeast, alternative splicing of some of these introns are specifically utilised to control developmental timing of during meiosis that takes place under conditions of limiting nutrients [76,77] so alternative splicing is used to control a complex stage in the lifecycle of this single celled organism. Taken as a whole, it is difficult to draw general correlations between overall frequencies of alternative splicing and organism sophistication.

Alternative splicing patterns can also evolve rapidly and sometimes differ between closely related species. For example, despite almost 99% identity in protein coding information, comparative transcriptome analyses have shown that 6%–8% of alternative spliced exons have different inclusion patterns between humans and chimps. These observations are consistent with alternative splicing contributing to species-species differences, and transcriptomes being more distinct between species than protein coding information [78]. Although the major conclusion from evolutionary comparisons is that much alternative splicing is not conserved between species, comparative genomics show some alternative exons have been highly conserved during evolution [79–81]. These include the highly conserved "AS4" alternative exon in the *Neurexin3* gene (abbreviated *Nrxn3*, Figure 4) that is conserved across the vertebrate lineage [51]. The *Neurexin* genes have been linked with autism and schizophrenia, and mice genetically engineered to be unable to regulate this AS4 alternative exon in the *Nrxn3* gene have different synaptic activity in the brain [82].

Post-genome analyses have also addressed the question what alternative splicing does. Protein sequences encoded by alternatively spliced exons are frequently involved in protein-protein interaction networks and contain signalling domains [83,84]. Some alternatively spliced mRNAs encode proteins with clearly different functional activity. For example, different mRNA splice isoforms encoding the FOXP1 transcription factor are made between neural stem cells and differentiated cells, and these encode proteins that activate different promoters [85]. Alternative splicing regulators have been implicated in human cognitive diseases like autism [86]. Different groups of genes are regulated by alternative splicing from those regulated by transcription [87].

Alternative splicing pathways have been shown to have important roles in controlling development, and some human diseases are caused by defects in alternative splicing including the multi-system disorder myotonic dystrophy [87]. However, individual RNA binding proteins likely regulate coordinated splicing programmes of many target exons, and each of these individual exons might only have somewhat subtle biological contributions. Furthermore, post-genome technologies are also starting to introduce a note of caution in the interpretation of high levels of alternative splicing. Some lower abundance splice variants might be non-functional isoforms that occur as a result of low frequency events mistakes in the splicing process if they are neither evolutionarily conserved nor protein-coding [21,88]. Hence the frequency of functional alternative splicing is likely to be lower than the total frequency of all alternative splicing events.

5. Human Genome Packaging into Chromatin Correlates with Its Intron/Exon Structure

Post genome analyses have shown that the genome and transcriptome are intimately linked. In particular the exon/intron structure of genes correlates with how the genome is packaged. Within the nucleus the genome is wrapped around protein complexes called nucleosomes to form chromatin. Each nucleosome is itself made up of eight positively charged histone molecules that can be modified by the addition of small chemical groups (typically methyl and acetyl groups) [89]. Packaging of the genome within chromatin is important to enable storage of chromosomes within the comparatively small space afforded by the nuclear volume.

After experimental treatment of chromatin with the enzyme DNAse I, the DNA sequences wrapped around nucleosomes remain protected, while the DNA linkages joining nucleosomes together become degraded. Genome-wide analysis of sequences protected from DNAse I digestion in humans and other species indicate a 1.5 fold enrichment of exon sequences over nucleosomes compared to intron sequences [90,91]. This means that in chromatin exons are preferentially (but not exclusively) associated with nucleosomes. This is likely to have a biochemical explanation: exon sequences are GC-enriched, while their flanking intron sequences are AT-rich. Nucleosomes interact more strongly with GC-rich sequences, which likely help anchor exons to nucleosomes. The association of exons with nucleosomes has in turn had important implications for genome evolution. A length of ~150 nucleotides of DNA is needed to wrap around a nucleosome, and this is also the average size of an exon. Hence nucleosome wrapping has placed an evolutionary constraint on exon size, while in contrast introns have been able to expand in size to thousands of nucleotides.

6. Most Splicing Occurs Co-Transcriptionally

Another important connection between the genome and the transcriptome is their physical proximity during important RNA processing steps. In several species including humans much pre-mRNA splicing has been shown to take place co-transcriptionally [92]. This means that "full length" pre-mRNA copies of genes are not made. Instead processing takes place as pre-mRNAs are produced on nascent pre-mRNAs still attached to RNA polymerases engaged in transcription. Deep sequencing of fractionated nascent RNA in human cell lines and total RNA in the brain show that exons located more upstream in genes are most likely to be spliced on nascent RNA [65,93,94]. Transcription of the genome also functionally depends on components more "traditionally" thought to be involved in splicing. Transcriptome-wide analysis of gene expression following depletion of U1 snRNP (a component of the spliceosome that recognises 5' splice sites) has shown that high nuclear concentrations of U1 snRNP are needed to prevent premature intragenic polyadenylation at sites upstream of the proper 3' boundary of genes [95].

The fact that splicing takes place on chromatin during ongoing transcription has important implications for alternative splicing patterns. Single molecule experiments have shown that nucleosomes slow down the progress of transcription [96,97]. This means that the time taken to traverse nucleosomes can provide pauses in RNA polymerase II elongation, allowing the spliceosome a window to assemble on nascent pre-mRNA. Nucleosomes have thus been described as "speed bumps" [98–101]. Interestingly only true exons, and not pseudoexons, are associated with nucleosomes [90,91]. Exon association with nucleosomes is likely to help in their recognition by the spliceosome. Pauses in transcriptional elongation on bona fide exons might facilitate spliceosome assembly before potentially competing splice sites in downstream pseudoexons can be transcribed. Changes in chromatin structure can locally speed up or slow down transcription within genes and be important for controlling alternative splicing [99]. Local transcriptional pauses would provide spliceosomes a longer "window" of time in which to assemble and sometimes even carry out splicing of an exon before a competing downstream splice site appears [102,103]. Faster elongating RNA polymerase II enzymes would give spliceosomes less time to assemble on nascent pre-mRNA before competing exons were transcribed, and so would favour exon skipping. In some cases changes in histone modification can recruit or stabilise RNA binding proteins which regulate splicing of the pre-mRNAs made from the gene [104]. The interactions between the different processes in gene expression have been recently reviewed in [105].

7. Conclusions

The human genome sequence has provided a catalyst for understanding the transcriptome. System-wide approaches of the transcriptome have led to a fuller appreciation of how genomes work including how the human genome operates with a finite gene number and providing a system wide view of RNA processing.

Acknowledgments

I thank Julian Venables for comments on the manuscript. This work was supported by the Wellcome Trust (grant numbers WT080368MA and WT089225/Z/09/Z to DJE) and the BBSRC (grant numbers BB/D013917/1 and BB/I006923/1 to DJE).

Author Contributions

DJE wrote the manuscript.

References

1. Lander, E.S.; Linton, L.M.; Birren, B.; Nusbaum, C.; Zody, M.C.; Baldwin, J.; Devon, K.; Dewar, K.; Doyle, M.; FitzHugh, W.; *et al.* Initial sequencing and analysis of the human genome. *Nature* **2001**, *409*, 860–921.
2. Venter, J.C.; Adams, M.D.; Myers, E.W.; Li, P.W.; Mural, R.J.; Sutton, G.G.; Smith, H.O.; Yandell, M.; Evans, C.A.; Holt, R.A.; *et al.* The sequence of the human genome. *Science* **2001**, *291*, 1304–1351.
3. Fatica, A.; Bozzoni, I. Long non-coding RNAs: New players in cell differentiation and development. *Nat. Rev. Genet.* **2014**, *15*, 7–21.
4. Berget, S.M.; Moore, C.; Sharp, P.A. Spliced segments at the 5' terminus of adenovirus 2 late mRNA. *Proc. Natl. Acad. Sci. USA* **1977**, *74*, 3171–3175.
5. Chow, L.T.; Gelinas, R.E.; Broker, T.R.; Roberts, R.J. An amazing sequence arrangement at the 5' ends of adenovirus 2 messenger RNA. *Cell* **1977**, *12*, 1–8.
6. Karolchik, D.; Barber, G.P.; Casper, J.; Clawson, H.; Cline, M.S.; Diekhans, M.; Dreszer, T.R.; Fujita, P.A.; Guruvadoo, L.; Haeussler, M.; *et al.* The UCSC Genome Browser database: 2014 update. *Nucleic Acids Res.* **2014**, *42*, D764–D770.
7. Gilbert, W. Why genes in pieces? *Nature* **1978**, *271*, 501.
8. Dunham, I.; Kundaje, A.; Aldred, S.F.; Collins, P.J.; Davis, C.; Doyle, F.; Epstein, C.B.; Frietze, S.; Harrow, J.; Kaul, R.; *et al.* An integrated encyclopedia of DNA elements in the human genome. *Nature* **2012**, *489*, 57–74.
9. Tennyson, C.N.; Klamut, H.J.; Worton, R.G. The human dystrophin gene requires 16 hours to be transcribed and is cotranscriptionally spliced. *Nat. Genet.* **1995**, *9*, 184–190.
10. Jackson, D.A.; Pombo, A.; Iborra, F. The balance sheet for transcription: An analysis of nuclear RNA metabolism in mammalian cells. *FASEB J.* **2000**, *14*, 242–254.
11. Okamura, K.; Hagen, J.W.; Duan, H.; Tyler, D.M.; Lai, E.C. The mirtron pathway generates microRNA-class regulatory RNAs in Drosophila. *Cell* **2007**, *130*, 89–100.
12. Ruby, J.G.; Jan, C.H.; Bartel, D.P. Intronic microRNA precursors that bypass Drosha processing. *Nature* **2007**, *448*, 83–86.
13. Jurica, M.S.; Moore, M.J. Pre-mRNA splicing: Awash in a sea of proteins. *Mol. Cell* **2003**, *12*, 5–14.
14. Wahl, M.C.; Will, C.L.; Luhrmann, R. The spliceosome: Design principles of a dynamic RNP machine. *Cell* **2009**, *136*, 701–718.
15. Brow, D.A. Allosteric cascade of spliceosome activation. *Annu. Rev. Genet.* **2002**, *36*, 333–360.
16. Mount, S.M. A catalogue of splice junction sequences. *Nucleic Acids Res.* **1982**, *10*, 459–472.
17. Lim, L.P.; Burge, C.B. A computational analysis of sequence features involved in recognition of short introns. *Proc. Natl. Acad. Sci. USA* **2001**, *98*, 11193–11198.

18. Patel, A.A.; Steitz, J.A. Splicing double: Insights from the second spliceosome. *Nat. Rev. Mol. Cell. Biol.* **2003**, *4*, 960–970.
19. Younis, I.; Dittmar, K.; Wang, W.; Foley, S.W.; Berg, M.G.; Hu, K.Y.; Wei, Z.; Wan, L.; Dreyfuss, G. Minor introns are embedded molecular switches regulated by highly unstable U6atac snRNA. *Elife* **2013**, *2*, e00780.
20. Fox-Walsh, K.L.; Hertel, K.J. Splice-site pairing is an intrinsically high fidelity process. *Proc. Natl. Acad. Sci. USA* **2009**, *106*, 1766–1771.
21. Pickrell, J.K.; Pai, A.A.; Gilad, Y.; Pritchard, J.K. Noisy splicing drives mRNA isoform diversity in human cells. *PLoS Genet.* **2010**, *6*, e1001236.
22. Berget, S.M. Exon recognition in vertebrate splicing. *J. Biol. Chem.* **1995**, *270*, 2411–2414.
23. Black, D.L. Finding splice sites within a wilderness of RNA. *RNA* **1995**, *1*, 763–771.
24. Robberson, B.L.; Cote, G.J.; Berget, S.M. Exon definition may facilitate splice site selection in RNAs with multiple exons. *Mol. Cell. Biol.* **1990**, *10*, 84–94.
25. Mardon, H.J.; Sebastio, G.; Baralle, F.E. A role for exon sequences in alternative splicing of the human fibronectin gene. *Nucleic Acids Res.* **1987**, *15*, 7725–7733.
26. Reed, R.; Maniatis, T. A role for exon sequences and splice-site proximity in splice-site selection. *Cell* **1986**, *46*, 681–690.
27. Black, D.L. Protein diversity from alternative splicing: A challenge for bioinformatics and post-genome biology. *Cell* **2000**, *103*, 367–370.
28. Maniatis, T.; Tasic, B. Alternative pre-mRNA splicing and proteome expansion in metazoans. *Nature* **2002**, *418*, 236–243.
29. Wang, Z.; Burge, C.B. Splicing regulation: From a parts list of regulatory elements to an integrated splicing code. *RNA* **2008**, *14*, 802–813.
30. Barash, Y.; Calarco, J.A.; Gao, W.; Pan, Q.; Wang, X.; Shai, O.; Blencowe, B.J.; Frey, B.J. Deciphering the splicing code. *Nature* **2010**, *465*, 53–59.
31. Schwartz, S.; Hall, E.; Ast, G. SROOGLE: Webserver for integrative, user-friendly visualization of splicing signals. *Nucleic Acids Res.* **2009**, *37*, W189–W192.
32. Grellscheid, S.; Dalgliesh, C.; Storbeck, M.; Best, A.; Liu, Y.; Jakubik, M.; Mende, Y.; Ehrmann, I.; Curk, T.; Rossbach, K.; *et al.* Identification of evolutionarily conserved exons as regulated targets for the splicing activator tra2beta in development. *PLoS Genet.* **2011**, *7*, e1002390.
33. Paz, I.; Akerman, M.; Dror, I.; Kosti, I.; Mandel-Gutfreund, Y. SFmap: A web server for motif analysis and prediction of splicing factor binding sites. *Nucleic Acids Res.* **2010**, *38*, W281–W285.
34. Zhang, X.H.; Chasin, L.A. Computational definition of sequence motifs governing constitutive exon splicing. *Genes Dev.* **2004**, *18*, 1241–1250.
35. Caceres, E.F.; Hurst, L.D. The evolution, impact and properties of exonic splice enhancers. *Genome Biol.* **2013**, *14*, R143.
36. Waterston, R.H.; Lindblad-Toh, K.; Birney, E.; Rogers, J.; Abril, J.F.; Agarwal, P.; Agarwala, R.; Ainscough, R.; Alexandersson, M.; An, P.; *et al.* Initial sequencing and

comparative analysis of the mouse genome. *Nature* **2002**, *420*, 520–562.

37. Sorek, R.; Ast, G. Intronic sequences flanking alternatively spliced exons are conserved between human and mouse. *Genome Res.* **2003**, *13*, 1631–1637.
38. Sugnet, C.W.; Srinivasan, K.; Clark, T.A.; O'Brien, G.; Cline, M.S.; Wang, H.; Williams, A.; Kulp, D.; Blume, J.E.; Haussler, D.; *et al.* Unusual intron conservation near tissue-regulated exons found by splicing microarrays. *PLoS Comput. Biol.* **2006**, *2*, e4.
39. Lareau, L.F.; Inada, M.; Green, R.E.; Wengrod, J.C.; Brenner, S.E. Unproductive splicing of SR genes associated with highly conserved and ultraconserved DNA elements. *Nature* **2007**, *446*, 926–929.
40. Ni, J.Z.; Grate, L.; Donohue, J.P.; Preston, C.; Nobida, N.; O'Brien, G.; Shiue, L.; Clark, T.A.; Blume, J.E.; Ares, M., Jr. Ultraconserved elements are associated with homeostatic control of splicing regulators by alternative splicing and nonsense-mediated decay. *Genes Dev.* **2007**, *21*, 708–718.
41. Buratti, E.; Baralle, M.; Baralle, F.E. From single splicing events to thousands: The ambiguous step forward in splicing research. *Brief. Funct. Genomics* **2013**, *12*, 3–12.
42. Licatalosi, D.D.; Mele, A.; Fak, J.J.; Ule, J.; Kayikci, M.; Chi, S.W.; Clark, T.A.; Schweitzer, A.C.; Blume, J.E.; Wang, X.; Darnell, J.C.; Darnell, R.B. HITS-CLIP yields genome-wide insights into brain alternative RNA processing. *Nature* **2008**, *456*, 464–469.
43. Konig, J.; Zarnack, K.; Rot, G.; Curk, T.; Kayikci, M.; Zupan, B.; Turner, D.J.; Luscombe, N.M.; Ule, J. iCLIP—Transcriptome-wide mapping of protein-RNA interactions with individual nucleotide resolution. *J. Vis. Exp.* **2011**, *50*, 2638.
44. Anko, M.L.; Neugebauer, K.M. RNA-protein interactions *in vivo*: Global gets specific. *Trends Biochem. Sci.* **2012**, *37*, 255–262.
45. Witten, J.T.; Ule, J. Understanding splicing regulation through RNA splicing maps. *Trends Genet.* **2011**, *27*, 89–97.
46. Johnson, J.M.; Castle, J.; Garrett-Engele, P.; Kan, Z.; Loerch, P.M.; Armour, C.D.; Santos, R.; Schadt, E.E.; Stoughton, R.; Shoemaker, D.D. Genome-wide survey of human alternative pre-mRNA splicing with exon junction microarrays. *Science* **2003**, *302*, 2141–2144.
47. Pan, Q.; Shai, O.; Misquitta, C.; Zhang, W.; Saltzman, A.L.; Mohammad, N.; Babak, T.; Siu, H.; Hughes, T.R.; Morris, Q.D.; *et al.* Revealing global regulatory features of mammalian alternative splicing using a quantitative microarray platform. *Mol. Cell* **2004**, *16*, 929–941.
48. Pan, Q.; Shai, O.; Lee, L.J.; Frey, B.J.; Blencowe, B.J. Deep surveying of alternative splicing complexity in the human transcriptome by high-throughput sequencing. *Nat. Genet.* **2008**, *40*, 1413–1415.
49. Sultan, M.; Schulz, M.H.; Richard, H.; Magen, A.; Klingenhoff, A.; Scherf, M.; Seifert, M.; Borodina, T.; Soldatov, A.; Parkhomchuk, D.; *et al.* A global view of gene activity and alternative splicing by deep sequencing of the human transcriptome. *Science* **2008**, *321*, 956–960.
50. Ule, J.; Ule, A.; Spencer, J.; Williams, A.; Hu, J.S.; Cline, M.; Wang, H.; Clark, T.; Fraser, C.; Ruggiu, M.; *et al.* Nova regulates brain-specific splicing to shape the synapse. *Nat. Genet.* **2005**, *37*, 844–852.

51. Ehrmann, I.; Dalgliesh, C.; Liu, Y.; Danilenko, M.; Crosier, M.; Overman, L.; Arthur, H.M.; Lindsay, S.; Clowry, G.J.; Venables, J.P.; *et al.* The tissue-specific RNA binding protein T-STAR controls regional splicing patterns of neurexin pre-mRNAs in the brain. *PLoS Genet.* **2013**, *9*, e1003474.
52. Iijima, T.; Wu, K.; Witte, H.; Hanno-Iijima, Y.; Glatter, T.; Richard, S.; Scheiffele, P. SAM68 regulates neuronal activity-dependent alternative splicing of neurexin-1. *Cell* **2011**, *147*, 1601–1614.
53. Brooks, A.N.; Yang, L.; Duff, M.O.; Hansen, K.D.; Park, J.W.; Dudoit, S.; Brenner, S.E.; Graveley, B.R. Conservation of an RNA regulatory map between Drosophila and mammals. *Genome Res.* **2011**, *21*, 193–202.
54. Llorian, M.; Schwartz, S.; Clark, T.A.; Hollander, D.; Tan, L.Y.; Spellman, R.; Gordon, A.; Schweitzer, A.C.; de la Grange, P.; Ast, G.; *et al.* Position-dependent alternative splicing activity revealed by global profiling of alternative splicing events regulated by PTB. *Nat. Struct. Mol. Biol.* **2010**, *17*, 1114–1123.
55. Xue, Y.; Zhou, Y.; Wu, T.; Zhu, T.; Ji, X.; Kwon, Y.S.; Zhang, C.; Yeo, G.; Black, D.L.; Sun, H.; *et al.* Genome-wide analysis of PTB-RNA interactions reveals a strategy used by the general splicing repressor to modulate exon inclusion or skipping. *Mol. Cell* **2009**, *36*, 996–1006.
56. Zhang, C.; Frias, M.A.; Mele, A.; Ruggiu, M.; Eom, T.; Marney, C.B.; Wang, H.; Licatalosi, D.D.; Fak, J.J.; Darnell, R.B. Integrative modeling defines the Nova splicing-regulatory network and its combinatorial controls. *Science* **2010**, *329*, 439–443.
57. Sorek, R.; Lev-Maor, G.; Reznik, M.; Dagan, T.; Belinky, F.; Graur, D.; Ast, G. Minimal conditions for exonization of intronic sequences: 5 ' splice site formation in Alu exons. *Mol. Cell* **2004**, *14*, 221–231.
58. Zarnack, K.; Konig, J.; Tajnik, M.; Martincorena, I.; Eustermann, S.; Stevant, I.; Reyes, A.; Anders, S.; Luscombe, N.M.; Ule, J. Direct competition between hnRNP C and U2AF65 protects the transcriptome from the exonization of Alu elements. *Cell* **2013**, *152*, 453–466.
59. Gerstein, M.B.; Bruce, C.; Rozowsky, J.S.; Zheng, D.; Du, J.; Korbel, J.O.; Emanuelsson, O.; Zhang, Z.D.; Weissman, S.; Snyder, M. What is a gene, post-ENCODE? History and updated definition. *Genome Res.* **2007**, *17*, 669–681.
60. Mudge, J.M.; Frankish, A.; Harrow, J. Functional transcriptomics in the post-ENCODE era. *Genome Res.* **2013**, *12*, 1961–1973.
61. Ast, G. How did alternative splicing evolve? *Nat. Rev. Genet.* **2004**, *5*, 773–782.
62. Brett, D.; Hanke, J.; Lehmann, G.; Haase, S.; Delbruck, S.; Krueger, S.; Reich, J.; Bork, P. EST comparison indicates 38% of human mRNAs contain possible alternative splice forms. *FEBS Lett.* **2000**, *474*, 83–86.
63. Mironov, A.A.; Fickett, J.W.; Gelfand, M.S. Frequent alternative splicing of human genes. *Genome Res.* **1999**, *9*, 1288–1293.
64. Clark, T.A.; Schweitzer, A.C.; Chen, T.X.; Staples, M.K.; Lu, G.; Wang, H.; Williams, A.; Blume, J.E. Discovery of tissue-specific exons using comprehensive human exon microarrays. *Genome Biol.* **2007**, *8*, R64.

65. Djebali, S.; Davis, C.A.; Merkel, A.; Dobin, A.; Lassmann, T.; Mortazavi, A.; Tanzer, A.; Lagarde, J.; Lin, W.; Schlesinger, F.; *et al.* Landscape of transcription in human cells. *Nature* **2012**, *489*, 101–108.
66. Gonzalez-Porta, M.; Frankish, A.; Rung, J.; Harrow, J.; Brazma, A. Transcriptome analysis of human tissues and cell lines reveals one dominant transcript per gene. *Genome Biol.* **2013**, *14*, R70.
67. Wang, E.T.; Sandberg, R.; Luo, S.; Khrebtukova, I.; Zhang, L.; Mayr, C.; Kingsmore, S.F.; Schroth, G.P.; Burge, C.B. Alternative isoform regulation in human tissue transcriptomes. *Nature* **2008**, *456*, 470–476.
68. Harrison, P.M.; Kumar, A.; Lang, N.; Snyder, M.; Gerstein, M. A question of size: The eukaryotic proteome and the problems in defining it. *Nucleic Acids Res.* **2002**, *30*, 1083–1090.
69. Stamm, S. Signals and their transduction pathways regulating alternative splicing: A new dimension of the human genome. *Hum. Mol. Genet.* **2002**, *11*, 2409–2416.
70. Adams, M.D.; Celniker, S.E.; Holt, R.A.; Evans, C.A.; Gocayne, J.D.; Amanatides, P.G.; Scherer, S.E.; Li, P.W.; Hoskins, R.A.; Galle, R.F.; *et al.* The genome sequence of Drosophila melanogaster. *Science* **2000**, *287*, 2185–2195.
71. Fleischmann, R.D.; Adams, M.D.; White, O.; Clayton, R.A.; Kirkness, E.F.; Kerlavage, A.R.; Bult, C.J.; Tomb, J.F.; Dougherty, B.A.; Merrick, J.M.; *et al.* Whole-genome random sequencing and assembly of Haemophilus influenzae Rd. *Science* **1995**, *269*, 496–512.
72. Genome sequence of the nematode *C. elegans*: A platform for investigating biology. *Science* **1998**, *282*, 2012–2018.
73. Ramani, A.K.; Calarco, J.A.; Pan, Q.; Mavandadi, S.; Wang, Y.; Nelson, A.C.; Lee, L.J.; Morris, Q.; Blencowe, B.J.; Zhen, M.; *et al.* Genome-wide analysis of alternative splicing in Caenorhabditis elegans. *Genome Res.* **2011**, *21*, 342–348.
74. Graveley, B.R.; Brooks, A.N.; Carlson, J.W.; Duff, M.O.; Landolin, J.M.; Yang, L.; Artieri, C.G.; van Baren, M.J.; Boley, N.; Booth, B.W.; *et al.* The developmental transcriptome of Drosophila melanogaster. *Nature* **2011**, *471*, 473–479.
75. Roy, S.W. Intron-rich ancestors. *Trends Genet.* **2006**, *22*, 468–471.
76. Munding, E.M.; Igel, A.H.; Shiue, L.; Dorighi, K.M.; Trevino, L.R.; Ares, M., Jr. Integration of a splicing regulatory network within the meiotic gene expression program of Saccharomyces cerevisiae. *Genes Dev.* **2010**, *24*, 2693–2704.
77. Pleiss, J.A.; Whitworth, G.B.; Bergkessel, M.; Guthrie, C. Rapid, transcript-specific changes in splicing in response to environmental stress. *Mol. Cell* **2007**, *27*, 928–937.
78. Calarco, J.A.; Xing, Y.; Caceres, M.; Calarco, J.P.; Xiao, X.; Pan, Q.; Lee, C.; Preuss, T.M.; Blencowe, B.J. Global analysis of alternative splicing differences between humans and chimpanzees. *Genes Dev.* **2007**, *21*, 2963–2975.
79. Barbosa-Morais, N.L.; Irimia, M.; Pan, Q.; Xiong, H.Y.; Gueroussov, S.; Lee, L.J.; Slobodeniuc, V.; Kutter, C.; Watt, S.; Colak, R.; *et al.* The evolutionary landscape of alternative splicing in vertebrate species. *Science* **2012**, *338*, 1587–1593.
80. Venables, J.P.; Vignal, E.; Baghdiguian, S.; Fort, P.; Tazi, J. Tissue-specific alternative splicing of Tak1 is conserved in deuterostomes. *Mol. Biol. Evol.* **2012**, *29*, 261–269.
81. Merkin, J.; Russell, C.; Chen, P.; Burge, C.B. Evolutionary dynamics of gene and isoform regulation in Mammalian tissues. *Science* **2012**, *338*, 1593–1599.

82. Aoto, J.; Martinelli, D.C.; Malenka, R.C.; Tabuchi, K.; Sudhof, T.C. Presynaptic neurexin-3 alternative splicing trans-synaptically controls postsynaptic AMPA receptor trafficking. *Cell* **2013**, *154*, 75–88.
83. Colak, R.; Kim, T.; Michaut, M.; Sun, M.; Irimia, M.; Bellay, J.; Myers, C.L.; Blencowe, B.J.; Kim, P.M. Distinct types of disorder in the human proteome: Functional implications for alternative splicing. *PLoS Comput. Biol.* **2013**, *9*, e1003030.
84. Ellis, J.D.; Barrios-Rodiles, M.; Colak, R.; Irimia, M.; Kim, T.; Calarco, J.A.; Wang, X.; Pan, Q.; O'Hanlon, D.; Kim, P.M.; *et al.* Tissue-specific alternative splicing remodels protein-protein interaction networks. *Mol. Cell* **2012**, *46*, 884–892.
85. Gabut, M.; Samavarchi-Tehrani, P.; Wang, X.; Slobodeniuc, V.; O'Hanlon, D.; Sung, H.K.; Alvarez, M.; Talukder, S.; Pan, Q.; Mazzoni, E.O.; *et al.* An alternative splicing switch regulates embryonic stem cell pluripotency and reprogramming. *Cell* **2011**, *147*, 132–146.
86. Voineagu, I.; Wang, X.; Johnston, P.; Lowe, J.K.; Tian, Y.; Horvath, S.; Mill, J.; Cantor, R.M.; Blencowe, B.J.; Geschwind, D.H. Transcriptomic analysis of autistic brain reveals convergent molecular pathology. *Nature* **2011**, *474*, 380–384.
87. Kalsotra, A.; Cooper, T.A. Functional consequences of developmentally regulated alternative splicing. *Nat. Rev. Genet.* **2011**, *12*, 715–729.
88. Hsu, S.N.; Hertel, K.J. Spliceosomes walk the line: Splicing errors and their impact on cellular function. *RNA Biol.* **2009**, *6*, 526–530.
89. Luger, K.; Mader, A.W.; Richmond, R.K.; Sargent, D.F.; Richmond, T.J. Crystal structure of the nucleosome core particle at 2.8 A resolution. *Nature* **1997**, *389*, 251–260.
90. Schwartz, S.; Meshorer, E.; Ast, G. Chromatin organization marks exon-intron structure. *Nat. Struct. Mol. Biol.* **2009**, *16*, 990–995.
91. Tilgner, H.; Nikolaou, C.; Althammer, S.; Sammeth, M.; Beato, M.; Valcarcel, J.; Guigo, R. Nucleosome positioning as a determinant of exon recognition. *Nat. Struct. Mol. Biol.* **2009**, *16*, 996–1001.
92. Brugiolo, M.; Herzel, L.; Neugebauer, K.M. Counting on co-transcriptional splicing. *F1000Prime Rep.* **2013**, *5*, 9.
93. Tilgner, H.; Knowles, D.G.; Johnson, R.; Davis, C.A.; Chakrabortty, S.; Djebali, S.; Curado, J.; Snyder, M.; Gingeras, T.R.; Guigo, R. Deep sequencing of subcellular RNA fractions shows splicing to be predominantly co-transcriptional in the human genome but inefficient for lncRNAs. *Genome Res.* **2012**, *22*, 1616–1625.
94. Ameur, A.; Zaghlool, A.; Halvardson, J.; Wetterbom, A.; Gyllensten, U.; Cavelier, L.; Feuk, L. Total RNA sequencing reveals nascent transcription and widespread co-transcriptional splicing in the human brain. *Nat. Struct. Mol. Biol.* **2011**, *18*, U1435–U1157.
95. Kaida, D.; Berg, M.G.; Younis, I.; Kasim, M.; Singh, L.N.; Wan, L.; Dreyfuss, G. U1 snRNP protects pre-mRNAs from premature cleavage and polyadenylation. *Nature* **2010**, *468*, 664–668.
96. Hodges, C.; Bintu, L.; Lubkowska, L.; Kashlev, M.; Bustamante, C. Nucleosomal fluctuations govern the transcription dynamics of RNA polymerase II. *Science* **2009**, *325*, 626–628.
97. Kornblihtt, A.R.; Schor, I.E.; Allo, M.; Blencowe, B.J. When chromatin meets splicing. *Nat. Struct. Mol. Biol.* **2009**, *16*, 902–903.

98. Schwartz, S.; Ast, G. Chromatin density and splicing destiny: On the cross-talk between chromatin structure and splicing. *EMBO J.* **2010**, *29*, 1629–1636.
99. Carrillo Oesterreich, F.; Bieberstein, N.; Neugebauer, K.M. Pause locally, splice globally. *Trends Cell. Biol.* **2011**, *21*, 328–335.
100. Schor, I.E.; Allo, M.; Kornblihtt, A.R. Intragenic chromatin modifications: A new layer in alternative splicing regulation. *Epigenetics* **2010**, *5*, 174–179.
101. Tilgner, H.; Guigo, R. From chromatin to splicing: RNA-processing as a total artwork. *Epigenetics* **2010**, *5*, 180–184.
102. De la Mata, M.; Alonso, C.R.; Kadener, S.; Fededa, J.P.; Blaustein, M.; Pelisch, F.; Cramer, P.; Bentley, D.; Kornblihtt, A.R. A slow RNA polymerase II affects alternative splicing *in vivo*. *Mol. Cell* **2003**, *12*, 525–532.
103. Schor, I.E.; Rascovan, N.; Pelisch, F.; Allo, M.; Kornblihtt, A.R. Neuronal cell depolarization induces intragenic chromatin modifications affecting NCAM alternative splicing. *Proc. Natl. Acad. Sci. USA* **2009**, *106*, 4325–4330.
104. Luco, R.F.; Pan, Q.; Tominaga, K.; Blencowe, B.J.; Pereira-Smith, O.M.; Misteli, T. Regulation of alternative splicing by histone modifications. *Science* **2010**, *327*, 996–1000.
105. Braunschweig, U.; Gueroussov, S.; Plocik, A.M.; Graveley, B.R.; Blencowe, B.J. Dynamic integration of splicing within gene regulatory pathways. *Cell* **2013**, *152*, 1252–1269.

8

The Genomic Signature of Breast Cancer Prevention

Jose Russo *, Julia Santucci-Pereira and Irma H. Russo

The Irma H. Russo MD Breast Cancer Research Laboratory, Fox Chase Cancer Center, Temple University Health System, 333 Cottman Avenue, Philadelphia, PA 19111, USA; E-Mail: Julia.Pereira@fccc.edu

* Author to whom correspondence should be addressed; E-Mail: Jose.Russo@fccc.edu;

Abstract: The breast of parous postmenopausal women exhibits a specific signature that has been induced by a full term pregnancy. This signature is centered in chromatin remodeling and the epigenetic changes induced by methylation of specific genes which are important regulatory pathways induced by pregnancy. Through the analysis of the genes found to be differentially methylated between women of varying parity, multiple positions at which beta-catenin production and use is inhibited were recognized. The biological importance of the pathways identified in this specific population cannot be sufficiently emphasized because they could represent a safeguard mechanism mediating the protection of the breast conferred by full term pregnancy.

Keywords: normal breast; breast cancer; genomic signature; prevention; pregnancy; splicing mechanisms; methylation; chromatin remodeling; Lnc-RNA; beta-catenin

1. Introduction

More than 300 years ago, an excess in breast cancer mortality in nuns was reported, in whom the increased risk was attributed to their childlessness [1] until MacMahon *et al.* [2] found an almost linear relationship between a woman's risk and the age at which she bore her first child. This work confirmed that pregnancy had a protective effect that was evident from the early teen years and persisted until the middle twenties [1]. Other studies have reported that additional pregnancies and breastfeeding confer greater protection to young women, including a statistically significantly reduced risk of breast cancer in women with deleterious BRCA1 mutations who breast-fed for a cumulative total of more than one year [3,4]. Our studies, designed to unravel what specific changes occurred in the breast during pregnancy that confer a lifetime protection from developing cancer, led us to the discovery that endogenous endocrinological or environmental influences affecting breast development before the first full term pregnancy were important modulators of the susceptibility of the breast to undergo neoplastic transformation. The fact that exposure of the breast of young nulliparous females to environmental physical agents [5] or chemical toxicants [6,7] results in a greater rate of cell transformation suggests that the immature breast possesses a greater number of susceptible cells that can become the site of the origin of cancer, similarly to what has been reported in experimental animal models [8–11]. In these models, the initiation of cancer is prevented by the differentiation of the mammary gland induced by pregnancy [11,12]. The molecular changes involved in this phenomenon are just starting to be unraveled [13–18]. The protection conferred by pregnancy is age-specific since a delay in childbearing after age 24 progressively increases the risk of cancer development. Eventually, this risk becomes greater than that of nulliparous women when the first full term pregnancy (FFTP) occurs after 35 years of age [2]. The higher breast cancer risk which has been associated with early menarche further emphasizes the importance of the length of the susceptibility "window" that encompasses the period of breast development occurring between menarche and the first pregnancy, when the organ is more susceptible to undergo complete differentiation under physiological hormonal stimuli. Differentiation is a hallmark that protects the breast from developing cancer by lessening the risk of suffering genetic or epigenetic damages. This postulate is supported by our observations that the architectural pattern of lobular development in parous women with cancer differs from that of parous women without cancer; the former being similar to the architectural pattern of lobular development of nulliparous women with or without cancer. Thus, the higher breast cancer risk in parous women might have resulted from either a failure of the breast to fully differentiate under the influence of the hormones of pregnancy and/or proliferation of transformed cells initiated by early damage or genetic predisposition [18].

Numerous studies have been performed to understand how the dramatic modifications that occur during pregnancy in the pattern of lobular development and differentiation, cell proliferation, and steroid hormone receptor content of the breast influence cancer risk [18]. Studies at the molecular level using different platforms for global genome analysis have confirmed the universality of this phenomenon in various strains of rats and mice [13–21]. Studies in experimental animal models have been useful for uncovering the sequential genomic changes occurring in the mammary gland in response to multiple hormonal stimuli of pregnancy that lead to the imprinting of a permanent genomic signature. Our results support our hypothesis that post-menopausal parous women exhibit a genomic "signature" that differs from the expression present in the breast of nulliparous women, who traditionally represent a high breast cancer risk group.

2. Phenotypic Changes Induced by Pregnancy in the Human Breast

Our study has been done using core biopsies of nulliparous (NP) and parous (P) postmenopausal women [22,23]. The nulliparous group included both nulligravida nulliparous (NN) and gravida nulliparous (GN); both NN and GN women were considered within the NP as a single group for most analyses, unless indicated otherwise. Our previous studies have in great part clarified the role of pregnancy-induced breast differentiation in the reduction in breast cancer risk, as well as the identification of lobules type 1 (Lob 1) or the terminal ductal lobular unit (TDLU) as the site of origin of breast cancer [4,7,24]. The morphological, physiological and genomic changes resulting from pregnancy and hormonally-induced differentiation of the breast and their influence on breast cancer risk have been addressed in previous publications [4,7,24,25]. Our observations that during the post-menopausal years the breast of both parous and nulliparous women contains preponderantly Lob 1, and the fact that nulliparous women are at higher risk of developing breast cancer than parous women, indicate that Lob 1 in these two groups of women either differ biologically, or exhibit different susceptibility to carcinogenesis [25]. The breast tissues of the P and NP women contained ducts and Lob 1 [4,12,26].

The microscopic analysis of the breast tissue revealed that the population of luminal cells lining ducts and Lob 1 was composed of cells that were characterized by their nuclear appearance into two types: one that contained large and palely stained nuclei with prominent nucleoli and another consisting of small hyper chromatic nuclei [27]. The pale staining of the large former nuclei is a feature indicative of a high content of non-condensed euchromatin; these nuclei were called euchromatin-rich nuclei (EUN). The hyperchromasia observed in the latter nuclei was indicative of chromatin condensation and high content of heterochromatin; these nuclei were identified as heterochromatin-rich nucleus (HTN). The analysis of the distribution of HTN and EUN cells in histological sections of the breast core biopsies revealed that EUN were more abundant in the NP than in the P breast tissues, whereas the inverse was true for the HTN; these differences were statistically significant [27]. We have confirmed the differences between the HTN and EUN using a quantitative image analysis system [27]. The nuclear size (diameter, area and perimeter) of the EUN as a whole was significantly higher ($p < 0.05$) than that of the HTN in both nulliparous and parous women. Differences were also found to be statistically significant ($p < 0.05$) regarding the nuclear shape (nuclear feret ratio) in the breast of nulliparous women, indicating that in these breasts the nuclei of the HTN had a more elongated ellipsoidal shape than the EUN. The light absorbance (mean gray values/nucleus) was always greater for EUN than for HTN of both NP and P breasts, either considered as two groups or individually, an indication that under densitometric terms HTN were always more densely stained than EUN. Comparison of the EUN of nulliparous *vs.* parous breasts revealed significant differences in nuclear size, stainability and densitometric energy, leading us to conclude that epithelial cell nuclei were larger, less stainable and with smaller regions with uniform densitometric intensity in nulliparous breasts. Comparison of the HTN of nulliparous *vs.* parous breasts revealed significant differences in nuclear diameter, perimeter, shape and stainability; cell nuclei showed larger contours and more elongated ellipsoidal shape and they were more stainable in nulliparous breasts. These observations indicated that a shift of the EUN cell population to a more densely packed chromatin cell (HTN) had occurred in association with the history of pregnancy as a distinctive pattern of the postmenopausal parous breast [27].

Since chromatin condensation is part of the process of chromatin remodeling towards gene silencing that is highly regulated by methylation of histones, we verified this phenomenon by immunohistochemistry (IHC) incubating NP and P breast tissues with antibodies against histone 3 dimethylated at lysine 9 (H3K9me2) and trimethylated at lysine 27 (H3K27me3) [27]. The IHC stain revealed that methylation of H3 at both lysine 9 and 27 was increased in the heterochromatin condensed nuclei of epithelial cells of the parous breast when compared to the euchromatin rich nuclei of the nulliparous breast. In the nulliparous breast, the reactivity in individual cells was less intense and the number of positive cells was significantly lower. These variations in chromatin reorganization were supported by the upregulation of CBX3, CHD2, L3MBTL, and EZH2 genes controlling this process (Table 1) [27].

Table 1. Genes upregulated in the parous breast.

Symbol	Log Ratio	*P* value	Gene Name
		Apoptosis (GO:0006915; GO:0006917; GO:0008624; GO:0042981)	
CASP4	0.37	0.0003	caspase 4, apoptosis-related cysteine peptidase
RUNX3	0.36	0.0000	runt-related transcription factor 3
LUC7L3	0.34	0.0002	LUC7-like 3 (S. cerevisiae)
ELMO3	0.30	0.0003	engulfment and cell motility 3
		DNA repair (GO:0006281; GO:0006284)	
SFPQ	0.46	0.0002	splicing factor proline/glutamine-rich
MBD4	0.36	0.0003	methyl-CpG binding domain protein 4
RBBP8	0.32	0.0000	retinoblastoma binding protein 8
		Cell adhesion (GO:0007155; GO:0030155)	
NRXN1	0.60	0.0001	neurexin 1
DSC3	0.51	0.0000	desmocollin 3
COL27A1	0.44	0.0002	collagen, type XXVII, alpha 1
PNN	0.37	0.0001	pinin, desmosome associated protein
COL4A6	0.36	0.0008	collagen, type IV, alpha 6
LAMC2	0.34	0.0008	laminin, gamma 2
COL7A1	0.33	0.0002	collagen, type VII, alpha 1
COL16A1	0.31	0.0000	collagen, type XVI, alpha 1
LAMA3	0.30	0.0008	laminin, alpha 3
		Cell cycle (GO:0000075; GO:0007049; GO:0045786)	
SYCP2	0.45	0.0000	synaptonemal complex protein 2
PNN	0.37	0.0001	pinin, desmosome associated protein
RUNX3	0.36	0.0000	runt-related transcription factor 3
RBBP8	0.32	0.0000	retinoblastoma binding protein 8
		Cell differentiation (GO:0001709; GO:0030154; GO:0030216)	
MGP	0.53	0.0003	matrix Gla protein
KRT5	0.41	0.0002	keratin 5
GATA3	0.35	0.0009	GATA binding protein 3
LAMA3	0.30	0.0008	laminin, alpha 3
		Cell proliferation (GO:0008283; GO:0008284; GO:0008285; GO:0042127; GO:0050679; GO:0050680)	
PTN	0.67	0.0002	Pleiotrophin
KRT5	0.41	0.0002	keratin 5
RUNX3	0.36	0.0000	runt-related transcription factor 3
IL28RA	0.34	0.0003	interleukin 28 receptor, alpha (interferon, lambda receptor)
CDCA7	0.31	0.0005	cell division cycle associated 7
		Cell motility (GO:0006928; GO:0030334)	
DNALI1	0.37	0.0001	dynein, axonemal, light intermediate chain 1
LAMA3	0.30	0.0008	laminin, alpha 3
		G-protein coupled receptor pathway (GO:0007186)	
OXTR	0.54	0.0006	oxytocin receptor

Table 1. *Cont.*

Symbol	Log Ratio	*P* value	Gene Name
RNA metabolic process (GO:0000398; GO:0001510; GO:0006376; GO:0006396; GO:0006397; GO:0006401; GO:0008380)			
METTL3	0.69	0.0000	methyltransferase like 3
HNRPDL	0.65	0.0001	heterogeneous nuclear ribonucleoprotein D-like
HNRNPD	0.59	0.0003	heterogeneous nuclear ribonucleoprotein D (AU-rich element RNA binding protein 1, 37 kDa)
HNRNPA2B1	0.56	0.0003	heterogeneous nuclear ribonucleoprotein A2/B1
SFPQ	0.47	0.0006	splicing factor proline/glutamine-rich
RBM25	0.38	0.0009	RNA binding motif protein 25
RBMX	0.38	0.0000	RNA binding motif protein, X-linked
LUC7L3	0.34	0.0002	LUC7-like 3 (S. cerevisiae)
SFRS1	0.30	0.0001	splicing factor, arginine/serine-rich 1
RNA transport (GO:0050658)			
HNRNPA2B1	0.56	0.0003	heterogeneous nuclear ribonucleoprotein A2/B1
Transcription (GO:0006350; GO:0006355; GO:0006357; GO:0006366; GO:0016481; GO:0045449; GO:0045893; GO:0045941)			
HNRPDL	0.65	0.0001	heterogeneous nuclear ribonucleoprotein D-like
HNRNPD	0.59	0.0003	heterogeneous nuclear ribonucleoprotein D (AU-rich element RNA binding protein 1, 37 kDa)
CBX3	0.53	0.0003	chromobox homolog 3 (HP1 gamma homolog, Drosophila)
NFKBIZ	0.48	0.0001	nuclear factor of kappa light polypeptide gene enhancer in B-cells inhibitor, zeta
FUBP1	0.47	0.0002	far upstream element (FUSE) binding protein 1
SFPQ	0.47	0.0006	splicing factor proline/glutamine-rich
EZH2	0.44	0.0000	enhancer of zeste homolog 2 (Drosophila)
ZNF207	0.41	0.0007	zinc finger protein 207
ZNF711	0.41	0.0003	zinc finger protein 711
GATA3	0.38	0.0009	GATA binding protein 3
PNN	0.37	0.0003	pinin, desmosome associated protein
ZNF107	0.37	0.0001	zinc finger protein 107
RUNX3	0.36	0.0000	runt-related transcription factor 3
CCNL1	0.35	0.0009	cyclin L1
ZNF692	0.34	0.0000	zinc finger protein 692
CHD2	0.33	0.0001	chromodomain helicase DNA binding protein 2
RBBP8	0.32	0.0000	retinoblastoma binding protein 8
ZNF789	0.32	0.0005	zinc finger protein 789
CDCA7	0.31	0.0005	cell division cycle associated 7
Chromatin organization (GO:0006333; GO:0006338)			
CBX3	0.53	0.0003	chromobox homolog 3 (HP1 gamma homolog, Drosophila)
CHD2	0.33	0.0001	chromodomain helicase DNA binding protein 2
Cell division (GO:0051301)			
SYCP2	0.45	0.0000	synaptonemal complex protein 2
DNA metabolic process (GO:0006139; GO:0006260; GO:0006310; GO:0015074)			
METTL3	0.69	0.0000	methyltransferase like 3
SFPQ	0.46	0.0002	splicing factor proline/glutamine-rich
GOLGA2B	0.32	0.0001	golgin A2 family, member B
Lactation (GO:0007595)			
OXTR	0.54	0.0006	oxytocin receptor

3. Transcriptomic Differences Induced by Pregnancy

Analysis of P and NP gene expression microarrays revealed that there were 305 probe sets, corresponding to 208 distinct genes, differentially expressed between these two groups. Of the 305 probe sets, 267 were up- and 38 were down-regulated [22,23]. From these 267 up-regulated genes, we described biological processes that were representative of the transcriptomic differences between the parous and the nulliparous breasts. Using bioinformatics based analysis of microarray data, we found that the biological processes involving the splicing machinery and mRNA processing were prevalent in the parous breast and were represented by the following upregulated genes: LUC7L3, SFRS1, HNRNPA2B1, HNRNPD, RBM25, SFRS5, METTL3, HNRNPDL, and SFPQ (Table 1). Transcription regulation and chromatin organization were also highly represented in the parous breast by the upregulation of CBX3, EBF1, GATA3, RBBP8, CCNL1, CCNL2, CDCA7, EZH2, FUBP1, NFKBIZ, RUNX3, ZNF107, ZNF207, ZNF692, ZNF711, ZNF789, CDCA7, and ZNF692 (Table 1). The parous breast also expressed upregulation of six non-coding regions that included XIST, MALAT-1 (or NEAT2) and NEAT1 [27].

Genes that were down-regulated in the parous breast represented transcription regulation, encompassing CBL, FHL5, NFATC3, NCR3C1, TCF7L2, and a set of genes that were involved in IGF-like growth factor signaling, somatic stem cell maintenance, muscle cell differentiation and apoptosis, such as IGF1, RASD1, EBF1, SOX 1,SOX6, SOX 17, RALGAPA2 and ABHD5. In rodents, also was observed the reduction of expression of genes related to growth factors, such as Igfl [15]. The level of expression was confirmed to be differentially expressed between nulliparous and parous breast tissues by real time RT-PCR for the following genes: CREBZF, XIST, MALAT1, NEAT1, CCNL2, GATA3, DDX17, HNRPDL, SOX6, SNHG12, SOX 17 and C1orf168 [23]. In addition to the level of expression, the localization of the alternative splicing regulator cyclin L2 protein (CCNL2) [28], was verified by IHC. CCNL2 protein was expressed in the nucleus of epithelial cells in breast tissues from NP and P women, although the level of expression was significantly higher in Lob 1 in the parous breast when compared with similar structures found in the breast of nulliparous women. These observations confirmed the localization of this gene product in the splicing factor compartment (nuclear speckles) [29].

4. Shifting of the Cell Population in the Human Breast

We found a shift in the cell population of the postmenopausal breast as a manifestation of the reprogramming of the organ after pregnancy. These observations are in agreement with what is observed in the rat mammary gland, which also contains two types of luminal epithelial cells, designated dark (DC) and intermediate (IC) cells, in addition to the myoepithelial cells [30]. The DC and IC are equivalent to the HTN and EUN cells described in the present work. DCs increase after pregnancy and lactational involution; whereas the ICs significantly outnumber the DC in ductal hyperplasias and ductal carcinomas [30,31]. Our analysis of nuclear ultrastructural and morphometric parameters of rodent IC have allowed us to differentiate the mammary progenitor stem cell from the cancer stem cells [25,30,31]. Nuclear morphometric analysis of breast and ovarian carcinomas has confirmed the predictive value of nuclear grade on the progression of premalignant lesions to invasiveness [32–34]. Our findings of a significant decrease in the number of EUN with a subsequent increase in the number of HTN cells expressing specific biomarkers identified at the chromatin and transcriptional levels support the value of morphometric analysis as an adjuvant to molecular studies [27]. Our data clearly indicate that there are morphological indications of chromatin remodeling in the parous breast, such as the increase in the number of epithelial cells with condensed chromatin and increased reactivity with anti-H3K9me2 and H3K27me3 antibodies. Histone methylation is a major determinant for the formation of active and inactive regions of the genome and is crucial for the proper programming of the genome during development [35]. In the parous breast, there is upregulation of transcription factors and chromatin remodeling genes such as CHD2 or chromodomain helicase DNA binding protein 2 and the CBX3 or Chromobox homolog 3, whose products are required for controlling recruitment of protein/protein or DNA/protein interactions. CBX3 is involved in transcriptional silencing in heterochromatin-like complexes, and recognizes and binds H3 tails methylated at lysine 9, leading to epigenetic repression. Two other important genes related to the polycomb group (PcG) protein that are upregulated in the parous breast are the L3MBTL gene or l(3)mbt-like and the histone-lysine N-methyltransferase or EZH2. Members of the PcG form multimeric protein complexes that maintain the transcriptional repressive state of genes over successive cell generations (Table 1). EZH2 is an enzyme that acts mainly as a gene silencer, performing this role by the addition of three methyl groups to lysine 27 of histone 3, a modification that leads to chromatin condensation [30,36,37].

5. Methylation Changes in the DNA of Parous Women are Part of Chromatin Remodeling and the Genomic Signature of Pregnancy

The chromatin remodeling process is demonstrated not only by the shifting of the EUN to the HTN cells, but also confirmed by the increase in methylation of histones H3K9me2 and H3K27me3. This is an indication that methylation of other genes could also be involved in the process. Using the DNA from five nulliparous and five parous breast core biopsies and applying the MBD-cap sequencing methodology [38], we have identified 583 genes showing different levels of methylation between the parous and nulliparous breasts. From the 583 genes, 455 were hypermethylated in the parous while 128 were hypermethylated in the nulliparous breast, confirming the reprogramming of the chromatin to a more silenced or resting stage. To get a better understanding of the methylation profile of the 583 genes, we used Integrative Genomics Viewer (IGV) software [39,40]. IGV was utilized to identify the distinct areas, throughout the entire gene, where the methylation levels differed between the sample groups. The identification of these areas, known as differentially methylated regions (DMRs), is important because they are more likely to affect gene expression [41]. We performed the comparison between the nulliparous and parous methylation profiles against the human reference genome "hg 18" and against each other. For example, the gene COBRA 1, which is the cofactor of BRCA1 and has been shown to work in its regulatory pathway [42], was hypermethylated in the nulliparous breast. It is shown in Figure 1 that the methylation levels for each sample at each base pair that an area of higher methylation occurring in at least four of the samples of one group as compared to all members of the opposing group, that area was defined as a (DMR) (Figures 1 and 2). COBRA1 had a DMR near the end of the gene, which was marked in Figure 1 using the IGV's marking tool. When a differentially methylated area is found and marked, hovering over the red marker at the top of the sample area gives the exact chromosomal location. Every gene within the 583 gene list was closely examined for DMRs. The chromosomal locations at which these DMRs were found and marked were recorded in Tables 2 and 3.

Figure 1. Overview of how the DNA methylation levels appear in the Integrative Genomics Viewer (IGV). At the top of the figure is the ideogram of the chromosome given by IGV, with the area currently being examined marked in red. At the bottom is the overall shape of the gene containing exons and introns. Exons are shown as thicker blue sections on the overall gene. The gray bars represent the methylation levels of each volunteer at each base pair. They are created by combining each read resulting from the sequencing done on the samples. The higher they are, the higher the percentage of methylation is at any given base pair. When there was an area of higher methylation occurring in at least four of the members of one parity group as compared to all members of the opposing group, that area was defined as a differentially methylated region (DMR).

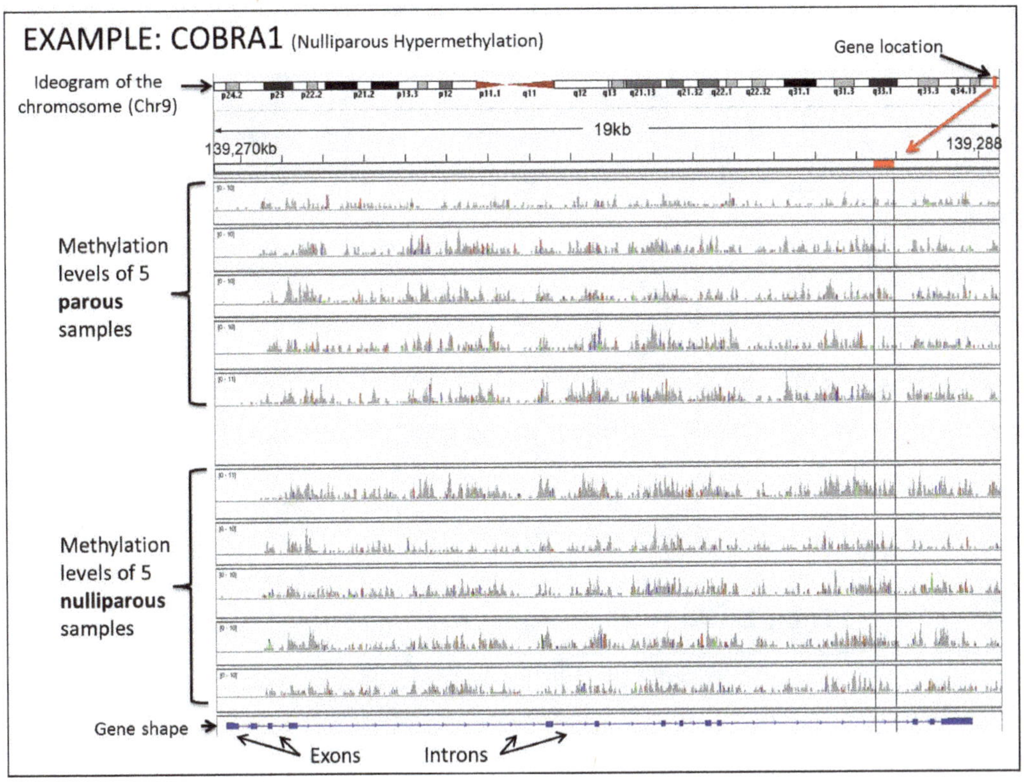

After analysis of the 583 genes using the IGV, we have identified the DMRs of 53 genes. Of the 455 parous hypermethylated genes, 41 had DMRs. These were NEGR1, NUF2, SYT14, POU4F1, FLRT2, ASAP2, DNAJC13, IFITM4P, ZNF292, SDK1, ELAVL4, DACT1, SPATA5L1, DYNC1I2, NLGN1, MAN1A1, AK5, DPYD, PROX1, PDE3A, NOVA1, SKAP1, ANKRD12, B4GALT5, CNTN4, ROBO1, GSK3B, INPP4B, FNIP2, IL6ST, TICAM2, PPP2CA, C6orf138, PRKAR2B, TTLL7, MAN1A2, CDC42BPA, OSBP, STIM2, NR3C2, and REV3L. The exact locations of these DMRs are recorded in Table 2. A point of interest within these genes is that DNAJC13 and GSK3B, while statistically given to be hypermethylated within parous women, had DMRs which suggested nulliparous hypermethylation. Because of this and for the scope of this experiment, those genes are treated as nulliparous hypermethylated. Of the 128 nulliparous hypermethylated genes, 12 had DMRs.

These were NHSL2, PTX4, LRRC37A3, C20orf166-AS1, TPPP, NELF, SAMD10, CELSR1, FZD1, TNFRSF18, SRMS, and COBRA1. The chromosomal locations of these DMRs can be seen in Table 3. Within this list only C20orf166-AS1 was found to have a DMR in the direction opposite to what the statistics showed. Visual examples of these differentially methylated areas are seen in Figure 2 and Supplementary Figures S1–S4.

Figure 2. DMRs for PRKAR2B. At the top we see the gene shape, with the red marked DMRs. Any colored locations within the gray bars indicate a nucleotide read which is different from the reference genome.

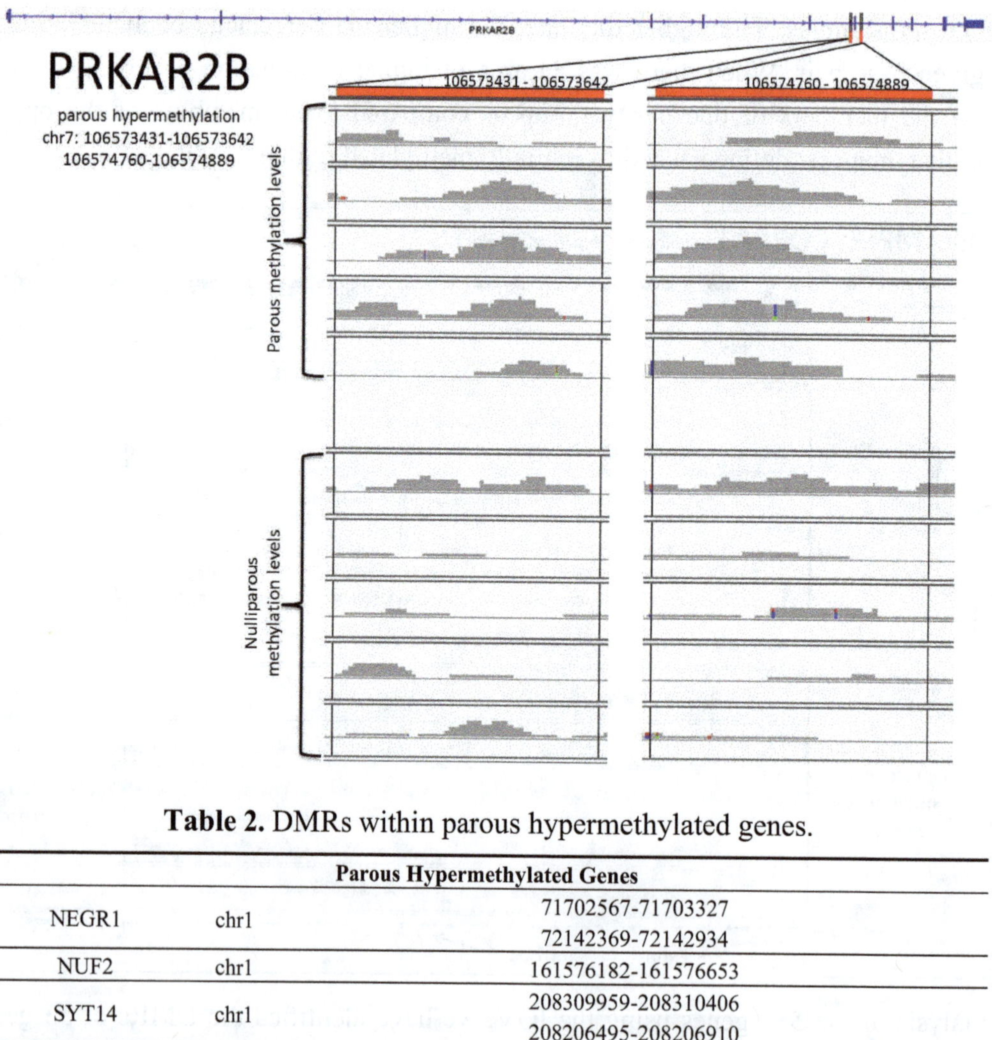

Table 2. DMRs within parous hypermethylated genes.

Parous Hypermethylated Genes		
NEGR1	chr1	71702567-71703327
		72142369-72142934
NUF2	chr1	161576182-161576653
SYT14	chr1	208309959-208310406
		208206495-208206910
POU4F1	chr13	78072725-78073146
FLRT2	chr14	85155301-85155789
ASAP2	chr2	9266977-9267464
		9432659-9433115
DNAJC13	Chr3	133712540-133712930
IFITM4P	Chr6	29826792-29827266
ZNF292	Chr6	88022117-88022631
SDK1	Chr7	4121961-4122279
		4230104-4230384
ELAVL4	Chr1	50387715-50388146
DACT1	Chr14	58182547-58182717
SPATA5L1	Chr15	43494615-43495210

Table 2. *Cont.*

Parous Hypermethylated Genes		
DYNC1I2	Chr2	172279940-172280462
NLGN1	Chr3	175147546-175148159 175156296-175156626 175277928-175278476
MAN1A1	Chr6	119623891-119624320
AK5	Chr1	77616541-77616886 77655265-77655548
DPYD	Chr1	98153997-98154252
PROX1	Chr1	212267523-212267905
PDE3A	Chr12	20432463-20432808
NOVA1	Chr14	26015695-26016215
SKAP1	Chr17	43591761-43592022
ANKRD12	Chr18	9168269-9168654
B4GALT5	Chr20	47704095-47704520
CNTN4	Chr3	2572819-2573349
ROBO1	Chr3	79026030-79023709
GSK3B	Chr3	121258375-121258501
INPP4B	Chr4	143292977-143293319 143347212-143347585 143966478-143966985
FNIP2	Chr4	159911129-159911596 160015288-160015809
IL6ST	Chr5	55271135-55271466
TICAM2	Chr5	114955685-114955992 114956473-114956938
PPP2CA	Chr5	133567556-133567871
C6orf138	Chr6	48025616-48025836 48067151-48067418
PRKAR2B	Chr7	106573431-106573642 106574760-106574889
TTLL7	Chr1	84185339-84185660
MAN1A2	Chr1	117816180-117816444
CDC42BPA	Chr1	225520202-225520399
OSBP	Chr11	59121100-59121437 59121927-59122155
STIM2	Chr4	26572404-26572775
NR3C2	Chr4	149367631-149368052
REV3L	Chr6	111804054-111804285

Table 3. DMRs within nulliparous hypermethylated genes.

Nulliparous Hypermethylated Genes		
NHSL2	chrX	71270541-71271527
C16orf38 (PTX4)	Chr16	1476600-1476773
LRRC37A3	Chr17	60311872-60311982
C20orf200 (C20orf166-AS1)	Chr20	60557111-60557421
TPPP	Chr5	742334-742618
NELF	Chr9	139471353-139471653 (HYPO) 139471653-139471895
SAMD10	Chr20	62077471-62077661
CELSR1	Chr22	45272965-45273071
FZD1	Chr7	90733372-90733621
TNFRSF18	Chr1	1130349-1130634
SRMS	Chr20	61646714-61647041
COBRA1	Chr9	139285424-139285977

Analysis and research into the functions of these 53 genes identified seven which interacted with each other in either the Wnt signaling pathway or its controlling PI3K/AKT/mTOR pathways. The DMRs of these genes (DACT1, PPP2CA, GSK3B, ROBO1, INPP4B, IL6ST, FZD1) are shown in Supplementary Figures S1–S4. An overview of the involvement in the canonical Wnt pathway is shown in Figure 3. The interworking of these genes with each other and with other genes within the statistically methylated 583 can be seen in Figure 4.

Figure 3. Canonical WNT/β-catenin signaling genes marked in green are hypermethylated in parous women (suggesting down-regulation of the gene in parous women). Genes in red are hypermethylated within nulliparous women. Genes marked with (*) were observed differentially expressed the microarray data. This canonical pathway was generated through the use of IPA (Ingenuity® Systems) [43].

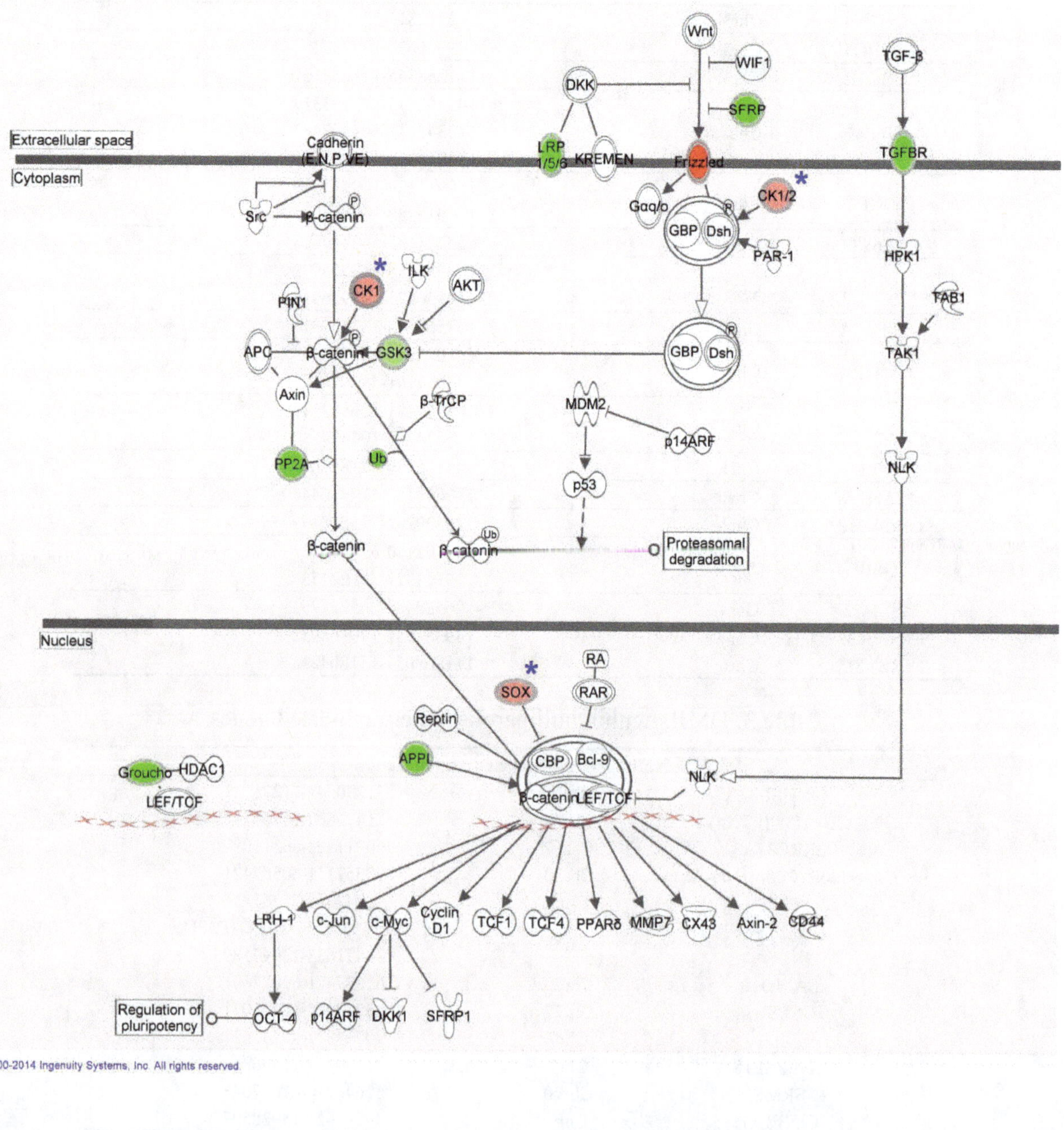

Figure 4. Interaction of target genes in Wnt/β-catenin signaling. The green genes are statistically parous hypermethylated, while the ones colored red are statistically nulliparous hypermethylated. The darker genes have recorded DMRs, and this is to the exception of GSK3B, which was first found statistically significant hypermethylated in the parous breast, but its DMR is hypermethylated in the nulliparous samples. This network was generated through the use of IPA (Ingenuity® Systems) [43].

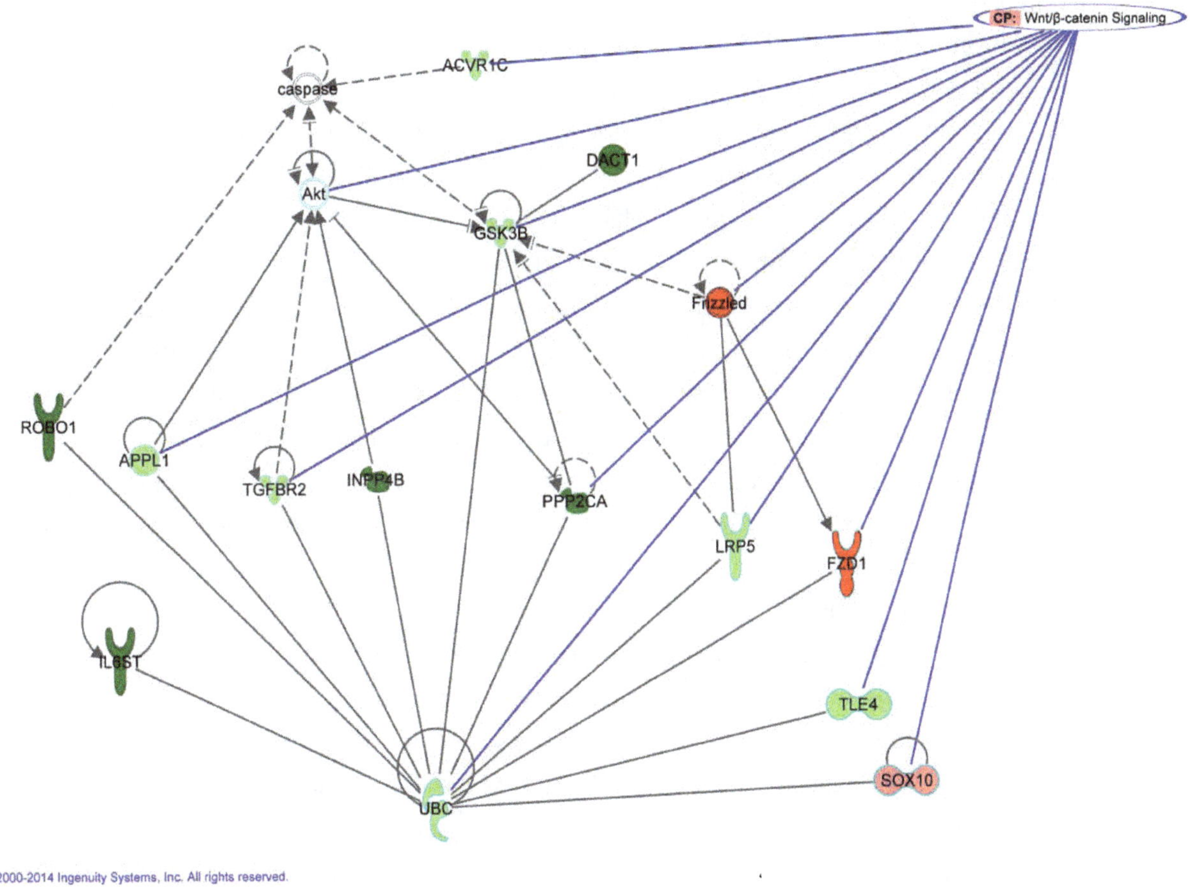

Of the seven genes with DMRs which we have shown to work together in the Wnt pathway or its controllers, three worked directly in canonical Wnt signaling. Interestingly, when we analyzed the genes differentially expressed between parous and nulliparous [23], we found genes that also participate in the Wnt pathway, such as CSNK1A1 and SOX family (Figure 3). FZD1, which is the hypermethylated in the nulliparous breast, codes for the Frizzled receptor. When activated, this receptor directly activates Disheveled (Dsh) in the cytosol to begin the Wnt signaling cascade [44]. GSK3B, which also contains DMRs hypermethylated in the nulliparous women, has as main rule to decrease beta-catenin levels in the Wnt signaling pathway [45]. PPP2CA (PP2A) is suggested to work both upstream and downstream of beta-catenin to assist in its stabilization [46]. DACT1 assists in Wnt signaling by up-regulating GSK3B [47]. ROBO1, INPP4B and IL6ST genes are active in PI3K dependent AKT signaling [48–50].

The potential significance of the Wnt signaling pathway is rooted in an experiment performed in 1982 to find which genes would mutate in mice injected with mouse mammary tumor virus locating int1, a proto-oncogene [51]. Int1 was soon found to be highly conserved across multiple species, including drosophila and humans. Int1 was discovered to be the mammalian homologue of the drosophila Wingless (Wg), a gene previously found to be a segment polarity gene in embryonic development. The Wnt signaling pathway was given its name from the combination of Wg and int1, and has always had a close relationship to both differentiation and breast cancer.

Mammary development requires complex, reciprocal epithelial mesenchymal interactions. During embryonic development, Wnt signaling is involved in the initiation and early formation of mammary buds [52]. Then, during pregnancy, the pathway is activated to help the differentiation of mammary ducts in preparation for lactation. It does this by increasing beta-catenin levels in the cytosol and the nucleus, which in turn increases epithelial-mesenchymal transition and aids in transcription. After weaning, the mammary glands go through involution and the E-cadherin binding domain for beta-catenin is truncated [53]. This decreases cellular adhesion and signal epithelial apoptosis. The result is a lessened need for beta-catenin. In fact, overexpression of beta-catenin during involution results in a lack of complete involution [54]. This suggests that lowered beta-catenin expression is essential for proper mammary involution. Studies in mouse model systems clearly demonstrate that activated Wnt signaling leads to mammary tumorigenesis [55]. Misra *et al.* observed alteration in Fzd4 and Wnt2 expression in rats after full term pregnancy [20]. Other studies have shown an increase in cytosolic/nuclear beta-catenin in about 60% of breast cancers. This is usually explained by the pathway's ability to aid in epithelial-mesenchymal transition and cell proliferation, two things incredibly important in the progression of cancer. Recently, the Wnt signaling pathway has been directly implicated in the parity induced protective effect against breast cancer [56]. It was revealed that parity induces differentiation and down-regulates the Wnt/Notch signaling ratio of basal stem/progenitor cells in mice. The down-regulation was attributed to a reduced expression of Wnt4, a necessary ligand in the activation stages of the Wnt pathway, in the mammary cells of parous mice [56].

The nulliparous hypermethylation of FZD1 suggests an up-regulation of the Frizzled family receptors and through this an up-regulation of all three types of Wnt signaling, indeed, we observed a slight overexpression of this gene in the parous women (not statistically significant). Increased Wnt signaling is associated with an increase in EMT in both development and cancer [57,58]. However, despite the Wnt signaling pathways being seemingly up-regulated, key genes within the pathways appear within our data to be down-regulated, thus changing the outcome of the signals sent through the Frizzled receptors. Signals sent through the Fz receptors activate the phosphoprotein Disheveled (Dsh). Dsh has three highly conserved protein domains, which interact differently depending on which Wnt pathway it is interacting with [44]. An up-regulation of FZD1 assumes an overall up-regulation of Dsh activation, and thus an increase in all three Wnt pathways. The three pathways are the canonical Wnt/beta-catenin pathway, the noncanonical planar cell polarity (PCP) pathway, and the noncanonical Wnt/calcium pathway.

The canonical pathway is the only one to involve beta-catenin, which is the TCF/LEF binding protein responsible for increased transcription and EMT [57,58]. Intracellular beta-catenin levels are maintained through constant creation and destruction, the processes of which are suggested to be regulated differently between our parity groups.

The canonical Wnt pathway contains the beta-catenin destruction complex, which is usually down-regulated or disrupted after the activation of Wnt signaling. The most effective way this occurs is through the binding of Fz to LRP5/6, which will disrupt the destruction complex before it can begin [59]. Our analysis showed an increased methylation of LRP5 within parous women, which suggests a decreased expression of LRP5/6 and a decreased cellular capability to stop the beta-catenin destruction complex in this way. The beta-catenin destruction complex begins with the binding of GSK3 to Axin, which leaves GSK3's active site open to phosphorylate beta-catenin. Once phosphorylated, beta-catenin is ubiquitinated and sent to the proteasome for removal [59]. It is suggested that initial tumor development requires rapid and effective repression of GSK3B [58]. In our analysis through IGV, GSK3B was found to have a DMR hypermethylated in the nulliparous samples. This suggests an increase in expression of GSK3 within parous women and subsequently an increase in the activity of the beta-catenin destruction complex.

PPP2CA, found to be hypermethylated within parous women, is also closely involved in canonical Wnt signaling. While the effect of PPP2CA in this context is still unclear, research leans toward a positive ability to stabilize beta-catenin [59]. The parous hypermethylation of PPP2CA, which suggests a lower expression in parous women, supports the idea of decreased beta-catenin.

The noncanonical Wnt/calcium pathway, which is also found to be up-regulated in parous women as a result of increased FZD1 expression, occurs independently of beta-catenin. However, the noncanonical Wnt/calcium pathway is an inhibitor of canonical Wnt/beta-catenin signaling further along the line by stopping the transcriptional efforts of beta-catenin in the nucleus [60]. This inhibition occurs in one of two ways. The first uses the CaMKII-TAK1-NLK pathway, which inhibits beta-catenin-TCF-dependent transcription through the phosphorylation of TCF. The second uses NFAT-mediated transcriptional regulation to suppress beta-catenin-dependent-transcription.

Whereas more mechanistic studies need to be done in human breast cells, the data analyzed thus far indicate that the methylation of genes involved in Wnt signaling pathway could be another path involved in the protective effect of pregnancy in the human breast.

6. Conclusions

Our work [22,23,27] clearly demonstrates that the breast of parous postmenopausal women exhibits a specific signature that has been induced by a full term pregnancy. This signature reveals for the first time that the differentiation process is centered in chromatin remodeling and the epigenetic changes induced by methylation of specific genes, that are important regulatory pathways induced by pregnancy. Through the analysis of the genes found to be differentially methylated between women of varying parity, multiple positions at which beta-catenin production and use is inhibited were recognized. First, the ability of the Fz receptor to bind to LRP5/6 and disrupt the beta-catenin destruction complex was down-regulated by a decrease in LRP5. Then, an increase in GSK3B suggests a strong up-regulation of the beta-catenin destruction complex, wherein GSK3B is responsible for marking beta-catenin for deletion. Third, a decrease in PPP2CA lowers its ability to stabilize beta-catenin. All of these transpire to decrease the amount of beta-catenin able to make it through the cytosol and into the nucleus. Once in the nucleus, however, the increased expression of the noncanonical Wnt/calcium signaling pathway interferes with the ability to beta-catenin to bind to TCF and help in transcription and EMT. The added effect of all of these differential methylations leans toward the conclusion that beta-catenin, especially as it pertains to the Wnt signaling pathway, is regulated differently between parous and nulliparous women. The decrease in beta-catenin production and accumulation may be a leftover effect from mammary involution, which would have been the last process of remodeling the mammary glands had undergone. This suggests that the decreased capacity for beta-catenin accumulation caused by involution is what causes the protective effect of pregnancy against breast cancer. The biological importance of the pathways identified in this specific population cannot be sufficiently emphasized due to the fact that they could represent another safeguard mechanism besides the ones discussed earlier [27], mediating the protection of the breast conferred by full term pregnancy.

Acknowledgments

This work was supported by grant 02-2008-034 from the Avon Foundation for Women Breast Cancer Research Program, NIH core grant CA06927 to Fox Chase Cancer Center and an appropriation from the Commonwealth of Pennsylvania. The authors thank the women of Norrbotten County, Sweden, for their willing contribution to the project and the staff of the Mammography Department, Sunderby Hospital, Luleå, Sweden.

Author Contributions

Designed the experiments: JR, IHR, JS-P. Analyzed the data: JR, JS-P. Wrote the paper: JR, JS-P, IHR.

References

1. Clarke, C.A.; Purdie, D.M.; Glaser, S.L. Population attributable risk of breast cancer in white women associated with immediately modifiable risk factors. *BMC Cancer* **2006**, *6*, 170–181.
2. MacMahon, B.; Cole, P.; Lin, T.M.; Lowe, C.R.; Mirra, A.P.; Ravnihar, B.; Salber, E.J.; Valaoras, V.G.; Yuasa, S. Age at first birth and breast cancer risk. *Bull. World Health Organ.* **1970**, *43*, 209–221.
3. Jemal, A.; Siegel, R.; Ward, E.; Murray, T.; Xu, J.; Thun, M.J. Cancer statistics, 2007. *CA Cancer J. Clin.* **2007**, *57*, 43–66.
4. Russo, J.; Balogh, G.A.; Russo, I.H. Full-term pregnancy induces a specific genomic signature in the human breast. *Cancer Epidemiol. Biomarkers Prev.* **2008**, *17*, 51–66.
5. Russo, J.; Russo, I.H. Influence of differentiation and cell kinetics on the susceptibility of the rat mammary gland to carcinogenesis. *Cancer Res.* **1980**, *40*, 2677–2687.
6. Tay, L.K.; Russo, J. Formation and removal of 7,12-dimethylbenz[a]anthracene–nucleic acid adducts in rat mammary epithelial cells with different susceptibility to carcinogenesis. *Carcinogenesis* **1981**, *2*, 1327–1333.
7. Russo, I.H.; Koszalka, M.; Russo, J. Comparative study of the influence of pregnancy and hormonal treatment on mammary carcinogenesis. *Br. J. Cancer* **1991**, *64*, 481–484.
8. Sinha, D.K.; Pazik, J.E.; Dao, T.L. Prevention of mammary carcinogenesis in rats by pregnancy: Effect of full-term and interrupted pregnancy. *Br. J. Cancer* **1988**, *57*, 390–394.
9. Srivastava, P.; Russo, J.; Mgbonyebi, O.P.; Russo, I.H. Growth inhibition and activation of apoptotic gene expression by human chorionic gonadotropin in human breast epithelial cells. *Anticancer Res.* **1998**, *18*, 4003–4010.
10. Thomas, D.B.; Rosenblatt, K.A.; Ray, R.M. Re: "Breastfeeding and reduced risk of breast cancer in an Icelandic cohort study". *Am. J. Epidemiol.* **2001**, *154*, 975–977.

11. Russo, J.; Moral, R.; Balogh, G.A.; Mailo, D.; Russo, I.H. The protective role of pregnancy in breast cancer. *Breast Cancer Res*. **2005**, *7*, 131–142.
12. Russo, J.; Russo, I.H. Role of differentiation in the pathogenesis and prevention of breast cancer. *Endocr. Relat. Cancer* **1997**, *4*, 7–21.
13. Srivastava, P.; Russo, J.; Russo, I.H. Chorionic gonadotropin inhibits rat mammary carcinogenesis through activation of programmed cell death. *Carcinogenesis* **1997**, *18*, 1799–1808.
14. Ginger, M.R.; Gonzalez-Rimbau, M.F.; Gay, J.P.; Rosen, J.M. Persistent changes in gene expression induced by estrogen and progesterone in the rat mammary gland. *Mol. Endocrinol.* **2001**, *15*, 1993–2009.
15. D'Cruz, C.M.; Moody, S.E.; Master, S.R.; Hartman, J.L.; Keiper, E.A.; Imielinski, M.B.; Cox, J.D.; Wang, J.Y.; Ha, S.I.; Keister, B.A.; *et al*. Persistent parity-induced changes in growth factors, TGF-beta3, and differentiation in the rodent mammary gland. *Mol. Endocrinol.* **2002**, *16*, 2034–2051.
16. Henry, M.D.; Triplett, A.A.; Oh, K.B.; Smith, G.H.; Wagner, K.U. Parity-induced mammary epithelial cells facilitate tumorigenesis in MMTV-neu transgenic mice. *Oncogene* **2004**, *23*, 6980–6985.
17. Medina, D. Breast cancer: The protective effect of pregnancy. *Clin. Cancer Res*. **2004**, *10*, 380S–384S.
18. Endocrine control of breast development. In *Molecular Basis of Breast Cancer: Prevention and Treatment*, Russo, J., Russo, I.H., Eds.; Springer, Berlin, Germany, 2004; pp. 64–67.
19. Ginger, M.R.; Rosen, J.M. Pregnancy-induced changes in cell-fate in the mammary gland. *Breast Cancer Res*. **2003**, *5*, 192–197.
20. Misra, Y.; Bentley, P.A.; Bond, J.P.; Tighe, S.; Hunter, T.; Zhao, F.Q. Mammary gland morphological and gene expression changes underlying pregnancy protection of breast cancer tumorigenesis. *Physiol. Genomics* **2012**, *44*, 76–88.
21. Santucci-Pereira, J.; George, C.; Armiss, D.; Russo, I.; Vanegas, J.; Sheriff, F.; Lopez de Cicco, R.; Su, Y.; Russo, R.; Bidinotto, L.; *et al*. Mimicking pregnancy as a strategy for breast cancer prevention. *Breast Cancer Manag*. **2013**, *2*, 283–294.
22. Belitskaya-Levy, I.; Zeleniuch-Jacquotte, A.; Russo, J.; Russo, I.H.; Bordas, P.; Ahman, J.; Afanasyeva, Y.; Johansson, R.; Lenner, P.; Li, X.; *et al*. Characterization of a genomic signature of pregnancy identified in the breast. *Cancer Prev. Res*. **2011**, *4*, 1457–1464.
23. Peri, S.; de Cicco, R.L.; Santucci-Pereira, J.; Slifker, M.; Ross, E.A.; Russo, I.H.; Russo, P.A.; Arslan, A.A.; Belitskaya-Levy, I.; Zeleniuch-Jacquotte, A.; *et al*. Defining the genomic signature of the parous breast. *BMC Med. Genomics* **2012**, *5*, 46–57.
24. Russo, J.; Rivera, R.; Russo, I.H. Influence of age and parity on the development of the human breast. *Breast Cancer Res. Treat*. **1992**, *23*, 211–218.
25. Russo, I.H.; Russo, J. Pregnancy-induced changes in breast cancer risk. *J. Mammary Gland Biol. Neoplasia* **2011**, *16*, 221–233.
26. *Molecular Basis of Breast Cancer: Prevention and Treatment*; Russo, J., Russo, I.H., Eds.; Springer-Verlag: Berlin, Germany, 2004; p. 447.
27. Russo, J.; Santucci-Pereira, J.; de Cicco, R.L.; Sheriff, F.; Russo, P.A.; Peri, S.; Slifker, M.; Ross, E.; Mello, M.L.; Vidal, B.C.; *et al*. Pregnancy-induced chromatin remodeling in the breast

of postmenopausal women. *Int. J. Cancer* **2012**, *131*, 1059–1070.

28. Long, J.C.; Caceres, J.F. The SR protein family of splicing factors: Master regulators of gene expression. *Biochem. J.* **2009**, *417*, 15–27.
29. Herrmann, A.; Fleischer, K.; Czajkowska, H.; Muller-Newen, G.; Becker, W. Characterization of cyclin L1 as an immobile component of the splicing factor compartment. *FASEB J.* **2007**, *21*, 3142–3152.
30. Russo, I.H.; Russo, J. Mammary gland neoplasia in long-term rodent studies. *Environ. Health Perspect.* **1996**, *104*, 938–967.
31. Russo, J.; Tait, L.; Russo, I.H. Susceptibility of the mammary gland to carcinogenesis. III. The cell of origin of rat mammary carcinoma. *Am. J. Pathol.* **1983**, *113*, 50–66.
32. Tan, P.H.; Goh, B.B.; Chiang, G.; Bay, B.H. Correlation of nuclear morphometry with pathologic parameters in ductal carcinoma *in situ* of the breast. *Mod. Pathol.* **2001**, *14*, 937–941.
33. Bussolati, G.; Marchio, C.; Gaetano, L.; Lupo, R.; Sapino, A. Pleomorphism of the nuclear envelope in breast cancer: A new approach to an old problem. *J. Cell Mol. Med.* **2008**, *12*, 209–218.
34. Palmer, J.E.; Sant Cassia, L.J.; Irwin, C.J.; Morris, A.G.; Rollason, T.P. The prognostic value of nuclear morphometric analysis in serous ovarian carcinoma. *Int. J. Gynecol. Cancer* **2008**, *18*, 692–701.
35. Cao, R.; Wang, L.; Wang, H.; Xia, L.; Erdjument-Bromage, H.; Tempst, P.; Jones, R.S.; Zhang, Y. Role of histone H3 lysine 27 methylation in Polycomb-group silencing. *Science* **2002**, *298*, 1039–1043.
36. Kubicek, S.; Schotta, G.; Lachner, M.; Sengupta, R.; Kohlmaier, A.; Perez-Burgos, L.; Linderson, Y.; Martens, J.H.; O'Sullivan, R.J.; Fodor, B.D.; *et al.* The role of histone modifications in epigenetic transitions during normal and perturbed development. *Ernst. Schering Res. Found. Workshop* **2006**, *57*, 1–27.
37. Lin, W.; Dent, S.Y. Functions of histone-modifying enzymes in development. *Curr. Opin. Genet. Dev.* **2006**, *16*, 137–142.
38. Zuo, T.; Tycko, B.; Liu, T.M.; Lin, H.J.; Huang, T.H. Methods in DNA methylation profiling. *Epigenomics* **2009**, *1*, 331–345.
39. Robinson, J.T.; Thorvaldsdottir, H.; Winckler, W.; Guttman, M.; Lander, E.S.; Getz, G.; Mesirov, J.P. Integrative genomics viewer. *Nat. Biotechnol.* **2011**, *29*, 24–26.
40. Thorvaldsdottir, H.; Robinson, J.T.; Mesirov, J.P. Integrative Genomics Viewer (IGV): High-performance genomics data visualization and exploration. *Brief Bioinform.* **2013**, *14*, 178–192.
41. Rakyan, V.K.; Down, T.A.; Thorne, N.P.; Flicek, P.; Kulesha, E.; Graf, S.; Tomazou, E.M.; Backdahl, L.; Johnson, N.; Herberth, M.; *et al.* An integrated resource for genome-wide identification and analysis of human tissue-specific differentially methylated regions (tDMRs). *Genome Res.* **2008**, *18*, 1518–1529.
42. Aiyar, S.E.; Cho, H.; Lee, J.; Li, R. Concerted transcriptional regulation by BRCA1 and COBRA1 in breast cancer cells. *Int. J. Biol. Sci.* **2007**, *3*, 486–492.
43. Ingenuity® Pathway Analysis. v.18030641. Available online: http://www.ingenuity.com/ (accessed on 24 January 2014).
44. Habas, R.; Dawid, I.B. Dishevelled and Wnt signaling: Is the nucleus the final frontier? *J. Biol.*

2005, *4*, 2.

45. Wang, X.; Goode, E.L.; Fredericksen, Z.S.; Vierkant, R.A.; Pankratz, V.S.; Liu-Mares, W.; Rider, D.N.; Vachon, C.M.; Cerhan, J.R.; Olson, J.E.; *et al.* Association of genetic variation in genes implicated in the beta-catenin destruction complex with risk of breast cancer. *Cancer Epidemiol. Biomarkers Prev.* **2008**, *17*, 2101–2108.
46. Ratcliffe, M.J.; Itoh, K.; Sokol, S.Y. A positive role for the PP2A catalytic subunit in Wnt signal transduction. *J. Biol. Chem.* **2000**, *275*, 35680–35683.
47. Cheyette, B.N.; Waxman, J.S.; Miller, J.R.; Takemaru, K.; Sheldahl, L.C.; Khlebtsova, N.; Fox, E.P.; Earnest, T.; Moon, R.T. Dapper, a Dishevelled-associated antagonist of beta-catenin and JNK signaling, is required for notochord formation. *Dev. Cell* **2002**, *2*, 449–461.
48. Zhao, L.; Hart, S.; Cheng, J.; Melenhorst, J.J.; Bierie, B.; Ernst, M.; Stewart, C.; Schaper, F.; Heinrich, P.C.; Ullrich, A.; *et al.* Mammary gland remodeling depends on gp130 signaling through Stat3 and MAPK. *J. Biol. Chem.* **2004**, *279*, 44093–44100.
49. Chang, P.H.; Hwang-Verslues, W.W.; Chang, Y.C.; Chen, C.C.; Hsiao, M.; Jeng, Y.M.; Chang, K.J.; Lee, E.Y.; Shew, J.Y.; Lee, W.H. Activation of Robo1 signaling of breast cancer cells by Slit2 from stromal fibroblast restrains tumorigenesis via blocking PI3K/Akt/beta-catenin pathway. *Cancer Res.* **2012**, *72*, 4652–4661.
50. Bertucci, M.C.; Mitchell, C.A. Phosphoinositide 3-kinase and INPP4B in human breast cancer. *Ann. N. Y. Acad. Sci.* **2013**, *1280*, 1–5.
51. Nusse, R.; Varmus, H. Three decades of Wnts: A personal perspective on how a scientific field developed. *EMBO J.* **2012**, *31*, 2670–2684.
52. Turashvili, G.; Bouchal, J.; Burkadze, G.; Kolar, Z. Wnt signaling pathway in mammary gland development and carcinogenesis. *Pathobiology* **2006**, *73*, 213–223.
53. Vallorosi, C.J.; Day, K.C.; Zhao, X.; Rashid, M.G.; Rubin, M.A.; Johnson, K.R.; Wheelock, M.J.; Day, M.L. Truncation of the beta-catenin binding domain of E-cadherin precedes epithelial apoptosis during prostate and mammary involution. *J. Biol. Chem.* **2000**, *275*, 3328–3334.
54. Imbert, A.; Eelkema, R.; Jordan, S.; Feiner, H.; Cowin, P. Delta N89 beta-catenin induces precocious development, differentiation, and neoplasia in mammary gland. *J. Cell Biol.* **2001**, *153*, 555–568.
55. Howe, L.R.; Brown, A.M. Wnt signaling and breast cancer. *Cancer Biol. Ther.* **2004**, *3*, 36–41.
56. Meier-Abt, F.; Milani, E.; Roloff, T.; Brinkhaus, H.; Duss, S.; Meyer, D.S.; Klebba, I.; Balwierz, P.J.; van Nimwegen, E.; Bentires-Alj, M. Parity induces differentiation and reduces Wnt/Notch signaling ratio and proliferation potential of basal stem/progenitor cells isolated from mouse mammary epithelium. *Breast Cancer Res.* **2013**, *15*, R36.
57. Prasad, C.P.; Rath, G.; Mathur, S.; Bhatnagar, D.; Parshad, R.; Ralhan, R. Expression analysis of E-cadherin, Slug and GSK3beta in invasive ductal carcinoma of breast. *BMC Cancer* **2009**, *9*, 325–335.
58. Logullo, A.F.; Nonogaki, S.; Pasini, F.S.; Osorio, C.A.; Soares, F.A.; Brentani, M.M. Concomitant expression of epithelial-mesenchymal transition biomarkers in breast ductal carcinoma: Association with progression. *Oncol. Rep.* **2010**, *23*, 313–320.

59. Kimelman, D.; Xu, W. Beta-catenin destruction complex: Insights and questions from a structural perspective. *Oncogene* **2006**, *25*, 7482–7491.
60. Sugimura, R.; Li, L. Noncanonical Wnt signaling in vertebrate development, stem cells, and diseases. *Birth Defects Res. C Embryo Today* **2010**, *90*, 243–256.

9

The Past, Present and Future of Human Centromere Genomics

Megan E. Aldrup-MacDonald [1,2] and Beth A. Sullivan [1,2,*]

[1] Department of Molecular Genetics and Microbiology, Duke University Medical Center, Durham, NC 27710, USA; E-Mail: megan.aldrup@duke.edu

[2] Division of Human Genetics, Duke University, Durham, NC 27710, USA

* Author to whom correspondence should be addressed; E-Mail: beth.sullivan@duke.edu;

Abstract: The centromere is the chromosomal locus essential for chromosome inheritance and genome stability. Human centromeres are located at repetitive alpha satellite DNA arrays that compose approximately 5% of the genome. Contiguous alpha satellite DNA sequence is absent from the assembled reference genome, limiting current understanding of centromere organization and function. Here, we review the progress in centromere genomics spanning the discovery of the sequence to its molecular characterization and the work done during the Human Genome Project era to elucidate alpha satellite structure and sequence variation. We discuss exciting recent advances in alpha satellite sequence assembly that have provided important insight into the abundance and complex organization of this sequence on human chromosomes. In light of these new findings, we offer perspectives for future studies of human centromere assembly and function.

Keywords: alpha satellite; higher order repeat; CENP; heterochromatin; human artificial chromosome; dicentric; chromosome truncation; transcription; tet operon

---

1. Introduction

The centromere is the chromosomal locus that controls chromosome segregation during cell division. Visually, the centromere appears on metaphase chromosomes, at least in metazoans that have excellent cytology, as a primary constriction. This is also the site of kinetochore assembly, the multi-protein structure that forms to coordinate attachment to and movement of chromosomes along microtubules. The proteins associated with centromeres are conserved among species, consistent with the functional significance of the locus. A surprising feature of centromeres is that the DNA sequences present at the locus are dissimilar, not only among organisms but often within the same organism. However, protein components of centromeres are shared among species, suggesting an epigenetic basis for centromere assembly. Such centromere proteins (CENPs) include CENP-A, CENP-C, and CENP-T that are important for structural and functional aspects of the centromere and kinetochore. CENP-A is of particular significance since it is a histone H3 variant that contributes to specialized chromatin at centromeres. The Holliday Junction Recognition Protein HJURP and its fungal homolog Scm3 are chaperones that direct the loading of CENP-A into chromatin primed by the Mis18 complex and ensure propagation of epigenetically marked centromeric nucleosomes (reviewed by [1,2]).

Despite the lack of sequence identity, many centromeres are located in regions of repetitive DNA or satellites. In humans, repetitive alpha satellite DNA defines the centromere. The sequence basis of centromere identity is widely debated, since variant centromeres have been identified in humans and other organisms. These unusual centromeres include neocentromeres, new centromeres that are formed on unique or non-centromeric DNA sequences [3,4]. Dicentric human chromosomes, those chromosomes that are formed by fusion or translocation, have two regions of centromeric DNA, but often only one is the site of kinetochore formation. In these instances, the alpha satellite DNA appears to be neither necessary nor sufficient for centromere function. Nevertheless, other evidence exists that supports the importance of DNA sequence in centromere formation in humans, particularly *de novo* centromere assembly. In this review, we will discuss advances in our understanding of human centromeric DNA, from the discovery of human centromeric sequences through integration of physical and genetic maps of centromeres during the Human Genome Project era to the first centromeric genome assemblies that are only now emerging.

2. Alpha Satellite DNA: Discovery, Organization, and Variation

Human centromeres, and in fact most primate centromeres, are composed of alpha satellite DNA [5]. This sequence is thought to be important for centromere function since it is present at the primary constriction of all human chromosomes. It comprises up to 5% of the genome. Alpha satellite is based on a 171 bp monomer arranged in a tandem, head-to-tail fashion. Individual monomers share 50%–70% sequence identity. An integral number of monomers give rise to a higher order repeat (HOR) unit that is itself repeated in a largely uninterrupted fashion so that within a given centromeric locus, the alpha satellite array can span from 250 to 5,000 kb. Such re-iteration of the HOR gives rise to a homogenized alpha satellite array in which the HORs differ in sequence by only a few percent (Figure 1), even though the constituent monomers show much less sequence similarity [6,7]. Some monomers within the HORs contain a 17 bp sequence called the CENP-B box, a motif that is recognized by the DNA-binding centromere protein CENP-B [8]. Outside of the higher order arrays, monomers are randomly arranged and span the region between the homogeneous array and the chromosome arm [9].

Figure 1. The genomic organization of human centromeres. The primary sequence at human centromeres is alpha satellite DNA that is based on 171 bp monomers (colored arrows) organized in a tandem head-to tail fashion. The monomeric sequences differ by as much as 40%. A set number of monomers give rise to a higher order repeat (colored bars with black arrowhead) and confer chromosome-specificity. Higher order repeats are themselves reiterated hundreds to thousands of times, so that the alpha satellite arrays are highly homogenous and span several hundred kilobases to several megabases. Unordered monomeric alpha satellite DNA flanks the higher order arrays, becoming progressively more divergent farther away from centromeric core.

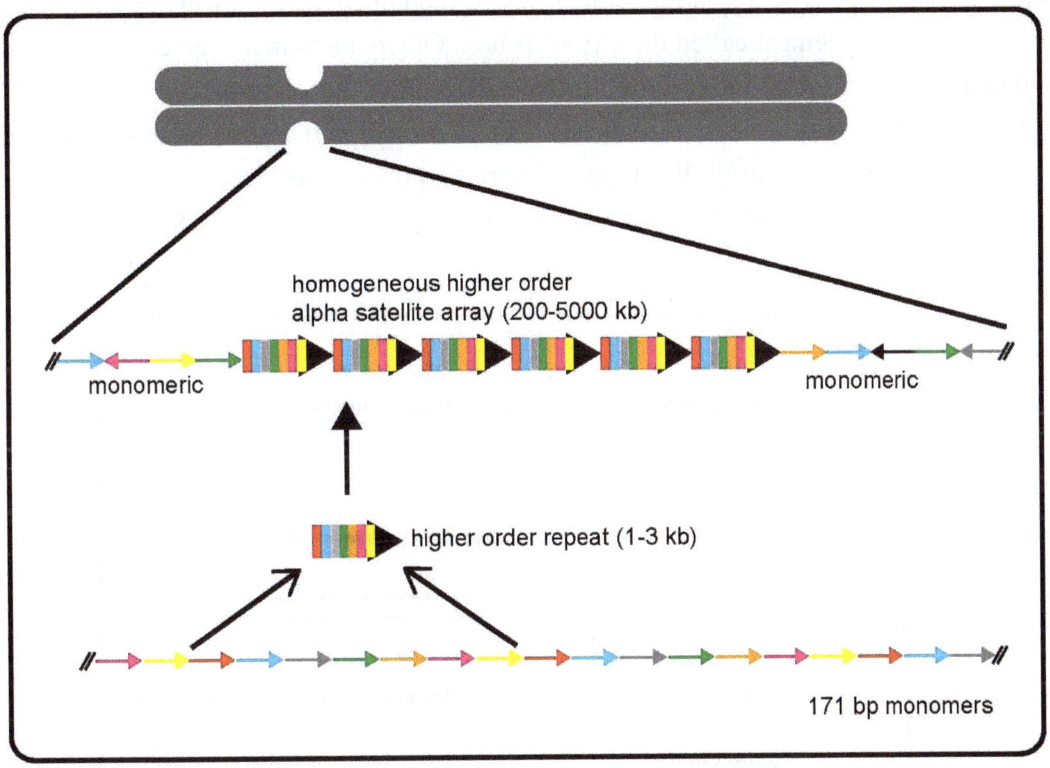

Variation within the alpha satellite is common and complex. Each chromosome type is defined by an alpha satellite array in which the multimers of the HOR contain a particular number of tandem monomers [7,10,11]. The homogeneity of HORs of the same monomer number makes the alpha satellite array chromosome-specific and distinguishable from related sequences at other centromeres. Certain chromosomes share greater homology of HORs based on monomer subtypes and organization, allowing them to be classified into one of three suprachromosomal families [12]. Diverged monomeric alpha satellite falls into two additional suprachromosomal families [13]. On a given chromosome type, the number of times the HOR is re-iterated is heterogeneous, spanning hundreds to thousands of copies. Consequently, total array size on a given chromosome varies between homologs and among individuals (Figure 2) [14–18]. Although array sizes can be as small as a few hundred kilobases or as large as five megabases [16,19,20], the range appears to be less extensive on a particular chromosome type [21]. Array size polymorphisms are largely stable through meiosis since they can be efficiently tracked through multigenerational families [17]. These polymorphisms make alpha satellite a useful centromeric marker for tracking inheritance of individual chromosomes.

Figure 2. Heterogeneity of alpha satellite DNA. The alpha satellite DNA at centromeres exhibits several types of polymorphism. (**A**) Total array size, defined by the number of higher order repeats (HOR; gray arrows), varies between homologues and among individuals; (**B**) The same alpha satellite array from a given chromosome type can contain HORs of different sizes. In addition, the number of each HOR variant can vary. For example, an alpha satellite array can contain a mixture of 10-mers and 6-mers, with a greater number of 10-mers. Another array from the same chromosome in a different individual might have an equal number of 10-mers and 6-mers or, alternatively, more 6-mers than 10-mers; (**C**) Alpha satellite DNA can also vary at the level of monomer (black arrowheads) type and arrangement. Some monomers (gray arrowheads) contain a specific sequence element called the CENP-B box. Others monomers can contain identical nucleotide changes or SNPs (yellow arrowheads) within the same array. Multiple SNPs (hot pink, orange, gray, yellow arrowheads) can be present in the same HOR or distributed across an alpha satellite array. Each type of variation (array size, HOR size, SNPs) is not mutually exclusive and all contribute to the heterogeneity of alpha satellite DNA in the human population.

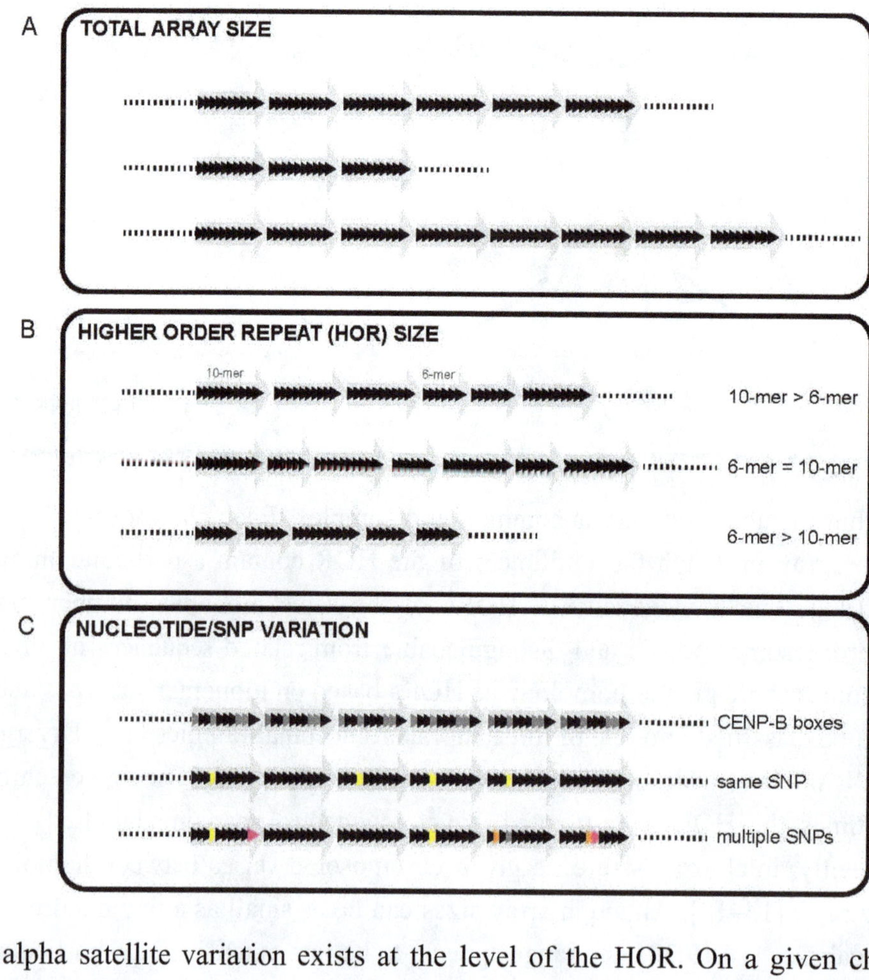

Additional alpha satellite variation exists at the level of the HOR. On a given chromosome, the primary HOR unit can exhibit size polymorphisms that are most likely the result of deletions caused by unequal exchange [22,23]. Human chromosome 17 is a good example of HOR polymorphism within the D17Z1 alpha satellite array. The predominant HOR on D17Z1 is a 16-monomer (16-mer) [18,22].

However, less prevalent 15-mer and 14-mer HORs are present on many D17Z1 arrays, as well as 13-mers and 12-mers [22,24]. Within this group, the 13-mer is the most abundant after the 16-mer. These size polymorphisms create centromeric haplotypes, with the 16-/15-/14-mer comprising a haplotype found on 65% of chromosome 17 s and the additional 13-mer present on 35% of chromosome 17 s [22,25]. A recent study that evaluated centromere assembly on multiple chromosome 17 s suggested that HOR variants might have different functional capacities [26]. This possibility, however, remains to be formally tested in an independent functional assay.

3. Functional Studies that Have Defined Genomic Centromeres

The strongest evidence for implicating alpha satellite DNA in human centromere function came from two lines of chromosome engineering experiments that took "bottom up" and "top down" approaches (Figure 3). In the "top down" strategy, telomere-mediated chromosomal truncation was used to modify the X chromosome or Y chromosome, some of which had been transferred to either rodent somatic cell hybrids or DT40 chicken cells (Figure 3B) [27,28]. Because DT40 cells are proficient in homologous recombination, targeted seeding of the telomere truncation constructs accelerated the deletion process. Multiple rounds of telomere truncation generated a series of deleted chromosomes, each containing less X or Y chromosome material (Figure 3B). The stability of the minichromosomes was monitored and those that maintained the least amount of the original chromosome but were still mitotically stable were concluded to contain the minimal sequence(s) necessary for centromere function. In both truncated X and Y chromosomes, minichromosomes containing alpha satellite DNA arrays DXZ1 and DYZ3, respectively, equated with the most stable linear minichromosomes. These studies strongly implicated alpha satellite DNA as the sequence that corresponds to centromere function and chromosome stability.

However, it could be argued that in the top-down studies the centromere, once established on any sequence, stays at that sequence, and does not shift with truncation. At the same time, pioneering experiments were being developed by two groups to take a "bottom up" approach to define the sequences required for centromere function. In these studies, alpha satellite sequences were introduced into linear yeast artificial chromosome (YAC) or circular bacterial artificial chromosome (BAC) vectors (Figure 3A). Hunt Willard's group created first generation artificial chromosomes from synthetic alpha satellite arrays [29]. One higher order repeat from D17Z1 (chromosome 17) or DYZ3 (chromosome Y) was amplified through successive rounds of directional cloning to yield a 1Mb array that was inserted into a BAC vector. Introduction of these artificial chromosome assembly constructs by liposome-mediated transfection into the HT1080 cell line yielded clones that contained a microchromosome or human artificial chromosome (HAC). The HACs recruited centromere proteins and were stable in mitosis for at least 6 months. Careful analysis of the HACs showed that the D17Z1 HACs were completely derived from the input construct. However, the DYZ3-derived HAC had acquired additional alpha satellite sequences from host chromosomes. The functional significance of the inability of DYZ3 to form a functional HAC containing only Y centromere sequences was not fully appreciated at the time. Subsequent studies shed light on the correlation between DYZ3 sequence and its competence for *de novo* centromere assembly (see below).

Figure 3. Minichromosome-based assays defining alpha satellite as the functional human centromere. (**A**) In the late 1990s, human artificial chromosome (HAC) assays (bottom up approach) were developed to test the ability of alpha satellite DNA to form *de novo* centromeres. Synthetic or clone arrays of alpha satellite DNA, such as D17Z1 from human chromosome 17 (green), were cloned into bacterial or plasmid (P1) artificial chromosome (BAC/PAC) vectors containing selectable marker genes (SM). The chromosome assembly constructs were introduced by transfection into human cells. In approximately half of the cells, an autonomous *de novo* chromosome (arrowhead) was produced, consisting of the same alpha satellite DNA (D17Z1, green, as shown) as the parental chromosome (arrow). Inset shows DAPI (DNA) staining of HAC. In the other proportion of transfected clones, the alpha satellite assembly BAC/PAC vector does not make a HAC, but integrates once or multiple times (as shown) into one or more chromosomes. In these instances, the alpha satellite DNA does not recruit any, or all, centromere proteins and is not a functional centromere. Inset shows DAPI (DNA) stained chromosome that contains multiple insertions of D17Z1. (**B**) In a complementary top-down approach, existing chromosomes (X and Y) were systematically deleted using plasmid constructs containing mammalian telomeric sequence (yellow arrowheads). These experiments yielded partially deleted chromosomes with integrated telomeres (red-orange-yellow rectangles) that were progressively deleted. Mitotic chromosome segregation of these minichromosomes was used as a measure of chromosome stability. Based on the molecular composition of the stable minichromosomes that were recovered, alpha satellite DNA (pink oval) was defined as the minimal sequence required for centromere function.

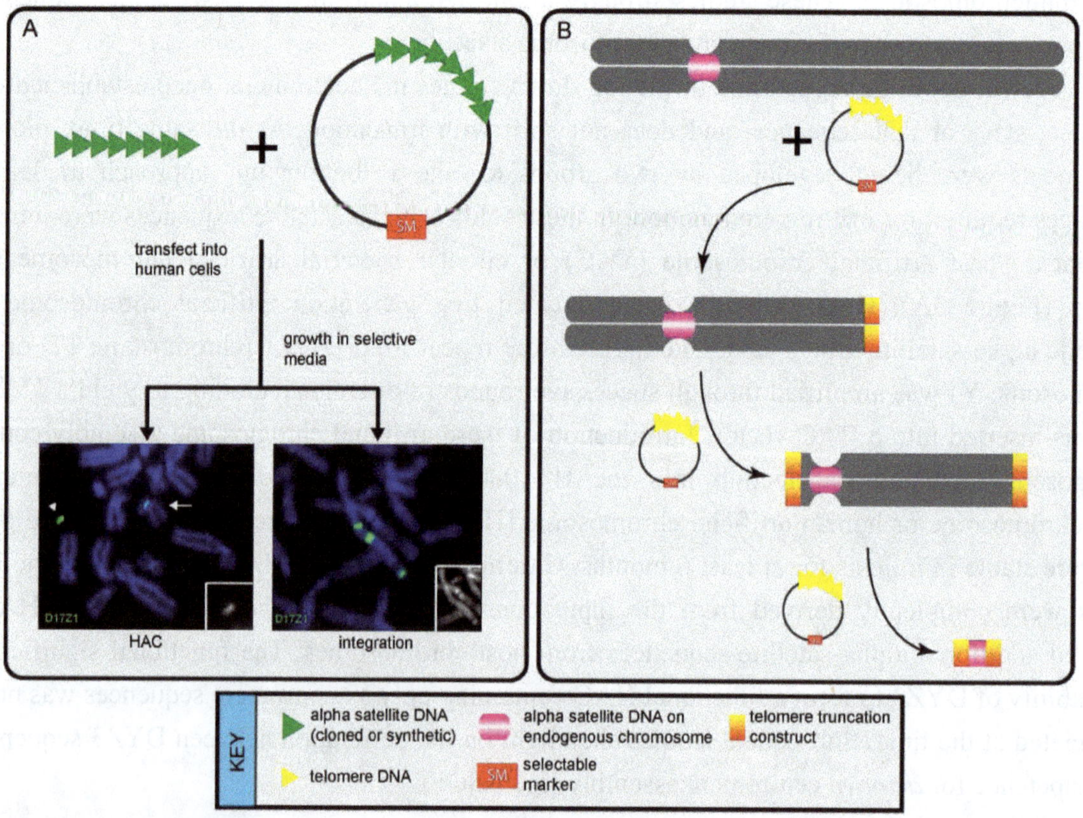

At the same time that the Willard group was creating HACs from synthetic alpha satellite DNA, Howard Cooke's and Hiroshi Masumoto's groups were collaborating to clone large alpha satellite arrays from human chromosome 21 into linear YAC vectors. In their studies, the higher order array α21-I and the unordered monomeric array α21-II were introduced into HT1080 cells and compared for *de novo* centromere competency [30]. Only YACs containing the α21-I HOR array were capable of forming mitotically stable HACs that properly assembled centromere proteins. These innovative studies complemented those of the Willard group, and contributed important structure-function information that implicated HOR alpha satellite as a preferred substrate for *de novo* centromere assembly. In the time that has elapsed since these groundbreaking experiments, additional studies have established HACs as models for testing the genomic (and epigenetic) requirements for *de novo* centromere assembly and function. Circular BAC and PAC vectors, rather than linear YACs, are the most useful assembly vectors and are associated with high rates of HAC formation [31,32]. Not all alpha satellite arrays translate to HAC formation. Y chromosome alpha satellite DNA (DYZ3) lacks CENP-B boxes and is unable to efficiently form *de novo* centromeres on HACs [29,32]. Furthermore, arrays containing mutated CENP-B boxes cannot form *de novo* HACs [33]. Thus, the presence of CENP-B binding sites is required for centromere assembly. This has been a perplexing finding, given that the Y chromosome clearly assembles a functional centromere and recruits essential centromere proteins. These findings hint at key differences between *de novo* versus established centromere function that are not well understood.

Initial studies that tested the ability of alpha satellite to nucleate functional centromeres introduced cosmids containing human alpha satellite DNA from chromosome 17 into African green monkey (AGM) cells [34]. These experiments did not result in supernumerary chromosomes or HACs, but instead, integration of the alpha satellite construct into AGM chromosomes (Figure 3A). Indeed, up to 60% of HAC constructs introduced into human cells integrate into the genome rather than forming an independent chromosome. While some might point out that this argues against the case for sequence-dependent centromere assembly, another interpretation is that *de novo* chromosome assembly and *de novo* centromere formation are two different processes. Indeed, some integrated alpha satellite arrays recruit centromere proteins [34,35], although they may not retain some or all of the proteins long-term. At the very least, both integrated and free-lying HAC studies suggest that alpha satellite provides sequence information for some aspects of centromere function.

Contemporary studies are now using centromere-based chromosome engineering to create a new generation of HACs that contain alpha satellite in addition to tetracycline operator (tetO) sequences [36]. The tetO sequences are bound with high affinity by the tet repressor (tetR) that can be fused to different proteins in order to manipulate the chromatin or protein composition of the HAC [37,38]. In this way, centromere assembly on the alpha satellite can be enhanced or inhibited, the long-term stability of the HAC can be monitored by tethering tetR fluorescent protein fusions, and expression of genes included on the HAC can be tested [39].

4. Centromere Regions in the Human Genome Project Era

As the understanding of the relationship between alpha satellite DNA and centromere function emerged at the end of the 20th century, it led to a call for the identification and mapping of functional centromere sequences [40]. However, the nature of alpha satellite, with its megabase-scale regions of higher-order repetitive structure, made it highly refractory to sequencing and assembly [41]. As the Human Genome Project (HGP) rapidly increased the sequence information available for testing human genome function, gains were largely not seen at the pericentromeres and centromeres of most human chromosomes. A 1998 plan for the project that outlined the HGP's goal for a 2001 working draft and a 2003 final draft acknowledged that "the small proportion of highly repeated sequence represented by the centromeres and other constitutive heterochromatic regions of the genome" might not be included in the final reference assembly [42]. A contemporary perspective on the plan warned of the possibility that potentially important duplications and tandem repeats would be "swept under the carpet". There was a repeated call for at least some centromeric regions to be characterized in order to confirm that their structure was as homogenous as originally claimed [43]. But again, due to the computational complexity required to accurately assemble such highly repetitive regions, few labs attempted to close these sequence gaps [44–46]. A decade later, multi-megabase-sized gaps remain at the centromeres of most chromosome assemblies. This problem is not exclusive to the human genome, since centromere and pericentromere sequence gaps in other organisms such as mouse and Drosophila remain unclosed [47–50]. Only in the past year have advances in sequencing technologies and innovative computational efforts focused on elucidating alpha satellite structure helped to make a full understanding of the genome and some of its most critical elements a real possibility [51,52].

5. Linking Physical and Genetic Maps of Human Centromeres

By the late 1990s and early 2000s, several groups had pushed forward the centromere field by producing integrated physical and genetic maps of centromere regions including chromosomes X, 5, and 12 [53–55]. These studies used pulsed-field gel electrophoresis to estimate physical alpha satellite array sizes and either radiation hybrid or linkage analyses to estimate genetic distance across the centromere. In addition to confirming the repression of recombination across centromeres, the integrated maps that resulted allowed for the anchoring of alpha satellite regions to existing genomic maps, and sometimes identified unique pericentric sequences that had not been represented in the human genome drafts [55].

Of the sequence assemblies around the centromere that do exist, the pericentric regions are the best characterized. Within these regions, a high proportion of segmental duplications have accumulated [44,56]. Many pericentric duplications corresponding to unmapped regions of the genome were identified using monochromosomal somatic cell hybrids and PCR or FISH with known pericentric sequences and genomic BACs that recognized paralogous sequences across the genome [56]. Genome-wide analysis of the January 2001 draft assembly further revealed pericentromeric and subtelomeric enrichment for duplicated sequences, and showed that such sequences were frequently present in unmapped or misassembled segments [57]. The discordance between FISH and BLAST results in these analyses was much higher than the genome-wide rate reported in the same year [58]. Together, these studies demonstrated the importance of elucidating highly duplicated pericentric regions in order to accurately understand the Human Genome Project's results. More recent progress was made in assembling "inaccessible" regions by using linkage disequilibrium analysis of genetically distinct (admixed) genomes to map almost 20 Mb of sequence near centromeres [59]. As the number of admixed genomes available for analysis increases, this powerful technique is expected to reduce the gaps in the current reference assembly.

Figure 4. The detailed genomic organization of the human X centromere. The first contiguous genomic map of a human centromere (CEN) on the X chromosome was completed in 2001 and showed that the higher order array (large light gray arrays containing black monomer arrowheads) is flanked by unordered, monomeric alpha satellite DNA (multi-colored arrows). The regions between monomeric alpha satellite and the chromosome short (Xp) and long (Xq) arms contain other types of satellite DNA, such as gamma satellite and HSAT4. LINEs (red lollipops) and SINEs (purple lollipops) punctuate the repetitive DNA between the centromere and chromosome arms. The Xq pericentromere contains monomeric alpha satellite and a LINE element at the pericentromere-arm junction. Some of the monomers within the unordered Xq satellite contain CENP-B boxes (black asterisks). The functional significance of these monomers remains unclear.

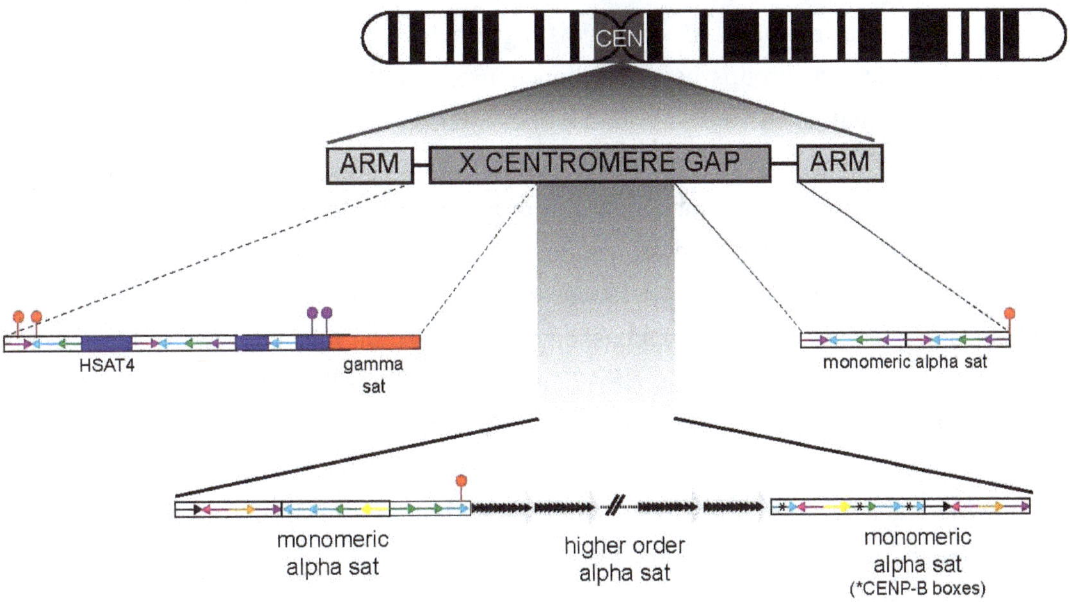

6. Correlating the Genetic and Functional Centromere

In 2001, a major breakthrough in reaching beyond the boundaries of alpha satellite occurred when chromosome X short arm (Xp) genomic clones were mapped into the homogenous higher order DXZ1 array [60]. This tour-de-force used combined *in silico* and high-stringency BAC clone screening to demonstrate that even in higher order alpha satellite, enough sequence variation existed to assemble a contig extending almost half a megabase from the satellite boundary towards the centromere core that is the location of the functional centromere (Figure 4). This study revealed that heterogeneity of alpha satellite DNA increased with more distance from the DXZ1 core. These studies permitted the definition of transitions between the higher order alpha satellite and flanking regions. Monomers of alpha satellite DNA that are not ordered into multi-monomer repeat units are located directly outside of the homogenous HOR domain [9]. These monomers exhibit enough sequence variation that they can be more easily assembled and in fact represent most of the alpha satellite that exists in the human reference assembly [60,61]. The monomeric alpha satellite regions show greater sequence dissimilarity and more interspersed elements, such as L1 sequences, as they approach the chromosome arms. Currently, HSAX and HSA8 are the only human chromosomes represented in the genome assembly with contiguous sequence from higher-order alpha satellite to both arms [62,63].

Subsequent to these findings, several groups began analyzing alpha satellite at increasing sequence depth, discovering new alpha satellite polymorphisms and repeat organization. Building on the work of previous decades, targeted sequencing of several well-characterized arrays was performed. The high copy number of alpha satellite HORs on each chromosome permitted analysis of intra-homolog SNPs in addition to inter-individual variation that was paired with restriction digestion for haplotype analysis [64]. These studies revisited the molecular basis for variation within alpha satellite by pinpointing where unequal exchange occurred to produce array homogenization.

7. The Computational Challenge of Alpha Satellite Genome Assemblies

The bottleneck in generating alpha satellite assemblies has undoubtedly been the sophistication of assembly tools that are required to order distinguishable monomeric sequences within highly homogenous arrays. Several groups have developed *in silico* tools for analyzing higher order alpha satellite sequence available in genome assemblies [65,66]. These computational and *in silico* approaches are most effective when combined with experimental approaches that mapped clones by FISH to verify their location in or near the higher order array. Indeed, such dry/wet approaches were used to map the region spanning the Xp centromere-arm junction and to characterize the centromere of human chromosome 17 [60,61,67]. In the latter instance, a novel higher-order array (D17Z1-B) was discovered on chromosome 17 [67], emphasizing the power of this integrative approach. Another novel HOR array, localized by BLAST to HSA22 and verified by FISH to hybridize to HSA14 and 22, was found by "rescuing" unassembled alpha satellite sequence information from whole genome sequencing (WGS) repositories [68]. These studies revealed that while challenging to assemble, repetitive satellite regions, particularly in the centromere, hold a wealth of complex genomic structure and potentially functional information.

8. Assembling Centromeres in the Present Day

Previous studies utilized traditional sequencing technologies that have the potential to contain several 171-bp monomers per read. Next generation short-read sequencing technology has enabled the recent increase in whole-genome sequencing and the amount of human sequence information available overall. Nevertheless, short reads present a particular challenge for assembling alpha satellite sequence. It appears that this obstacle of aligning short-read alpha satellite sequences can be overcome to utilize functional information gleaned from chromatin immunoprecipitation with centromeric protein antibodies and Illumina sequencing of the DNA that is captured [51]. This ChIP-sequencing (ChIP-seq) approach utilized the reference assembly as well as the HuRef genome, first by aligning the HuRef alpha satellite reads to the reference assembly. After this alignment, the reference alpha satellite was broken into sliding windows, and the alignment checked back onto the HuRef reads to determine the "mappability" of each window. This mappability information was then used for alignment of the short Illumina reads generated by ChIP-seq. It should be noted, however, that this study did not have the means to extend beyond the edges of the reference assembly into the homogenous centromere cores (see Future Perspectives). Another major discovery from the assembly annotation of this work was that many more chromosomes than previously thought contain two or more higher order alpha satellite arrays [51,61,69–71]. This finding has raised the complexity of centromeres to a new level and introduced the possibility that the location of centromere assembly may be quite variable in humans. This is indeed the case for human chromosome 17 on which the centromere can be assembled at either of the two higher order repeat arrays [26]. This new information suggests that in addition to alpha satellite haplotypes, there may a number of functional centromeric genotypes. How a functional genotype might affect long-term chromosome stability is an open question.

9. Future Perspectives

It is now 2014, so what can we expect from the centromere field in the next decade? Based on the foundation laid by the Human Genome Project era, the most exciting areas of centromere research are in some of the following areas.

9.1. Centromere Assemblies

Clearly, the most significant frontier that remains to be explored in centromere biology is complete genomic centromere assemblies. With the recent advances in the past two years alone using long-template sequencing and advanced computational approaches that have sampled, annotated, and assembled centromere sequences in multiple genomes, centromere reference sequences are a real possibility. Just recently, ordering of monomer sequences from whole-genome shotgun reads has produced the first linear characterization of centromeric assemblies for alpha satellite arrays from chromosomes X and Y [72]. Increasing read lengths offered by multiple platforms offer the potential to contain several multi-kilobase HORs in one read. In fact, long PacBio reads have already accelerated the discovery and mapping of centromeric tandem repeats in a variety of species [52]. These third generation sequencing techniques should enable longer alpha satellite sequence assemblies and better understanding of centromere structure and neighboring variant HORs. Completion of even a few centromere assemblies will undoubtedly be important, but given the amount of variation in alpha satellite organization and size, the ultimate goal would be to produce centromere assemblies for each individual. These personalized maps would be useful for defining the spectrum of sequences that correlate with functional competency. In addition, they will allow identification of other features—such as genes or non-coding elements—that are present within current centromere/pericentromere gaps. These sequences may require centromeric locations for proper function, similar to heterochromatic genes in *Drosophila* [48,73]. Indeed, a human muscle disorder has been mapped to the gene KCNJ18 that is located in an assembly gap on 17p11.2 [74]. It is possible that other genes or elements within centromere regions may be associated with diseases for which the molecular basis remains undefined.

9.2. Centromeric Variation and Functional Capacity

The ability to confidently assemble centromeric contigs should permit identification of the full range of variability in alpha satellite, including sequence and size variants [72]. Such variation will shed light on the molecular mechanisms that regulate alpha satellite homogenization, but also effects of fundamental processes such as DNA replication and DNA repair on alpha satellite stability. Ultimately, characterization of alpha satellite variation would reveal the range of sequences that are capable of supporting centromere function. HAC studies have taught us that not all alpha satellite sequences have the capacity to support *de novo* centromere assembly [29,32]. The reasons for this have been largely unexplored, and mostly attributed to the presence or absence of CENP-B boxes in alpha satellite [33,75]. One would expect that like a given complex human disease that is often associated with various SNPs, many types of sequence variation would be associated with diminished centromere function. Complete, personalized centromeric assemblies linked to functional centromere status would expedite experiments to compare efficiencies of various sequence variants in *de novo* centromere assembly and/or centromere maintenance.

9.3. Maps of Functional Centromeric Domains

The consensus in the centromere field is that centromere identity is specified by epigenetic mechanisms. However, without detailed genomic information, this theory is not irrefutable. Centromere proteins, such as the histone H3-like protein CENP-A, are assembled onto alpha satellite DNA to create a specialized type of nucleosome within unique chromatin that distinguishes the centromere from the rest of the genome [76,77]. CENP-A and other proteins create a complicated network of protein sub-complexes that link the chromatin to the structural kinetochore that interacts with spindle microtubules [78]. However, chromatin that contains CENP-A nucleosomes is only assembled on a portion of alpha satellite DNA [79,80]. How and why CENP-A is recruited to only a subset of alpha satellite HOR and/or monomers is unclear. Recent studies have revealed that CENP-A nucleosomes on the human X chromosome are positioned at monomers that do not contain CENP-B boxes [81]. One could speculate that distribution of CENP-B boxes within an alpha satellite array and sequence variation that interrupts the CENP-B box motif or makes the motif non-functional (not bound by CENP-B) might impact CENP-A chromatin assembly and centromere function. Complete centromeric assemblies of many human chromosomes will be important for addressing this possibility experimentally.

10. Conclusions

Since the discovery of alpha satellite DNA in the late 1970s, the field has moved from identification of centromeric sequences at every human centromere to a basic molecular understanding of the organization and structure of alpha satellite monomers into homogeneous higher order repetitive arrays (Figure 5). The Human Genome Project was essential in providing a rough and limited reference assembly for centromeres of three chromosomes (X, 8, 17). These fundamental studies of alpha satellite DNA paved the way for pioneering functional assays in which the sequence was tested in *de novo* centromere assembly in human artificial chromosome assays. HACs have been the gold standard for testing centromere assembly, but are now being used to explore chromosome stability and gene expression. The next challenge will be to complete genomic assemblies for all human centromeres in multiple individuals and populations and to develop the next generation of functional assays to test the role of alpha satellite variation in centromere function, chromosome stability, and disease association.

Figure 5. Timeline of major discoveries in human centromere genomics. Since the discovery of alpha satellite DNA in 1979, the understanding of the sequence, organization, and functional aspects of this sequence flourished during the Human Genome Project era. Recent years have shown the use of human artificial chromosomes (HACs) and the creation of the first database of alpha satellite sequences linked to their functional capacity.

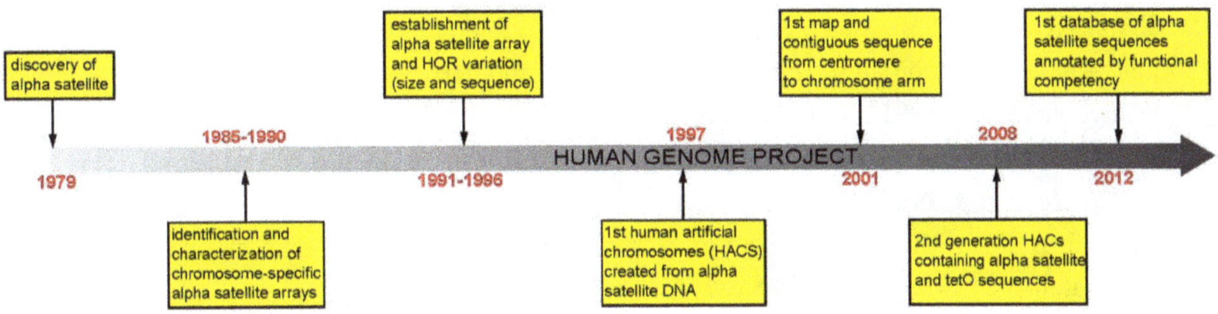

Acknowledgments

We apologize to our colleagues whose work on alpha satellite DNA and human centromeres could not be cited due to space constraints. Research in the Sullivan lab is supported in part by R01 GM098500 (NIH) and Gene Discovery and Translational Research Grant #1-FY13-517 (March of Dimes Foundation).

Author Contributions

Wrote the paper: MEAM, BAS.

References

1. Panchenko, T.; Black, B.E. The epigenetic basis for centromere identity. *Prog. Mol. Subcell. Biol.* **2009**, *48*, 1–32.
2. Valente, L.P.; Silva, M.C.; Jansen, L.E. Temporal control of epigenetic centromere specification. *Chromosome Res.* **2012**, *20*, 481–492.
3. Choo, K.H. Domain organization at the centromere and neocentromere. *Dev. Cell* **2001**, *1*, 165–177.
4. Warburton, P.E. Chromosomal dynamics of human neocentromere formation. *Chromosome Res.* **2004**, *12*, 617–626.
5. Manuelidis, L.; Wu, J.C. Homology between human and simian repeated DNA. *Nature* **1978**, *276*, 92–94.
6. Waye, J.S.; Willard, H.F. Nucleotide sequence heterogeneity of alpha satellite repetitive DNA: A survey of alphoid sequences from different human chromosomes. *Nucleic Acids Res.* **1987**, *15*, 7549–7569.

7. Willard, H.F. Chromosome-specific organization of human alpha satellite DNA. *Am. J. Hum. Genet.* **1985**, *37*, 524–532.
8. Muro, Y.; Masumoto, H.; Yoda, K.; Nozaki, N.; Ohashi, M.; Okazaki, T. Centromere protein B assembles human centromeric alpha-satellite DNA at the 17-bp sequence, CENP-B box. *J. Cell Biol.* **1992**, *116*, 585–596.
9. Schueler, M.G.; Sullivan, B.A. Structural and functional dynamics of human centromeric chromatin. *Annu. Rev. Genomics Hum. Genet.* **2006**, *7*, 301–313.
10. Choo, K.H.; Vissel, B.; Nagy, A.; Earle, E.; Kalitsis, P. A survey of the genomic distribution of alpha satellite DNA on all the human chromosomes, and derivation of a new consensus sequence. *Nucleic Acids Res.* **1991**, *19*, 1179–1182.
11. Vissel, B.; Choo, K.H. Human alpha satellite DNA—Consensus sequence and conserved regions. *Nucleic Acids Res.* **1987**, *15*, 6751–6752.
12. Alexandrov, I.A.; Mitkevich, S.P.; Yurov, Y.B. The phylogeny of human chromosome specific alpha satellites. *Chromosoma* **1988**, *96*, 443–453.
13. Alexandrov, I.; Kazakov, A.; Tumeneva, I.; Shepelev, V.; Yurov, Y. Alpha-satellite DNA of primates: Old and new families. *Chromosoma* **2001**, *110*, 253–266.
14. Devilee, P.; Kievits, T.; Waye, J.S.; Pearson, P.L.; Willard, H.F. Chromosome-specific alpha satellite DNA: Isolation and mapping of a polymorphic alphoid repeat from human chromosome 10. *Genomics* **1988**, *3*, 1–7.
15. Mahtani, M.M.; Willard, H.F. Pulsed-field gel analysis of alpha-satellite DNA at the human X chromosome centromere: High-frequency polymorphisms and array size estimate. *Genomics* **1990**, *7*, 607–613.
16. Greig, G.M.; Parikh, S.; George, J.; Powers, V.E.; Willard, H.F. Molecular cytogenetics of alpha satellite DNA from chromosome 12: Fluorescence in situ hybridization and description of DNA and array length polymorphisms. *Cytogenet. Cell Genet.* **1991**, *56*, 144–148.
17. Wevrick, R.; Willard, H.F. Long-range organization of tandem arrays of alpha satellite DNA at the centromeres of human chromosomes: High-frequency array-length polymorphism and meiotic stability. *Proc. Natl. Acad. Sci. USA* **1989**, *86*, 9394–9398.
18. Willard, H.F.; Waye, J.S.; Skolnick, M.H.; Schwartz, C.E.; Powers, V.E.; England, S.B. Detection of restriction fragment length polymorphisms at the centromeres of human chromosomes by using chromosome-specific alpha satellite DNA probes: Implications for development of centromere-based genetic linkage maps. *Proc. Natl. Acad. Sci. USA* **1986**, *83*, 5611–5615.
19. Abruzzo, M.A.; Griffin, D.K.; Millie, E.A.; Sheean, L.A.; Hassold, T.J. The effect of Y-chromosome alpha-satellite array length on the rate of sex chromosome disomy in human sperm. *Hum. Genet.* **1996**, *97*, 819–823.
20. Oakey, R.; Tyler-Smith, C. Y chromosome DNA haplotyping suggests that most European and Asian men are descended from one of two males. *Genomics* **1990**, *7*, 325–330.
21. Willard, H.F. Evolution of alpha satellite. *Curr. Opin. Genet. Dev.* **1991**, *1*, 509–514.
22. Waye, J.S.; Willard, H.F. Molecular analysis of a deletion polymorphism in alpha satellite of human chromosome 17: Evidence for homologous unequal crossing-over and subsequent fixation.

Nucleic Acids Res. **1986**, *14*, 6915–6927.
23. Waye, J.S.; Willard, H.F. Structure, organization, and sequence of alpha satellite DNA from human chromosome 17: Evidence for evolution by unequal crossing-over and an ancestral pentamer repeat shared with the human X chromosome. *Mol. Cell. Biol.* **1986**, *6*, 3156–3165.
24. Willard, H.F.; Greig, G.M.; Powers, V.E.; Waye, J.S. Molecular organization and haplotype analysis of centromeric DNA from human chromosome 17: Implications for linkage in neurofibromatosis. *Genomics* **1987**, *1*, 368–373.
25. Warburton, P.E.; Willard, H.F. Interhomologue sequence variation of alpha satellite DNA from human chromosome 17: Evidence for concerted evolution along haplotypic lineages. *J. Mol. Evol.* **1995**, *41*, 1006–1015.
26. Maloney, K.A.; Sullivan, L.L.; Matheny, J.E.; Strome, E.D.; Merrett, S.L.; Ferris, A.; Sullivan, B.A. Functional epialleles at an endogenous human centromere. *Proc. Natl. Acad. Sci. USA* **2012**, *109*, 13704–13709.
27. Brown, K.E.; Barnett, M.A.; Burgtorf, C.; Shaw, P.; Buckle, V.J.; Brown, W.R. Dissecting the centromere of the human Y chromosome with cloned telomeric DNA. *Hum. Mol. Genet.* **1994**, *3*, 1227–1237.
28. Farr, C.J.; Bayne, R.A.; Kipling, D.; Mills, W.; Critcher, R.; Cooke, H.J. Generation of a human X-derived minichromosome using telomere-associated chromosome fragmentation. *EMBO J.* **1995**, *14*, 5444–5454.
29. Harrington, J.J.; van Bokkelen, G.; Mays, R.W.; Gustashaw, K.; Willard, H.F. Formation of *de novo* centromeres and construction of first-generation human artificial microchromosomes. *Nat. Genet.* **1997**, *15*, 345–355.
30. Ikeno, M.; Grimes, B.R.; Okazaki, T.; Nakano, M.; Saitoh, K.; Hoshino, H.; McGill, N.I.; Cooke, H.; Masumoto, H. Construction of YAC-based mammalian artificial chromosomes. *Nat. Biotechnol.* **1998**, *16*, 431–439.
31. Grimes, B.R.; Babcock, J.; Rudd, M.K.; Chadwick, B.; Willard, H.F. Assembly and characterization of heterochromatin and euchromatin on human artificial chromosomes. *Genome Biol.* **2004**, *5*, R89.
32. Grimes, B.R.; Rhoades, A.A.; Willard, H.F. Alpha-satellite DNA and vector composition influence rates of human artificial chromosome formation. *Mol. Ther.* **2002**, *5*, 798–805.
33. Ohzeki, J.; Nakano, M.; Okada, T.; Masumoto, H. CENP-B box is required for *de novo* centromere chromatin assembly on human alphoid DNA. *J. Cell Biol.* **2002**, *159*, 765–775.
34. Haaf, T.; Warburton, P.E.; Willard, H.F. Integration of human alpha-satellite DNA into simian chromosomes: Centromere protein binding and disruption of normal chromosome segregation. *Cell* **1992**, *70*, 681–696.
35. Nakashima, H.; Nakano, M.; Ohnishi, R.; Hiraoka, Y.; Kaneda, Y.; Sugino, A.; Masumoto, H. Assembly of additional heterochromatin distinct from centromere-kinetochore chromatin is required for *de novo* formation of human artificial chromosome. *J. Cell Sci.* **2005**, *118*, 5885–5898.
36. Nakano, M.; Cardinale, S.; Noskov, V.N.; Gassmann, R.; Vagnarelli, P.; Kandels-Lewis, S.; Larionov, V.; Earnshaw, W.C.; Masumoto, H. Inactivation of a human kinetochore by specific targeting of chromatin modifiers. *Dev. Cell* **2008**, *14*, 507–522.

37. Bergmann, J.H.; Rodriguez, M.G.; Martins, N.M.; Kimura, H.; Kelly, D.A.; Masumoto, H.; Larionov, V.; Jansen, L.E.; Earnshaw, W.C. Epigenetic engineering shows H3K4me2 is required for HJURP targeting and CENP-A assembly on a synthetic human kinetochore. *EMBO J.* **2011**, *30*, 328–340.
38. Cardinale, S.; Bergmann, J.H.; Kelly, D.; Nakano, M.; Valdivia, M.M.; Kimura, H.; Masumoto, H.; Larionov, V.; Earnshaw, W.C. Hierarchical inactivation of a synthetic human kinetochore by a chromatin modifier. *Mol. Biol. Cell* **2009**, *20*, 4194–4204.
39. Kononenko, A.V.; Lee, N.C.; Earnshaw, W.C.; Kouprina, N.; Larionov, V. Re-engineering an alphoid(tetO)-HAC-based vector to enable high-throughput analyses of gene function. *Nucleic Acids Res.* **2013**, *41*, e107.
40. Murphy, T.D.; Karpen, G.H. Centromeres take flight: Alpha satellite and the quest for the human centromere. *Cell* **1998**, *93*, 317–320.
41. Henikoff, S. Near the edge of a chromosome's "black hole". *Trends Genet.* **2002**, *18*, 165–167.
42. Collins, F.S.; Patrinos, A.; Jordan, E.; Chakravarti, A.; Gesteland, R.; Walters, L. New goals for the U.S. Human Genome Project: 1998–2003. *Science* **1998**, *282*, 682–689.
43. Eichler, E.E. Repetitive conundrums of centromere structure and function. *Hum. Mol. Genet.* **1999**, *8*, 151–155.
44. Horvath, J.E.; Bailey, J.A.; Locke, D.P.; Eichler, E.E. Lessons from the human genome: Transitions between euchromatin and heterochromatin. *Hum. Mol. Genet.* **2001**, *10*, 2215–2223.
45. Horvath, J.E.; Viggiano, L.; Loftus, B.J.; Adams, M.D.; Archidiacono, N.; Rocchi, M.; Eichler, E.E. Molecular structure and evolution of an alpha satellite/non-alpha satellite junction at 16p11. *Hum. Mol. Genet.* **2000**, *9*, 113–123.
46. She, X.; Horvath, J.E.; Jiang, Z.; Liu, G.; Furey, T.S.; Christ, L.; Clark, R.; Graves, T.; Gulden, C.L.; Alkan, C.; *et al.* The structure and evolution of centromeric transition regions within the human genome. *Nature* **2004**, *430*, 857–864.
47. Hoskins, R.A.; Carlson, J.W.; Kennedy, C.; Acevedo, D.; Evans-Holm, M.; Frise, E.; Wan, K.H.; Park, S.; Mendez-Lago, M.; Rossi, F.; *et al.* Sequence finishing and mapping of Drosophila melanogaster heterochromatin. *Science* **2007**, *316*, 1625–1628.
48. Smith, C.D.; Shu, S.; Mungall, C.J.; Karpen, G.H. The Release 5.1 annotation of Drosophila melanogaster heterochromatin. *Science* **2007**, *316*, 1586–1591.
49. Kalitsis, P.; Griffiths, B.; Choo, K.H. Mouse telocentric sequences reveal a high rate of homogenization and possible role in Robertsonian translocation. *Proc. Natl. Acad. Sci. USA* **2006**, *103*, 8786–8791.
50. Mouse Genome Sequencing Consortium; Waterston, R.H.; Lindblad-Toh, K.; Birney, E.; Rogers, J.; Abril, J.F.; Agarwal, P.; Agarwala, R.; Ainscough, R.; Alexandersson, M.; *et al.* Initial sequencing and comparative analysis of the mouse genome. *Nature* **2002**, *420*, 520–562.
51. Hayden, K.E.; Strome, E.D.; Merrett, S.E.; Lee, H.R.; Rudd, M.K.; Willard, H.F. Sequences associated with centromere competency in the human genome. *Mol. Cell. Biol.* **2012**, *33*, 763–772.
52. Melters, D.P.; Bradnam, K.R.; Young, H.A.; Telis, N.; May, M.R.; Ruby, J.G.; Sebra, R.; Peluso, P.; Eid, J.; Rank, D.; *et al.* Comparative analysis of tandem repeats from hundreds of species reveals unique insights into centromere evolution. *Genome Biol.* **2013**, *14*, R10.

53. Mahtani, M.M.; Willard, H.F. Physical and genetic mapping of the human X chromosome centromere: Repression of recombination. *Genome Res.* **1998**, *8*, 100–110.
54. Puechberty, J.; Laurent, A.M.; Gimenez, S.; Billault, A.; Brun-Laurent, M.E.; Calenda, A.; Marçais, B.; Prades, C.; Ioannou, P.; Yurov, Y.; *et al.* Genetic and physical analyses of the centromeric and pericentromeric regions of human chromosome 5: Recombination across 5cen. *Genomics* **1999**, *56*, 274–287.
55. Vermeesch, J.R.; Duhamel, H.; Raeymaekers, P.; van Zand, K.; Verhasselt, P.; Fryns, J.P.; Marynen, P. A physical map of the chromosome 12 centromere. *Cytogenet. Genome Res.* **2003**, *103*, 63–73.
56. Horvath, J.E.; Schwartz, S.; Eichler, E.E. The mosaic structure of human pericentromeric DNA: A strategy for characterizing complex regions of the human genome. *Genome Res.* **2000**, *10*, 839–852.
57. Bailey, J.A.; Yavor, A.M.; Massa, H.F.; Trask, B.J.; Eichler, E.E. Segmental duplications: Organization and impact within the current human genome project assembly. *Genome Res.* **2001**, *11*, 1005–1017.
58. Cheung, V.G.; Nowak, N.; Jang, W.; Kirsch, I.R.; Zhao, S.; Chen, X.N.; Furey, T.S.; Kim, U.J.; Kuo, W.L.; Olivier, M.; *et al.* Integration of cytogenetic landmarks into the draft sequence of the human genome. *Nature* **2001**, *409*, 953–958.
59. Genovese, G.; Handsaker, R.E.; Li, H.; Kenny, E.E.; McCarroll, S.A. Mapping the human reference genome's missing sequence by three-way admixture in Latino genomes. *Am. J. Hum. Genet.* **2013**, *93*, 411–421.
60. Schueler, M.G.; Higgins, A.W.; Rudd, M.K.; Gustashaw, K.; Willard, H.F. Genomic and genetic definition of a functional human centromere. *Science* **2001**, *294*, 109–115.
61. Rudd, M.K.; Willard, H.F. Analysis of the centromeric regions of the human genome assembly. *Trends Genet.* **2004**, *20*, 529–533.
62. Nusbaum, C.; Mikkelsen, T.S.; Zody, M.C.; Asakawa, S.; Taudien, S.; Garber, M.; Kodira, C.D.; Schueler, M.G.; Shimizu, A.; Whittaker, C.A.; *et al.* DNA sequence and analysis of human chromosome 8. *Nature* **2006**, *439*, 331–335.
63. Ross, M.T.; Grafham, D.V.; Coffey, A.J.; Scherer, S.; McLay, K.; Muzny, D.; Platzer, M.; Howell, G.R.; Burrows, C.; Bird, C.P.; *et al.* The DNA sequence of the human X chromosome. *Nature* **2005**, *434*, 325–337.
64. Roizes, G. Human centromeric alphoid domains are periodically homogenized so that they vary substantially between homologues. Mechanism and implications for centromere functioning. *Nucleic Acids Res.* **2006**, *34*, 1912–1924.
65. Paar, V.; Pavin, N.; Rosandic, M.; Gluncic, M.; Basar, I.; Pezer, R.; Zinic, S.D. ColorHOR—Novel graphical algorithm for fast scan of alpha satellite higher-order repeats and HOR annotation for GenBank sequence of human genome. *Bioinformatics* **2005**, *21*, 846–852.
66. Rosandic, M.; Paar, V.; Gluncic, M.; Basar, I.; Pavin, N. Key-string algorithm—Novel approach to computational analysis of repetitive sequences in human centromeric DNA. *Croat. Med. J.* **2003**, *44*, 386–406.
67. Rudd, M.K.; Schueler, M.G.; Willard, H.F. Sequence organization and functional annotation of human centromeres. *Cold Spring Harb. Symp. Quant. Biol.* **2003**, *68*, 141–149.

68. Alkan, C.; Ventura, M.; Archidiacono, N.; Rocchi, M.; Sahinalp, S.C.; Eichler, E.E. Organization and evolution of primate centromeric DNA from whole-genome shotgun sequence data. *PLoS Comput. Biol.* **2007**, *3*, 1807–1818.
69. Alexandrov, I.A.; Mashkova, T.D.; Akopian, T.A.; Medvedev, L.I.; Kisselev, L.L.; Mitkevich, S.P.; Yurov, Y.B. Chromosome-specific alpha satellites: Two distinct families on human chromosome 18. *Genomics* **1991**, *11*, 15–23.
70. Choo, K.H.; Earle, E.; Vissel, B.; Filby, R.G. Identification of two distinct subfamilies of alpha satellite DNA that are highly specific for human chromosome 15. *Genomics* **1990**, *7*, 143–151.
71. Wevrick, R.; Willard, H.F. Physical map of the centromeric region of human chromosome 7: Relationship between two distinct alpha satellite arrays. *Nucleic Acids Res.* **1991**, *19*, 2295–2301.
72. Miga, K.H.; Newton, Y.; Jain, M.; Altemose, N.; Willard, H.F.; Kent, W.J. Centromere reference models for human chromosomes X and Y satellite arrays. *arXiv* **2013**, arXiv:1307.0035v3 [q-bio.GN].
73. Schulze, S.; Sinclair, D.A.; Silva, E.; Fitzpatrick, K.A.; Singh, M.; Lloyd, V.K.; Morin, K.A.; Kim, J.; Holm, D.G.; Kennison, J.A.; *et al.* Essential genes in proximal 3L heterochromatin of Drosophila melanogaster. *Mol. Gen. Genet.* **2001**, *264*, 782–789.
74. Ryan, D.P.; da Silva, M.R.; Soong, T.W.; Fontaine, B.; Donaldson, M.R.; Kung, A.W.; Jongjaroenprasert, W.; Liang, M.C.; Khoo, D.H.; Cheah, J.S.; *et al.* Mutations in potassium channel Kir2.6 cause susceptibility to thyrotoxic hypokalemic periodic paralysis. *Cell* **2010**, *140*, 88–98.
75. Masumoto, H.; Nakano, M.; Ohzeki, J. The role of CENP-B and alpha-satellite DNA: *De novo* assembly and epigenetic maintenance of human centromeres. *Chromosome Res.* **2004**, *12*, 543–556.
76. Blower, M.D.; Sullivan, B.A.; Karpen, G.H. Conserved organization of centromeric chromatin in flies and humans. *Dev. Cell* **2002**, *2*, 319–330.
77. Sullivan, B.A.; Karpen, G.H. Centromeric chromatin exhibits a histone modification pattern that is distinct from both euchromatin and heterochromatin. *Nat. Struct. Mol. Biol.* **2004**, *11*, 1076–1083.
78. Hori, T.; Fukagawa, T. Establishment of the vertebrate kinetochores. *Chromosome Res.* **2012**, *20*, 547–561.
79. Spence, J.M.; Critcher, R.; Ebersole, T.A.; Valdivia, M.M.; Earnshaw, W.C.; Fukagawa, T.; Farr, C.J. Co-localization of centromere activity, proteins and topoisomerase II within a subdomain of the major human X alpha-satellite array. *EMBO J.* **2002**, *21*, 5269–5280.
80. Sullivan, L.L.; Boivin, C.D.; Mravinac, B.; Song, I.Y.; Sullivan, B.A. Genomic size of CENP-A domain is proportional to total alpha satellite array size at human centromeres and expands in cancer cells. *Chromosome Res.* **2011**, *19*, 457–470.
81. Hasson, D.; Panchenko, T.; Salimian, K.J.; Salman, M.U.; Sekulic, N.; Alonso, A.; Warburton, P.E.; Black, B.E. The octamer is the major form of CENP-A nucleosomes at human centromeres. *Nat. Struct. Mol. Biol.* **2013**, *20*, 687–695.

10

Mechanisms of Base Substitution Mutagenesis in Cancer Genomes

Albino Bacolla [1], David N. Cooper [2] and Karen M. Vasquez [1,*]

1 Dell Pediatric Research Institute, Division of Pharmacology and Toxicology, College of Pharmacy, The University of Texas at Austin, 1400 Barbara Jordan Blvd., Austin, TX 78723, USA; E-Mail: albino.bacolla@austin.utexas.edu

2 Institute of Medical Genetics, School of Medicine, Cardiff University, Cardiff CF14 4XN, UK; E-Mail: cooperDN@cardiff.ac.uk

* Author to whom all correspondence should be addressed; E-Mail: karen.vasquez@austin.utexas.edu;

Abstract: Cancer genome sequence data provide an invaluable resource for inferring the key mechanisms by which mutations arise in cancer cells, favoring their survival, proliferation and invasiveness. Here we examine recent advances in understanding the molecular mechanisms responsible for the predominant type of genetic alteration found in cancer cells, somatic single base substitutions (SBSs). Cytosine methylation, demethylation and deamination, charge transfer reactions in DNA, DNA replication timing, chromatin status and altered DNA proofreading activities are all now known to contribute to the mechanisms leading to base substitution mutagenesis. We review current hypotheses as to the major processes that give rise to SBSs and evaluate their relative relevance in the light of knowledge acquired from cancer genome sequencing projects and the study of base modifications, DNA repair and lesion bypass. Although gene expression data on APOBEC3B enzymes provide support for a role in cancer mutagenesis through U:G mismatch intermediates, the enzyme preference for single-stranded DNA may limit its activity genome-wide. For SBSs at both CG:CG and YC:GR sites, we outline evidence for a prominent role of damage by charge transfer reactions that follow interactions of the DNA with reactive oxygen species (ROS) and other endogenous or exogenous electron-abstracting molecules.

Keywords: genetic alterations; cancer etiology; cancer genomes; functional genomics; human genome sequence; single base substitutions; DNA repair; translesion synthesis; oxidative damage; cytosine deamination

1. Introduction

The explosion of cancer genome sequencing projects over the past few years has revealed the complexity of the processes whose alterations are associated with, and are often causative of, various types of cancer [1,2]. These include mutational mechanisms that give rise to tissue-specific mutation rates, variations in mutation frequencies in distinct compartments of chromosomes, sequence context-dependent mutation spectra, and regional hypermutation, often termed kataegis (thunderstorm) [3]. Cancer genome studies have also served to catalogue the extent to which translocations, chromosomal gains and losses and focal copy-number alterations take place, often mediated by catastrophic chromosomal shattering events, as in the case of some bone and pediatric cancers [4–8]. From a mechanistic and therapeutic perspective, the arsenal of gene classes and pathways that are frequently altered, such as signal transduction, metabolism, DNA repair, transcription, epigenetics, RNA splicing and protein homeostasis, has also greatly expanded [1,2,9–11]. Attempts to address key questions concerning the causes leading to the mutational events that characterize and contribute to driving a normal cell towards tumorigenesis have also burgeoned [12–14]. These attempts are, however, necessarily indirect since the only material available for analysis are the catalogues of mutations that survived and accumulated, in most cases, over long periods of time. With few exceptions, such as mutation patterns observed in the lung cancers of heavy smokers and in skin cancers following UV exposure, most mutational patterns have remained enigmatic.

The goal of this review is to examine two prominent single base substitution (SBS) patterns observed in cancer genomes, both of which display sequence context-dependent signatures: C→T transitions at CG:CG (the colon separates complementary bases written in a 5'→3' direction) dinucleotide sequences and substitutions at C:G base-pairs in the context of Y<u>C</u>:<u>G</u>R (Y, pyrimidine; R, purine) motifs. Whereas spontaneous deamination of 5-methylcytosine has been proposed to account for the first pattern, two mechanisms have been recently suggested for the latter pattern: over-activity by the APOBEC family of cytosine deaminases and electron transfer following oxidative damage. After considering several factors associated with SBSs, such as regional variations in mutation frequencies, mechanisms leading to base modification, and DNA repair systems, we conclude that for both SBS patterns, oxidative base damage from ROS and other electron-abstracting molecules appears to play a more significant role than previously anticipated.

2. Meta-Analyses of Cancer Genomes

2.1. Mutational Signatures in Cancer Genomes

The large number of sequenced cancer genomes now available has made it possible to address the issue of mutational spectra and relative mutation frequencies, both exome-wide and genome-wide across different cancer types. Some of the largest meta-analyses have included 4,938,362 somatic substitutions and small insertions/deletions (indels) from 7,042 primary cancers of 30 different classes [13], ~1,000,000 somatic exome mutations from 4,800 tumors representing 19 different cancers [15], 617,354 somatic mutations in 3,281 tumors from 12 cancer types [16], 533,482 somatic SBSs from 1,149 cancer samples and 2 cell lines representing 14 different tissues [17], and 373,909 non-silent coding mutations in 3,083 tumor-normal pairs across 27 tumor types [18]. The prevalence of SBSs was highly variable, both between and within cancer classes, ranging from ~0.001 per megabase (Mb) to >400 per Mb, with childhood cancers generally carrying the fewest mutations, acute myeloid leukemia exhibiting a very low median mutation frequency (~0.28/Mb), and cancers associated with chronic mutagen exposure, such as lung (tobacco smoking) and malignant melanoma (UV light) displaying the highest mutation frequencies (8.15/Mb for lung squamous cell carcinoma) [13,15,16].

With regard to mutational spectra, the most consistent and frequent mutational signature across cancer types has been noted at CG:CG (we identify both nucleosides and nucleotides by their base) dinucleotides, with 25/30 cancer types in [13] and 13/14 in [17], and gastrointestinal tumors displaying <u>C</u>G:C<u>G</u>→TG:CA (target base underlined) among the highest fractions [13,16,17] (Table 1A). The preponderance of C→T transitions at CG:CG sequences has been attributed to high rates of spontaneous deamination of 5-methylcytosine (5mC) as compared to unmethylated cytosine [13,16,17,19]; such deamination events yield T:G mismatches and, subsequently, G→A transitions at the next round of DNA replication. A second prominent mutational signature found across several cancer types, including breast, ovary, bladder, head and neck, cervix, liver and lung [13,15–17,20,21], has been noted at C:G base-pairs in the context of T<u>C</u>:<u>G</u>A dinucleotides (C→T, C→G and C→A) (Table 1A); this has been attributed either to over-activity of members of the apolipoprotein B mRNA-editing catalytic polypeptide (APOBEC) cytosine deaminases [13,15,18,20], or to electron transfer reactions following oxidative damage [17].

Cancer type-specific mutational signatures have also been identified, particularly in cancers of the lung and skin. For example, in a cohort of 17 non-small cell lung cancer patients, the total number of somatic mutations was ~10-fold higher in smokers (median 15,659, range 7,424–26,202) than in never-smokers (median 888, range 842–1,268) [22], consistent with other reports that smoking-associated lung cancer is distinguished by a significantly high number of mutations per Mb [23–28]. Tumors from smokers were also characterized by high fractions (up to 46%) of C:G→A:T transversions [22,27–30] (Table 1B), a signature of exposure to alkylating nitrosamines and polycyclic aromatic hydrocarbons (PAHs) present in tobacco smoke, which yield miscoding G adducts [31–34]. This conclusion is further supported by a recent whole-genome sequencing analysis of an arsenic exposure-related lung squamous cell carcinoma, which was instead characterized by a high fraction (16.3%) of T:A→G:C transversions but a low fraction (~6.1%) of C:G→A:T transversions [35] (Table 1B). In melanomas, up to 87% of all mutations were represented by C:G→T:A transitions, mostly at pyrimidine dimers [36–42], consistent with DNA translesion synthesis across UV-induced covalently linked pyrimidine dimers [43]. Mutations in skin cancer at pyrimidine dimers, particularly CC→TT transitions on the non-transcribed strand of expressed genes, have also been attributed to APOBEC3A, a member of the APOBEC family of cytosine deaminases which are active mostly on single-stranded DNA, and expressed in skin keratinocytes [44] (Table 1B). Other prominent patterns included T→C transitions in hepatocellular carcinomas, which have been attributed to bulky DNA adducts on adenine, A→T transversions in the

TA:TA context, particularly in leukemia samples [13,18], a 2-fold increase in mutations at NGRA relative to NGRB (B = C or G or T) in melanomas [17] (Table 1B) and several others, for which the underlying mechanisms remain unknown.

Table 1. Mutational signatures in cancer genomes.

A. Main mutational signatures revealed from meta-analyses of cancer genomes				
Total number of mutations	**Total number of cancer types**	**Major SBS signature (% cancer types)**	**Sequence context**	**References**
4,938,362	30	C:G→T:A (80%)	NCG:CGN	[13]
		C:G→T:A or G:C (50%)	TCN:NGA	
1,000,000	19	C:G→any subst. (32%)	TCN:NGA	[15]
617,354	12	C:G→T:A (33%)	CG:CG	[16]
		C:G→G:C (25%)	TC:GA	
533,482	14	C:G→any subst. (93%)	NNCG:CGNN	[17]
		C:G→any subst. (36%)	NYCH:DGRN	
373,909	27	C:G→T:A (30%)	CG:CG	[18]
		C:G→any subst. (11%)	TC:GA	

B. Main cancer type-specific mutational signatures				
Cancer type	**Putative cause**	**SBS signature**	**Sequence context**	**References**
Lung cancer	tobacco smoke	C:G→A:T	none	[22,27–30]
	arsenic exposure	T:A→G:C	none	[35]
Melanoma	UV, APOBEC3A	C:G→T:A	pyrimidine dimers	[36–42,44]
	unknown	G:C→any subst.	NGRA:TYCN	[17]
Liver carcinoma	carcinogens	T:A→C:G	none	[13,18]
Leukemia	unknown	A:T→T:A	TA:TA	[13,18]
Endometrial cancer	$POLE^{P286R}$	G:C→T:A	AGA:TCT	

N, any nucleotide; *Y*, C or T; *R*, A or G; *D*, A or G or T; *H*, A or C or T.

2.2. Mismatch Repair and DNA Replicative Polymerase Proofreading Genes

The high fidelity of human DNA replication achieves nucleotide incorporation error rates of ~10^{-9}–10^{-10} in part through the proofreading (3'→5' exonuclease) activities of replicative polymerases, Pol ε and Pol δ, and postreplicative mismatch repair (MMR), which decrease the rates of misincorporation on the newly synthesized daughter strands by 100–1,000-fold each [45–48]. The proofreading domains reside in the large 261 and 125 kDa POLE and POLD1 catalytic subunits of the Pol ε and Pol δ holoenzymes, respectively, which perform their specific function predominantly during leading-strand (Pol ε) and lagging-strand (Pol δ) DNA synthesis in S phase [49]. The MMR pathway comprises 6 genes (*MSH2*, *MSH3*, *MSH6*, *MLH1*, *PMS1* and *PMS2*) whose products yield 4 types of heterodimeric complexes [MutSα (MSH2/6), MutSβ (MSH2/3), MutLα (MLH1/PMS2) and MutLβ (MLH1/PMS1)], active on mismatches, bulges, small loops, and a number of DNA lesions [50,51].

More than 1,000 constitutional gene variants in *MLH1*, *MSH2*, *MSH6* and *PMS2* have been classified as pathogenic or likely pathogenic in patients affected by Lynch syndrome [52], an autosomal dominant condition also known as hereditary non-polyposis colorectal cancer (HNPCC) and characterized by increased susceptibility to colorectal (25%–70%), endometrial (30%–70%), and other types of cancer [53]. Such MMR defects have also been known to lead to an accumulation of mutations, mostly in the form of microsatellite length changes (microsatellite instability, MSI) [53]. The patterns of SBSs in cancers showing MSI have recently been addressed from the exome and whole-genome sequencing data of two large cohorts of colorectal [54] and endometrial [55] cancer patients (224 and 373 tumors, respectively), and the reconstructed whole-genomes from two gastric cancer patients [56]. In general, there was no relationship between MSI status (high/low) and SBS mutation rates. By contrast, most samples with elevated SBS mutation rates also displayed somatic mutations in the proofreading *POLE* domain [57]. In addition, *POLE*-mutated samples could be classified into two distinct groups based on *MLH1* status: group 1, with low SBS mutations rates, *MLH1* inactivation and MSI-high; and group 2, with high SBS mutation rates, functional *MLH1* and MSI-low. Thus, paradoxically, concomitant *POLE* and *MLH1* mutations do not appear to act synergistically on SBS mutation rates in most patients [57].

Mutations in the proofreading domains of *POLE* and *POLD1* were also reported in two human colorectal cancer cell lines (DLD-1 and LoVo) and 1/76 colorectal cancer patients, all three samples exhibiting MMR deficiency [58]. More recently, two recurrent germline mutations in the proofreading domains of *POLE* (L424V) and *POLD1* (S478N) have been found in a cohort of 3,805 colorectal cancer patients selected for family history of colorectal tumors and multiple adenomas [59]. In the 62 tumors analyzed, no MSI was found; rather, loss of heterozygosity, chromosomal instability and driver mutations in known cancer genes, including *KRAS*, *BRAF*, *APC*, *PIK3CA*, *FBXW7*, but not *CTNNB1*, were revealed, suggesting that mutant $POLE^{L424V}$ and $POLD1^{S478N}$ may promote tumor formation by increasing the rates of SBS, without any apparent bias for a predominant type of base substitution [59]. Twelve missense somatic mutations predicted to affect the proofreading domain of *POLE*, and all associated with microsatellite stability, have also been identified in a study of 173 endometrial cancers, P286R being the most commonly represented (6 times) POLE mutation [60]. Using a panel of 75 cancer genes, *POLE*-mutated tumors exhibited an ~6-fold increase in mutations relative to non-*POLE*-mutated tumors, with a prevalence of G:C→T:A transversions, particularly at G:C base-pairs flanked 5' and 3' by an A:T base-pair [60]. Further validation of the putative role for *POLE*, and to a lesser extent *POLD1*, mutations in endometrial cancer has been obtained from an analysis of unpublished TCGA genomic data, where 21 (8.5%) and 1 (0.4%) tumors out of 248 samples were found to harbor *POLE* and *POLD1* mutations, respectively, including 8 cases of P286R and 5 cases of V411L changes in POLE [60]. *POLE/POLD1*-mutated cancers displayed high SBS rates, ranging from 227 to 14,695 exonic events, compared to a range of 22 to 2,014 in cancers lacking *POLE/POLD1* mutations. Likewise, cancers carrying the $POLE^{P286R}$ allele exhibited an overrepresentation of G:C→T:A substitutions at A<u>G</u>A:T<u>C</u>T motifs (Table 1B), supporting a DNA sequence-specific proofreading defect for this particular POLE mutation [60]. In summary, germline and somatic mutations in the *POLE* and *POLD1* genes appear to predispose to, or promote, colorectal and endometrial cancers in part by increasing the rates of SBSs [61]. Less direct investigations from cell culture nuclear extracts suggest that defects in DNA replication fidelity might also be associated with ovarian cancers [62].

2.3. The APOBEC Family of Cytosine Deaminases

The family of APOBEC cytosine deaminases comprises eleven members with distinct functions: activation–induced deaminase (AID), a B cell-specific enzyme required for both somatic hypermutation

(SHM) and class-switch recombination (CSR); APOBEC1, which is expressed primarily in the gastrointestinal compartment and is active in the transcript sequence editing of the apolipoprotein B mRNA; APOBEC2, which is expressed in heart and skeletal muscles and which appears to be essential for muscle development; APOBEC3s (A, B, C, D, F, G and H), which are active against exogenous viruses and endogenous retroelements and hence important for innate immunity; and APOBEC4, which is mostly expressed in the testes but whose function remains unknown [63]. APOBECs have in common a zinc-dependent cytidine deaminase domain (ZDD), which catalyzes the conversion of cytosine and deoxycytidine to uracil and deoxyuracil [64] in single-stranded RNA and DNA, often in a sequence-dependent context.

Editing activities of APOBEC3s play a critical role in restricting viral infectivity, and have also been postulated to have counteracted the actions against genome stability, mostly in terms of integration, exerted by both non-LTR (long terminal repeats) and LTR retrotransposons during evolutionary time [63,65]. For example, in Δvif (virion infectivity factor) HIV-1 particles, APOBEC3G proteins interact with the nucleocapsid domain of viral Gag to form nucleoprotein complexes with several Pol-II and Pol-III transcribed RNAs, which are then encapsulated into virions. During HIV-1 reverse transcription, up to 10% of cytosines can be deaminated to uracil on the minus strand of the viral complementary DNA (cDNA), thereby promoting loss of genetic information and the production of large populations of defective virions [63]. Vif antagonizes the activity of APOBEC3G (and other APOBEC3s) by binding and targeting the enzyme for polyubiquitination and subsequent degradation [63,66,67]. Other viruses targeted by APOBEC3 enzymes include human T-cell lymphotropic virus (HTLV), hepatitis B virus (HBV), hepatitis C virus (HCV), human papillomavirus (HPV) and human herpesviruses (HHV). In addition to deaminase activity, APOBEC enzymes restrict exogeneous viruses and endogenous retroelements through editing-independent activities. These include inhibition of viral replication, for example by interfering with tRNA priming and the initiation of DNA replication during HIV-1 reverse transcription, binding with positive regulators of viral gene expression, as in the case of heterogeneous nuclear ribonucleoprotein K (hnRNP) for HBV [63], and other less well-characterized mechanisms aimed at restricting non-LTR retrotransposon transcription, DNA synthesis and integration [65].

At least three recent reports have suggested the involvement of aberrant APOBEC3B deaminase activity as a frequent cause of SBSs in cancer. For example, APOBEC3B mRNA was found to be upregulated relative to controls in 28/38 established breast cancer cell lines and to be expressed in the nucleus [68]. In selected nuclear extracts, C→U editing activity was detected on synthetic DNA substrates specifically at T<u>C</u>:<u>G</u>A dinucleotides when treated with a control shRNA, whereas no activity was evident upon treatment with short hairpin RNA (shRNA) targeting APOBEC3B mRNA. In these cell lines, treatment with anti-APOBEC3B shRNA also led to a decrease in genomic uracil loads from ~100,000 to ~60,000 per haploid genome by HPLC-ESI-MS/MS and, as expected, test amplicons displayed a decrease in C→T transition mutation frequency. The involvement of APOBEC3B activity in cancer was further supported by data on Ref-seq APOBEC3B expression, which was shown to be high in several tumor types, including breast, uterus, bladder, head and neck and lung (both adenocarcinoma and squamous cell carcinoma) [15,20]. The top five cancer types with the majority of mutations at C:G base-pairs were also among the top six datasets in terms of APOBEC3B mRNA expression, and a positive correlation between the proportion of mutations at C:G base-pairs and median APOBEC3B levels was observed. Bladder, cervical, lung squamous cell carcinoma, lung adenocarcinoma, head and neck, and breast cancers shared a strong bias for TCN (N = A or C or G or T) mutation signatures, as observed for the recombinant APOBEC3B protein. Interestingly, a significant enrichment of strand-coordinated and clustered (2 or more per 10 kb) C→T and C→G mutations at TCW (W = A or T) motifs were discovered, a number of which were in close proximity to chromosomal rearrangement breakpoints, particularly in bladder, cervical, head and neck, breast and lung tumors [15,20], a phenomenon termed kataegis [3]. In summary, these analyses are consistent with the possibility that aberrant APOBEC deaminase activity, particularly at TC:GA sites, may represent a general endogenous mutagen that contributes to several different types of human cancer.

2.4. Electron Transfer in DNA Oxidation

Mutation spectra analyses of SBSs arising spontaneously, both in cell culture and in whole animals, have indicated the frequent occurrence of sequence context-dependent mutations. For example, an analysis of 837 spontaneous SBSs in the *supF* tRNA gene in 18 cell lines and 2 transgenic mouse models indicated that the most mutable regions involved guanine and cytosine tracts [69]. In human osteosarcoma cells, shRNA knock-down of the *WRN* helicase gene, mutations in which are associated with the progeroid Werner syndrome, led to a doubling in genomic 8-oxo-7,8-dihydro-2'-deoxyguanosine (8-oxoG) content and an increase in SBSs in the *supF* reporter gene, consisting largely of G→C (49%), G→A (28%) and G→T (23%) substitutions at GA:TC dinucleotides within the *supF* reporter gene [70]. Because 8-oxoG is a well-recognized marker of oxidative damage and guanine oxidation depends upon flanking sequence (*i.e.*, GR > GY; R = A or G; Y = T or C) as a result of stacking-induced electron transfer [71,72], these data suggested a role for oxidative damage in sequence context-dependent mutagenesis.

A recent study addressed the question as to whether electron transfer might also cause sequence context-dependent SBSs in cancer [17]. The analysis compared the fractions of mutations occurring at G:C base-pairs in the context of all 64 possible combinations of NGNN:NNCN motifs (NGNN for simplicity), and included 21 cancer datasets representing 14 tissues comprising 1,149 patient samples and 2 cell lines for a total of 533,482 SBSs. With the exception of two melanoma datasets, CGNN sequences were more frequently mutated than DGNN (D = A or G or T), consistent with the CG:CG dinucleotide, the most prominent substrate for cytosine methylation, being a common mutation hotspot (Table 1A). In 7 cancer datasets, including lung, head and neck and melanoma, for which association with exposure to either cigarette smoke or sunlight was documented, G followed by a 3' purine was associated with increased mutations as compared to a 3' pyrimidine, *i.e.*, DGRN > DGYN (Table 1A). Notably, significant correlations were observed between the fractions of mutated DGNN motifs and the sequence-dependent free energies of base stacking along the DGNN motifs for 5 of the 7 cancer datasets. Significant correlations were also observed between the fractions of mutated DGNN motifs and the energies required to abstract an electron from the target guanines, as assessed from the values of vertical ionization energies computed for all G-centered trimer motifs. These results are consistent with the conclusion that DNA oxidation may be a source of sequence context-dependent SBSs in cancer as a result of electron transfer, as postulated from model sequences *in vitro* [73,74].

2.5. DNA Replication Timing

During eukaryotic DNA replication, more than 20,000 pre-replicative complexes comprising the origin recognition complex (ORC), Cdc6 and Cdt1 load inactive minichromosome maintenance (MCM) helicase complexes to generate "licensed" replication origins. Initiation of DNA synthesis is controlled by Dbf4-dependent kinase (DDK), which recruits Cdc45 and Sld3, and cyclin-dependent kinase (CDK), which by phosphorylating Sld3 and Sld2 recruit GINS and additional factors, including DNA polymerases. Origin firing is controlled both temporally and spatially. Chromatin correlating positively with gene expression, G + C-richness and active chromatin marks is replicated in early S phase in the nuclear interior, whereas chromatin associated with gene-poor regions, A + T-richness and repressive chromatin marks is preferentially replicated during late S phase at the nuclear periphery [75,76].

In cancer, late replicating chromatin has been shown to harbor higher relative fractions of SBSs than early replicating chromatin. For example, the fractions of SBSs from several completely sequenced cancer genomes (melanoma, prostate cancer, small cell lung cancer, chronic lymphocytic leukemia and colorectal cancer) were found to be significantly more extensive in the constant late than in the constant early replicating zones of neutrally evolving regions [77]. Such genomic regions were those remaining after sequences from centromeres, telomeres, the Y chromosome, genes, promoters, repetitive elements and ultra-conserved regions had been excluded. Analyses of the mutation spectra indicated that, with the exception of A→T (T→A) transversions, which occurred more often in the constant late replicating regions in all five cancer types, the relative proportions of substitutions were very similar between constant early and constant late replicating regions, even though mutation frequencies were higher in the latter. Similarly, a large study of SBSs in exomes from 3,083 tumor-normal pairs representing 27 different cancer types also found that the average mutation fraction was higher (~2.9-fold) in the latest- as compared to the earliest-replicating percentiles [18].

The increased mutation frequency in late as compared to early replicating regions does not appear to be a unique property of cancer cells. For example, a comparison of 1-Mb non-overlapping regions containing SBSs between human and chimpanzee and the pooled cancer data [77] indicated that most regions harboring human-chimpanzee SBSs also harbored SBSs in cancer. Similarly, in human populations, late-replicating regions of the human genome have been shown to be characterized by a greater density of single nucleotide polymorphisms (SNPs) than early replicating regions [78]. Finally, deep-sequencing of human lymphoblastoid cell lines from father-mother-offspring trios revealed that transition mutations were >2-fold more abundant in late-replicating than in early-replicating regions of the genome, whereas transversion mutations were increased >6-fold [79]. In summary, these analyses suggest that the increased mutation rate in late- *versus* early-DNA replicating regions, which has been noted in both population and cancer genome studies, are potentially caused by mechanisms that share some commonalities between the germline and the soma.

2.6. Chromatin Organization

Chromatin structure has been found to strongly correlate with regional SBS rates along chromosomes in cancer genomes. Chromatin organization is regulated by many factors, including epigenetics, *i.e.*, reversible changes both at the level of DNA and involving histone tail amino acids. Chromatin condensation, or heterochromatin, is associated with reduced gene expression and involves the accumulation of specific histone marks, including methylation of lysines 9 and 27 on histone 3 (H3K9me2 and H3K27me3) by histone methyltransferases (HMT) such as G9a, GLP and SETBD1, and DNA methylation at C5 of cytosine, mainly at CG:CG dinucleotides, by DNA (cytosine-5)-methyltransferases (DNMT1, DNMT3a and DNMT3b). Among other critical interactions, H3K9me2 serves as a high-affinity binding site for the recruitment of heterochromatin protein 1 (HP1), a platform protein that collapses chromatin into higher-order fibers as a result of dimerization between nucleosomes. By contrast, promoter DNA demethylation and acetylation of H3K9 maintains open chromatin structure, or euchromatin, and supports active gene transcription [80]. These represent only a few epigenetic marks that are known to influence chromatin structure.

A total of 84,879 unique SBS positions from leukemia, melanoma, small cell lung cancer and prostate cancer genomes were used to identify potential sources of mutation rate variation across the genome, by correlating site variation with a set of 46 diverse genetic and epigenetic features that had been mapped genome-wide in human cells [81]. On a megabase scale, cancer SBS density correlated with many features of somatic cell chromatin organization, with the highest positive correlations being represented by the repressive histone modification H3K9me3, followed by H3K9me2 and H4K20me3. In fact, >55% of the variance in cancer SBS regional variation could be accounted for by combining features, with H3K9me3 alone associating with more than 40% of the observed variance in SBS density. Interestingly, the associations remained strong when only non-genic or only genic regions of the genome were considered, suggesting that transcription or transcription-coupled nucleotide excision repair may have played a comparatively minor role as compared to epigenetic modifications. We should note that, in a separate study, levels of gene transcription were found to correlate inversely with mutation rates [18], although the process of transcription is known to be a source of genetic instability [82,83]. The use of a metric that employed data on physical contacts between regions through three-dimensional folding of chromosomes, thereby distinguishing between densely packed chromatin with strong short-range interactions and accessible euchromatin with more diverse interactions, also revealed an anti-correlation pattern with somatic SBS density [81]. These findings led the authors to propose that chromatin organization is a major determinant of variation in regional mutation rates in cancer. Specifically, SBS rates in cancer cells appear to be highest in inaccessible, heterochromatin-like regions and lowest in accessible euchromatin-like domains. Reversible histone acetylation and deacetylation events in the cell are also critical for mutation avoidance. For example, failure to deacetylate H3K59ac marks following the S-phase in *hst3*Δ *hst4*Δ double-mutant yeast cells increased the rates of SBS ~10-fold and the rates of gross chromosomal rearrangements 15,600-fold, whereas lack of H3K59 acetylation in *rtt109*Δ cells led to a ~10-fold increase in complex mutations [84]. These effects were synergistic with mutations in MMR and the proofreading activities of replicative Pols δ and ε, suggesting a model in which cyclic acetylation and deacetylation of chromatin is critical for replication fidelity.

3. Mechanisms of Base Modification

3.1. Cytosine Methylation and Demethylation

The ability of site-specific DNA (cytosine-5)-methyltransferases to transfer the methyl group from *S*-adenosylmethionine to the C5 position of cytosine in the context (mainly) of CG:CG dinucleotides in mammals and the role of DNA methylation in transcriptional regulation, genomic imprinting and silencing of repetitive DNA have been well reviewed [85,86]. During the past few years, it has become evident that DNA methylation can be reversed by a group of enzymes belonging to the ten-eleven translocation (TET1, 2, and 3) family of iron and α-ketoglutarate (α-KG)-dependent dioxygenases, which utilize molecular oxygen to transfer a hydroxyl group to 5mC to form 5-hydroxymethylcytosine (5hmC) (Figure 1, top). Whereas approximately 4% of all cytosines (70%–80% of CG:CGs) are estimated to be methylated, only ~0.1%–0.7% total appear to be marked by hydroxymethylation [87]. TET enzymes can further oxidize 5hmC sequentially to yield 5-formylcytosine (5fC) and 5-carboxycytosine (5caC), and evidence is increasing for a role of thymine DNA glycosylase (TDG), a member of the base excision repair pathway, in cleaving 5fC and 5caC thereby yielding abasic sites which are then replaced with unmodified cytosines [88,89] (Figure 1, top).

A critical step in the TET-dependent oxidation reactions is the role played by α-KG, which binds to a His-His-Asp-coordinated Fe(II) cluster in the enzymatic active site to effect the transfer of an oxygen to the substrate 5mC and release 5hmC. α-KG is synthesized from isocitrate in a fully reversible reaction by different isoforms of $NADP^+$-dependent isocitrate dehydrogenases (IDH), with IDH2 and IDH3 acting in the mitochondria as part of the tricarboxylic acid (TCA) cycle (Krebs cycle), and IDH1 providing a source of NADPH in the cytoplasm for lipid biosynthesis and protection from oxidative stress [90].

Figure 1. (Top) Cytosine methylation and demethylation pathways; **(Bottom)** Products of cytosine and C5-substituted cytosine deamination.

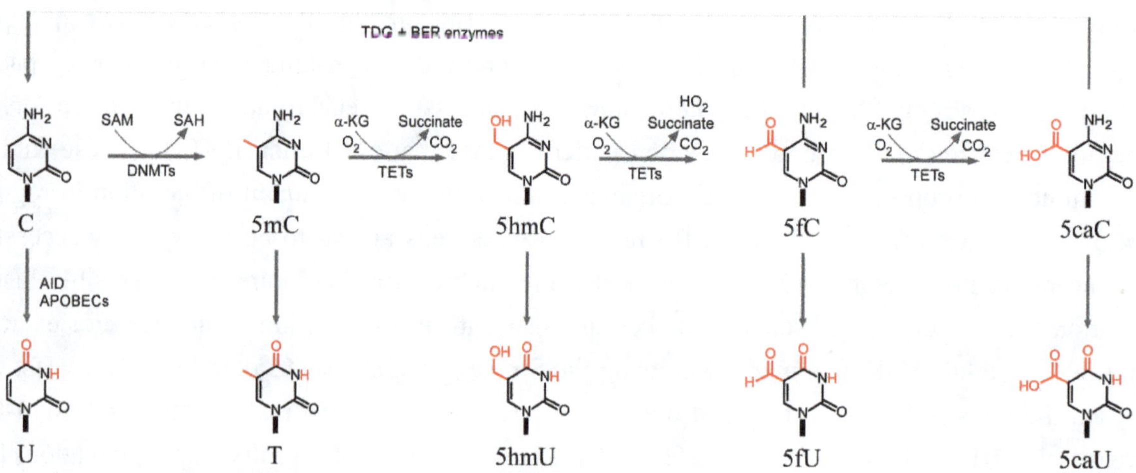

Recurrent gain-of-function mutations in *IDH1* and/or *IDH2* typify ~70% of sporadic high-grade gliomas and secondary glioblastomas, ~10% of acute myeloid leukemias [16] and colangiocarcinomas [91], and have been reported in patients with acute lymphoblastic leukemia, chondrosarcomas, angioimmunoblastic T-cell lymphoma, cholangiocarcinoma and pancreatic cancers [90,92], where they have been found to induce DNA hypermethylation at CG:CG islands and shores in a tissue-specific manner [91]. The mutations, which occur at the active site of IDHs, alter the reaction order of the enzymes such that high concentrations of D-2-hydroxyglutarate (2-HG), rather than α-KG, are released [92] (Figure 2). Thus, 2-HG binding to both TET enzymes as well as other cellular dioxygenases, including histone demethylases and propyl hydroxylases, effectively inhibits their activities.

Figure 2. Conversion of isocitrate to α-ketoglutarate (α-KG) by isocitrate dehydrogenases (IDH) enzymes and conversion of α-KG to D-2-hydroxyglutarate (2-HG) by gain-of-function mutations in *IDH1* or *IDH2*.

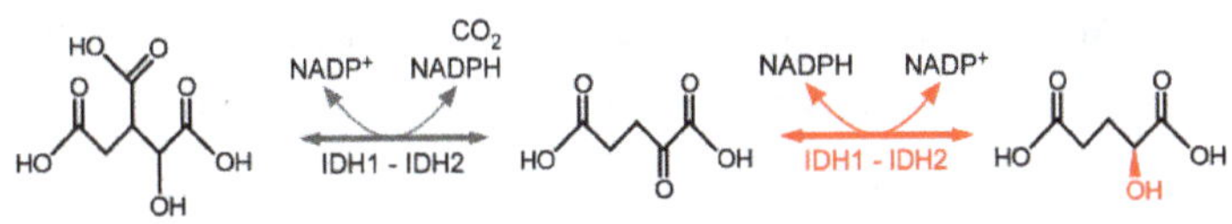

In melanoma, loss of 5mC through oxidation to 5hmC has been observed both as a result of *IDH1* or *IDH2* neomorphic mutations as well as the downregulation of *TET* and *IDH2* genes [93]. A comparison between benign nevi and melanoma further supported the selective loss of 5hmC in melanoma; in addition, the extent of 5hmC loss in melanoma correlated directly with Breslow depth, a predictor of prognosis, pathological stage and, most significantly, Kaplan-Meier survival curves [93]. Large losses of 5hmC peaks and higher levels of 5mC in melanoma *versus* nevi were detected within gene coding and flanking regions, including genes associated with adherens junctions, Wnt signaling, additional pathways in cancer, and melanogenesis pathways, implying a role for 5hmC in pathways that are fundamental to cellular differentiation and dedifferentiation. In mouse embryonic stem cells, 5fC and 5hmC were found to be enriched in intragenic regions, especially within exons and enhancers, where they colocalized with histone acetyltransferase p300 sites, DNaseI hypersensitive sites and CTCF-bound regions, specifically at poised enhancers, which are marked by H3K4me1[+] and H3K27ac[–], in comparison to active enhancers (H3K4me1[+] H3K27ac[+]), concomitantly with a decrease in 5mC [88]. Accordingly, in the absence of TDG, accumulation of 5fC correlated with increased binding of the transcriptional activator p300 at poised enhancers. These data support a role for 5mC and 5hmC oxidation in the regulation of the epigenetic state of functional enhancer elements in mammalian genomes. CG:CG hypermethylation at specific genes was also found to represent a marker for relapse-free survival time after surgery in a cohort of 444 patients with non-small cell lung cancer. Specifically, patients with zero to one methylated markers in the *HIST1H4F*, *PCDHGB6*, *NPBWR1*, *ALX1* and *HOXA9* genes were characterized by a longer relapse-free survival time than those with two or more hypermethylated markers; 48% from the enriched methylated group relapsed, as compared with only 18% of those in the less methylated group [94].

3.2. Deamination of Cytosine Bases

3.2.1. Spontaneous Deamination

The rate constants for the spontaneous deamination of cytosine (C) and protonated C to uracil (U), 5mC to T, 5hmC to 5hmU, 5fC to 5fU and 5caC to 5caU (Figure 1, bottom) have been determined from both the extrapolations of Arrhenius plots and genetic assays [95–100]. In double-stranded DNA, deamination rates for C and 5mC are extremely slow, of the order of 10^{-13} s^{-1} (Table 2), which translates into half-lives ranging between ~30,000 to ~85,000 years. Rates increase approximately 3-orders of magnitude in both single-stranded DNA and in isolated deoxyribonucleotides, supporting the view that hydrogen bonding plays a major role in shielding cytosines from spontaneous deamination. Comparison of the data given in Table 2 indicates that rates of deamination for 5mC are only marginally higher than for cytosine. Rates increase by an additional 3-orders of magnitude upon protonation; however, since cytosine protonation occurs at acidic pH (pKa = 2.4) within a C:G Watson-Crick base-pair [101], it is unlikely to play a significant role *in vivo*. That DNA melting is rate-limiting for cytosine deamination is further suggested by evolutionary studies, which indicate that CG:CGs embedded within G+C-rich areas (H isochores), and thus characterized by increased melting temperatures, have been depleted to a lesser extent than CG:CGs embedded in G+C-poor (L isochores) regions, which melt at lower temperatures [102]. Further support is provided by the finding that C→T transitions decrease gradually with increasing nucleosome occupancy score in comparative studies of *S. cerevisiae*, medaka (*Oryzias lapites*) and *C. elegans* genomes [103]. Analysis of the hydrolytic deamination reaction using density functional theory shed considerable light on the requirements for base unpairing and the effect of protonation [104]. Two pathways have been identified (Figure 3). In pathway A, upon protonation of N3 (Figure 3a), nucleophilic addition of a first water molecule to carbon C4, leads to the formation of a tetrahedral intermediate with the assistance of a second water molecule (Figure 3b). The C4-N4 bond is then broken and a proton transfer takes place from the hydroxyl group at C4 to NH_3, thereby forming thymine and an ammonium cation. In pathway B, nucleophilic addition of the first water molecule to C4 occurs on the neutral 5mC (Figure 3b), again with the assistance of a second water molecule, yielding a neutral tetrahedral intermediate (not shown). The exocyclic amino group is then protonated through an intermolecular proton transfer, after which the reaction proceeds as in pathway A. For both pathways, the nucleophilic addition is the rate-determining step; however, whereas nucleophilic addition to carbon C4 of 5mC is easier than to the N3-protonated form, the trend is reversed in the case of C and N3-protonated C. Thus, deamination of 5mC is more difficult than that of C in pathway A, whereas the opposite is seen in pathway B.

This study has several implications. First, nucleophilic attack and formation of the tetrahedral intermediate cannot occur on duplex DNA; second, only pathway B is compatible with the greater susceptibility of 5mC to deamination, compared with C; third, a protonated base is not required for pathway B; and fourth, for pathway B, the activation free energy for 5mC (134.1 kJ/mol in aqueous solution) is only 4.4 kJ/mol less than that associated with C (138.5 kJ/mol in aqueous solution), implying that the susceptibility to deamination of 5mC relative to C is no more than 4-5-fold. In summary, these investigations show that (1) spontaneous deamination of cytosine bases only occurs in single-stranded DNA; (2) deamination of 5mC is only marginally more efficient than for C; and (3) all C5 substituted bases, with the exception of 5caC, display detectable (and rather similar) rates of deamination.

3.2.2. ROS-Induced Deamination

ROS, such as the non-radical hydrogen peroxide (H_2O_2) and free superoxide radicals (O_2^-), are generated as a result of mitochondrial respiration, from the activation of growth factor receptors through NADPH oxidase, the arachidonic acid cascade and others, and play crucial roles as signal transduction molecules and neuroregulators [105,106]. H_2O_2 may also generate free radicals, including the hydroxyl radical (OH), the most potent oxidizing radical generated by the cell, through routes including ionizing radiation, interactions with O_2^- through the Haber-Weiss reaction and by interactions with transition metal ions [Fe(II) and Cu(I)/Cu(II)] through Fenton chemistry, as exemplified in reaction 1.

$$Cu(II) + H_2O_2 \rightarrow Cu(I) + H_2O + H^+;\ Cu(I) + H_2O_2 \rightarrow Cu(II) + OH + OH^- \quad (1)$$

Free radicals constitute an important endogenous source of damage to DNA and other molecules, and therefore strict homeostatic controls exist in the cell between the generation and neutralization of ROS species by catalase, superoxide dismutase 1 and 2 (SOD1 and 2), the glutathione peroxidase (GPX) and peroxiredoxin (PRX) families of detoxifying enzymes and other antioxidants, such as vitamins C and E [106,107]. A number of studies in leukemia, breast cancer, ovarian cancer, benign and malignant prostate cancer, non-small cell lung carcinoma, cervical squamous cell carcinoma, stomach cancer, and Hodgkin's disease patients all concur with the conclusion that levels of ROS detoxifying enzymes are generally lower in cancer than in surrounding normal tissue, leading to oxidative stress, *i.e.*, altered ROS homeostasis in favor of increased steady-state levels of ROS [107]. In addition, as in the case of *BCR-ABL1* translocations in myeloid leukemia, Rac2 activation has been shown to reduce the mitochondrial membrane potential ($\Delta\Psi_m$), thereby inducing electron leakage from the mitochondrial respiratory chain complexes I-III and II-III (MRC-cIII) and, as a consequence, a 2- to 6-fold increase in cellular ROS [108].

Treatment of duplex oligonucleotides containing methylated and unmethylated CG:CGs with Cu(II)/H_2O_2/ascorbate to effect Fenton-type reactions led to much more frequent modifications of 5mC than C [109]. One of the main products involved the saturation of the C5-C6 double-bond of 5mC, to yield 5-methyl-5,6-dihydroxy-5,6-dihydro-2'-deoxycytidine (5-methyl-2'-deoxycytidine glycol, 5mCg). Kinetic determinations of the spontaneous deamination of two stereoisomers of 5mCg, *i.e.*, 5mCg(*5S,6S*) and 5mCg(*5R,6R*) in duplex DNA, which affords 5,6-dihydroxy-5,6-dihydrothymidine (thymidine glycol, Tg) [110], indicated rates in the range of $\sim 10^{-6}\ s^{-1}$, similar to the values determined for isolated nucleotides. Thus, ROS-induced saturation of the C5-C6 double-bond in 5mC increases rates of deamination by ~4 orders of magnitude with respect to single-stranded DNA, and ~7 orders of magnitude with respect to unmodified 5mC in double-stranded DNA. The susceptibility of duplex DNA to damage by Fenton-type reactions has been assessed using 5S rDNA, either alone or upon reconstitution on nucleosome particles. The number of single base lesions was found to be 8-fold higher on nucleosomal DNA than on isolated DNA, implying that Fenton chemistry is not only unrestricted by chromatin compaction but actually appears to be facilitated [111]. Although the roles of histone octamers in permitting DNA damage are not fully understood, X-ray crystal structure studies revealed the presence of many divalent metal binding sites in nucleosome particles [112], and several peptide models of histones H2A, H2B, H3 and H4 have been shown to coordinate Cu(II), mostly through macrochelate rings involving histidine and carboxylate groups [113]. The extremely short (<1 ns) half-life of OH likely restricts Fenton chemistry at sites of OH generation. Thus, given that H_2O_2 is relatively stable and able to diffuse across cells, copper coordination within nucleosome particles might provide a suitable environment for oxidative DNA damage in chromatin. Determinations of copper concentrations by both atomic absorption spectroscopy and X-ray fluorescence in plasma and tumor samples from several types of cancer indicate that levels are usually higher (up to 2- to 3-fold) in cancer patients than in normal controls [107]. Indeed, high levels of copper appear to be required for tumor growth [114].

Table 2. Rates of spontaneous deamination for cytosines.

Deaminating base	Sequence context	Deaminated base product	Deamination rate at 37 °C (s^{-1})	References
C	Free nucleoside	U	$9.4 \pm 0.5 \times 10^{-10}$	[95]
3mC$^+$	Free nucleoside	3mU	5×10^{-7}	[100]
3mC$^+$	Free nucleotide	3mU	13×10^{-7}	[100]
5mC	Free nucleoside	T	$7.8 \pm 0.3 \times 10^{-10}$	[95]
5hmC	Free nucleoside	5hmU	$5.8 \pm 0.8 \times 10^{-10}$	[95]
5fC	Free nucleoside	5fU	$1.2 \pm 0.2 \times 10^{-9}$	[95]
5caC	Free nucleoside	5caU	not detected	[95]
5mCg(*5S,6S*)	Free nucleoside	Tg	1.1×10^{-5}	[115]
5mCg(*5R,6R*)	Free nucleoside	Tg	8.6×10^{-6}	[115]
C	ssDNA	U	2.1×10^{-10}	[96]
C	ssDNA	U	$\sim 1 \times 10^{-10}$	[98]
5mC	ssDNA	T	9.5×10^{-10}	[96]
C	dsDNA	U	2.6×10^{-13}	[97]
C	dsDNA	U	4×10^{-13}	[99]
C	dsDNA	U	$\sim 7 \times 10^{-13}$	[98]
5mC	dsDNA	T	5.8×10^{-13}	[97]
5mC	dsDNA	T	1.5×10^{-11}	[99]
5mCg(*5S,6S*)	dsDNA	Tg	5.2×10^{-6}	[109]
5mCg(*5R,6R*)	dsDNA	Tg	7.0×10^{-6}	[109]

C, cytosine; 5mC, 5-methylcytosine; 3mC$^+$, N^3-methylcytosine; 5hmC, 5-hydroxymethylcytosine; 5fC, 5-formylcytosine; 5caC, 5-carboxycytosine; 5mCg(*5S,6S*) and 5mCg(*5R,6R*), 5-methylcytosine glycol stereoisomers; U, uracil; 3mU, N^3-methyluracil; T, thymine; 5hmU, 5-hydroxymethyluracil; 5fU, 5-formyluracil; Tg, thymine glycols.

Figure 3. **(a)** Numbering scheme for cytosines; **(b)** Pathways for the spontaneous deamination of C and 5mC (only 5mC is shown).

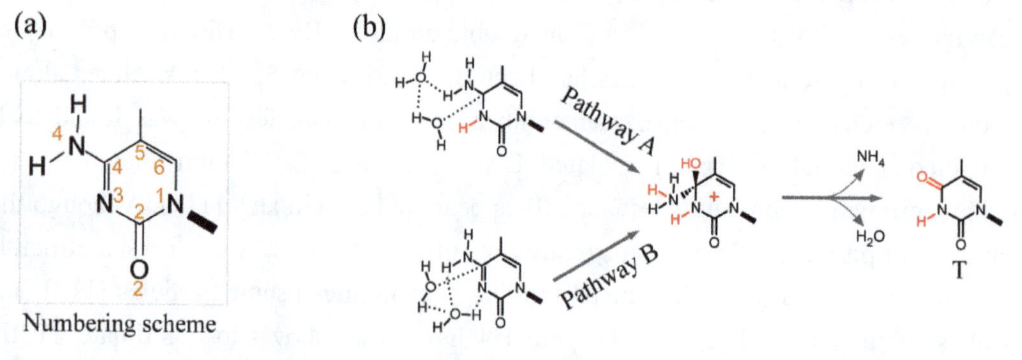

In summary, oxidation of 5mC by copper ions and ROS generate cytosine glycol intermediates, which deaminate at high rates to yield Tg:G mispairs. DNA oxidation by Fenton chemistry is enhanced in nucleosomal DNA, in which several coordination sites for copper and other metal ions have been identified. Both ROS and copper concentrations have been found to be enhanced in tumors, raising the possibility that ROS-dependent deamination at methylated CG:CG sites may contribute to mutation in cancer.

3.2.3. Enzymatic Deamination by Single-Stranded DNA-Specific AID/APOBECs

AID/APOBEC enzymes catalyze the deamination of cytosine to uracil on single-stranded DNA (Figure 1, bottom). The active site of APOBEC enzymes is recognized by a conserved motif (H-X-E-$X_{23–28}$-P-C-X-C), in which a coordinated zinc ion carries out the nucleophilic attack during the deamination reaction. In the high-resolution crystal structures of human APOBEC2 and the catalytic domain of APOBEC3G (aa 197–380), a water molecule serves as a hydrogen donor, whereas a conserved glutamate residue (E100 in APOBEC2 and E259 in APOBEC3G) functions as a proton shuffler during the hydrolytic cycle [116]. As with other DNA metabolizing enzymes, the target cytosine is flipped-out and inserted into the active site; because the flipping step appears to involve passive DNA breathing, it probably accounts for the greater enzymatic activities observed for deamination of single-stranded, as opposed to double-stranded, nucleic acids [117,118]. Whether AID/APOBEC enzymes deaminate cytosines modified at C5, including 5mC, 5hmC, 5fC and 5caC has been controversial, with recent studies reporting generally weak or no activities towards C5-substituted cytosines [118–120], and earlier work reporting strong activities on 5mC by human AID and rat APOBEC1 [121]. Experiments performed by scoring mutations either in viral DNA or *in vitro* with model sequences indicate strong effects on deamination rates by nucleotides flanking the target cytosine: optimal substrates include (C/G)T$\underline{\text{C}}$(A/G) for APOBEC3B [122,123], T(T/C)$\underline{\text{C}}$ for APOBEC3C [124], CC$\underline{\text{C}}$A for APOBEC3G [122,124–127], TT$\underline{\text{C}}$T for APOBEC3F [124–126], WR$\underline{\text{C}}$Y (W = A or T; R = A or G; Y = C or T) for AID [128], although comparative analyses using enzymatic kinetic constants awaits further work. In summary, APOBEC enzymes favor deamination of unmodified cytosine residues, to yield C-to-U modifications, in single-stranded DNA and in a sequence-dependent manner that is specific to each family member.

3.3. Sequence Context-Dependent Guanine Oxidation Products

3.3.1. Guanine

DNA oxidation plays a significant role in the pathophysiology of cancer, with epidemiological studies demonstrating a strong association between the generation of ROS and reactive nitrogen species (RNS) from chronic inflammation and increased cancer risk [129]. Guanine has the lowest redox potential of all DNA bases, and it has consistently been found to be a highly susceptible site for reactions with a variety of agents, including singlet oxygen, OH radicals, peroxynitrite, UV radiation with riboflavin and many others [130]. For example, in a study in which duplex DNA was subjected to 266 nm wavelength laser pulses as a source of photonic ionization, the quantum yield for the formation of 8-oxoG was much higher than that of oxidized nucleosides arising from the degradation of the other bases [131]. In addition, extensive experimental evidence supports a role for sequence context in terms of the chemistry and extent of DNA damage at guanine residues. Pioneering work in which DNA cleavage was induced by riboflavin as an electron-accepting photosensitizer in double-stranded 30-mers containing a target G in different sequence contexts (5'-TXGYT-3'), showed that the extent of cleavage at the target G depended upon DNA flanking sequence composition [73], implying that the ease of losing an electron by the target G was also dependent on flanking sequence. Indeed, computations of the ionization potentials (*i.e.*, the energy required to abstract an electron) (IPs) for the target G were found not only to differ with varying X and Y in a 5'-XGY-3' context, but also to correlate inversely with the extent of cleavage [73]. Specifically, the ionization potentials at a G followed by a 3' A or G were found to be up to 0.44 eV lower than when followed by a 3' T or C, whereas the base composition 5' of a G made little (<0.1 eV) contribution [74]. Hence, these and other investigations have together laid the foundation for the concept that a positive charge (a hole) inserted into DNA by abstracting an electron migrates through base stacking, either by hopping from one base to the next over long distances or by a tunneling mechanism over short distances (1–3 bases) [132], from the original location to sites of lowest IPs, *i.e.*, 5' G in GA and GG sequences.

3.3.2. 8-oxoG

In addition to unmodified bases, sequence context-dependent reactivity has also been observed for 8-oxoG, one of the main products of DNA exposed to ROS and RNS. 8-oxoG is several orders-of-magnitude more susceptible to further oxidation than G itself due to a lower ionization potential (6.93 *versus* 7.31 eV for unstacked 8-oxoG and G, respectively) [133], yielding more stable secondary oxidation products, including dehydroguanidinohydantoin (DGh), *N*-nitro-dehydroguanidinohydantoin (NO_2-DGh), 5-guanidinohydantoin (Gh), 2-imino-5,5'-spirodihydantoin (Sp), 2,5-diamino-4*H*-imidazol-4-one (imidazolone, Iz), its hydrolysis product 2,2,4-triamino-5(2*H*)-oxazolone (oxazolone, Oz) [134] and guanidinoformimine (Gf), the decarboxylated product of Oz [135] (Figure 4).

Earlier studies demonstrated that 8-oxoG reactivity to a variety of oxidants, including NiCR/$KHSO_5$, $IrCl_6^{2-}$, $IrBr_6^{2-}$, $Fe(CN)_6^{3-}$, SO_4^{-} and $^1O_2$, increased when located 5' to a G (8-oxoGG) compared to 3' to a G (G8-oxoG) [136], a trend that followed the computed sequence-dependent ionization potentials (6.38 eV for 8-oxoGG and 6.51 eV for G8-oxoG) [133]. More recent studies further established the sequence-dependent reactivity of 8-oxoG in duplex oligonucleotides to UVA-irradiated riboflavin to follow: C8-oxoGA ≈ A8-oxoGG > G8-oxoGG > C8-oxoGT > T8-oxoGC > A8-oxoGC, supporting a model whereby indiscriminate removal of electrons from all four nucleobases by riboflavin creates holes that migrate to sites of lower IPs (8-oxoG), with 8-oxoG reactivity modulated by sequence-dependent variations in the IPs by neighboring bases [137]. In addition to the extent of reactivity, also the types of products formed by riboflavin-oxidized 8-oxoG varied with flanking sequence composition. For example, although three main products were generally observed (Sp > Gh > Iz), at low riboflavin concentrations (<15 μM) oligonucleotides containing G8-oxoGG, C8-oxoGT and T8-oxoGC yielded relatively high levels of Sp that decreased as a function of increasing riboflavin concentration. By contrast, at riboflavin doses >30 μM, DGh was the most abundant species in some sequence contexts (G8-oxoGG and C8-oxoGA), with Iz matching DGh in the A8-oxoGG and G8-oxoGG sequence contexts. In contrast to riboflavin, nitrosoperoxycarbonate ($ONOOCO_2^-$), generated from macrophage-derived nitric oxide (NO) and superoxide (O_2^-), failed to yield sequence-dependent 8-oxoG reactivity and displayed a rather uniform spectrum of oxidation products, which were dominated by DGh > Oxaluric acid > NO_2-DGh [138]. Likewise, Gh and Sp have been established as the main products of G and 8-oxoG oxidation by peroxynitrite, peroxyl radicals, and hypochlorous acid, reactive species also released by macrophages during an inflammatory response [139]. Thus, as noted by Lim *et al.* [138], "the observation of strong sequence context effects on the final chemistry of DNA oxidation complicates our understanding of the mechanistic basis for both mutation frequency and mutational spectra caused by DNA damage *in vivo*", a task that is further complicated by sequence-dependent variations in the rates of DNA repair of individual DNA lesions. The analyses reported above clearly point to a critical role being played by charge (electron) transfer in the sequence context-dependent oxidation of DNA and the migration of the original sites of damage to distant sites of lower IPs, mostly G and 8-oxoG in the GA, GG, 8-oxoGA, and 8-oxoGG contexts.

Figure 4. Sequence context-dependent reaction products of 8-oxoG.

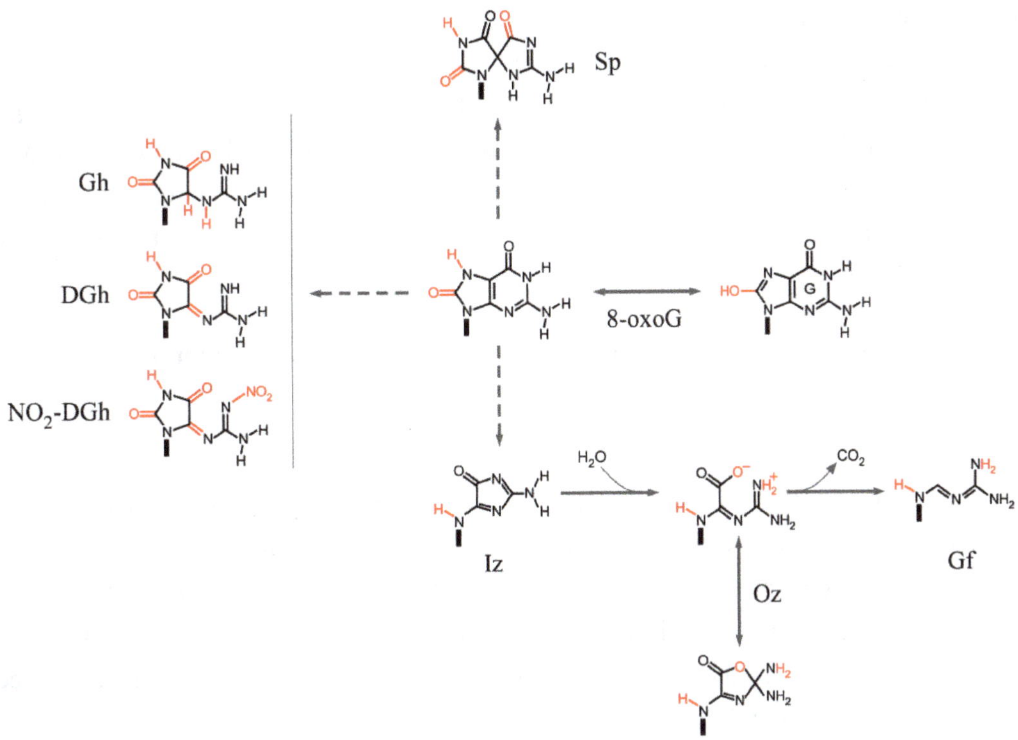

3.3.3. Charge Transfer in Nucleosomal DNA

Whereas bases in single-stranded DNA are generally more easily oxidized than in double-stranded DNA [138], the stability of stacking interactions in duplex DNA enables the charge transfer process towards guanine bases to take place more efficiently in duplex DNA than in single-stranded DNA. For example, increasing the ionic strength, which results in a more stable duplex, also enhanced the yield of 8-oxoG and, concomitantly, decreased the yield of thymine and adenine oxidation products upon 266 nm laser pulses in isolated DNA [131]. Charge transfer was also shown to occur more efficiently in chromatinized DNA than in naked DNA. A detailed study of the location and types of guanine oxidation products generated by UVA photodamage along a duplex DNA fragment wrapped around a nucleosome core particle (NCP) indicated that, whereas in naked DNA lesions were mostly localized at the distal sites, in nucleosomal DNA there was substantial enhancement of internally (*i.e.*, in contact with the NCP) damaged guanine sites [140]. Surprisingly, removing the histone tails from nucleosomes, most of which were in molecular contact with the packaged DNA, was sufficient to abrogate the effects of nucleosomal packing on long-range charge transfer, implying that weakening histone-DNA interactions also dampened the efficiency of charge transfer along DNA. In addition, a shift in the nature of guanine lesions was observed, from Oz mostly in the linker regions and in naked DNA to 8-oxoG for those sites in closest contact with the NCP. Because the guanine radical cation (G^{+}), a key intermediate in guanine oxidation, may react with either oxygen or water to yield Oz and 8-oxoG, respectively, these results clearly show that the NCP shields G^{+} from reacting with molecular oxygen. As the authors pointed out, the implications of this study are two-fold: first, "the enhancement of damage in the most tightly packaged nucleosomes could result in enhanced guanine oxidation in heterochromatin *versus* euchromatin"; and second "the distribution of guanine oxidation products is modulated by nucleosomal packaging. Therefore, the spectrum of guanine lesions generated by DNA oxidation could vary in different regions of chromatin".

In summary, sites of oxidation in DNA may migrate from their original location to sites of lower ionization potentials, e.g., predominantly G in the context of <u>G</u>A and <u>G</u>G sequences (charge/electron transfer); charge transfer also occurs towards 8-oxoG, which yields different oxidation products depending on the nature of the oxidizing agent and flanking sequence composition. Finally, charge transfer in DNA is favored in chromatin, where guanine oxidation products are modulated by their position along the NCP.

4. DNA Repair Pathways and Synthesis across Modified Bases and Mismatches

4.1. Base Excision Repair

Base excision repair (BER) is initiated by the activity of a DNA glycosylase that recognizes small perturbations in the DNA helical structure caused by base modifications or a mismatched base-pair [141]. The basic steps of BER, the distinction between short and long patch repair, the nature of monofunctional *versus* bifunctional enzymes and the involvement of Pol β in cancer have been thoroughly reviewed [141–143]. Herein, we shall focus on those BER enzymes that have been shown to process the modified bases and mismatches described above.

4.1.1. Modified CG:CG Sites

Two monofunctional DNA glycosylases display a preference for correcting T:G and U:G mismatches in the CG:CG sequence context in double-stranded DNA, methyl-CpG binding domain protein 4 (MBD4) and thymine DNA glycosylase (TDG). MBD4 contains an N-terminal methyl-CpG binding domain and a C-terminal DNA glycosylase domain that acts on T:G, 5hmU:G and U:G mismatches with relative rate constants of 0.5, 1.0 and 1.7 min^{-1}, respectively [144], and on Tg:G with half the efficiency observed for T:G. Thus, the enzyme is poised to recognize deamination products of 5mC:G, 5hmC:G and C:G within CG:CG sequences [145]. Consistent with these activities *in vitro*, *Mbd4*$^{-/-}$ mice display a ~3-fold increase in C→T transitions at CG:CG sites relative to wild-type littermates [146,147]; however, a direct role for *Mbd4*$^{-/-}$ in accelerating tumorigenesis has not been confirmed [148]. Competition experiments indicate that CG:CG methylation enhances MBD4 binding and that, whereas glycosylase activity is observed on reconstituted chromatin, activity is enhanced upon histone tail acetylation, consistent with increased accessibility of the target sites on a less compact chromatin environment [149]. TDG actively processes a number of lesions resulting from oxidation, alkylation and deamination of C, 5mC, 5hmC, T and A, with the strongest activities observed on U:G > T:G > Tg:G mismatches. The enzyme also cleaves the products of 5hmC oxidation, 5fC and 5caC. The *TDG* gene is expressed at high levels in the G2-M and G1 phases of the cell cycle, and then rapidly declines at the onset of S-phase. Loss of *TDG* expression is embryonic lethal in mice and, indeed, a number of investigations support a role for TDG in demethylation during embyogenesis, whereas interactions with transcription factors, transcriptional coregulators, DNMT3a, DNMT3b and others, suggest a scenario in which coordinated CG:CG methylation/demethylation and chromatin organization serve to regulate gene expression [145]. A striking preponderance (86%) of C→T transitions at mutated CG:CG sites, which are normally methylated, was recently reported in a mismatch-repair (*PMS2*) deficient 13-year-old colorectal cancer patient with a heterozygous germline missense mutation in *TDG* [150], in line with a potential role for TDG in repairing C5-substituted C deaminated products at CG:CG sites.

Because thymidine glycol may exist in four different configurations, $5R_{cis/trans}$ (Figure 5a) and $5S_{cis/trans}$ (Figure 5b) pairs, the efficiency of repair by DNA glycosylases will vary depending upon whether Tg isomers oppose G (Tg:G), which results from 5mC oxidation and deamination, or A (Tg:A), which results from T oxidation. Under conditions of single turnover, the stereoselectivity of nth endonuclease III-like 1 (NTHL1, a bifunctional enzyme with β-lyase activity) was similar for Tg(*5R*):A and Tg(*5R*):G, but the amount of Tg(*5R*) cleaved was ~13-fold higher than for the Tg(*5S*) due to stronger product inhibition by the latter. By contrast, for nei endonuclease VIII-like 1 (NEIL1, a bifunctional enzyme with β,δ-lyase activity), no stereoselectivity was detected; however, Tg:G was excised much more rapidly than Tg:A, suggesting that NEIL1 may be primarily involved in the repair of modified CG:CG sites [151]. $Nth1^{-/-}Neil1^{-/-}$ double mutant, but not single mutant, mice developed a high incidence of lung and liver tumors after the first year, implying overlapping roles in DNA repair [152].

Figure 5. (**a**) *cis-trans* stereoisomer pair of *5R* thymine glycol; (**b**) *cis-trans* stereoisomer pair of *5S* thymine glycol.

(a) (b)

cis-5R,6S *trans-5R,6R* *cis-5S,6R* *trans-5S,6S*

NEIL1 synthesis is activated during S-phase, and NEIL1 has been proposed to act at the replication fork to remove oxidative DNA lesions in a scheme involving: (1) damage recognition on the single-stranded template but cleavage inhibition by replication protein A (RPA); (2) fork reversal, which places the lesion back into duplex DNA; (3) base cleavage; and (4) resumption of the collapsed replication fork and DNA synthesis. Although NEIL2, which displays substrate specificities similar to NEIL1, was able to partially complement NEIL1 at the replication fork, its activity is believed to be more relevant during transcription [153]. A study on promoter methylation for 160 DNA repair genes in ~40 head and neck squamous cell carcinoma samples and controls identified *NEIL1* as the most prominently hypermethylated gene, with 81% samples (35/43) displaying significant hypermethylation relative to controls, suggesting a role for diminished DNA repair activity in cancer onset or progression [154].

4.1.2. Uracil

Two additional monofunctional DNA glycosylases, uracil-DNA glycosylase 2 (UNG2) and single-stranded selective monofunctional uracil-DNA glycosylase 1 (SMUG1), serve to remove uracil from nuclear DNA [141]. UNG2 is a single-stranded DNA-specific enzyme that plays an indispensable role in somatic hypermutation (SHM), which is part of the antigen-driven high-affinity antibody diversification program in follicular B cells, by removing uracil generated by AID at WRCY sequence hotspots. *In vitro*, SMUG1 is active either on single-stranded or double-stranded DNA, depending on salt concentrations. However, at physiological mono and divalent metal ion concentrations, SMUG1 is active only on double-stranded DNA, and therefore it is considered to be a double-stranded-specific DNA glycosylase. This distinction is crucial, since it places UNG2 as the sole enzyme acting on uracil in single-stranded DNA *in vivo*. Thus, in addition to sequence-specificity, the ability of RPA, a single-stranded DNA binding protein, to recruit UNG2 to single-stranded DNA has been proposed as a key feature that restricts UNG2 (rather than SMUG1, TDG or MBD4) activity to SHM [155]. In mice, during SHM, UNG2-generated AP sites are "copied" by error-prone translesion synthesis (TLS) polymerases during DNA replication, including Rev1 and Pol η, yielding C→T, C→G and C→A mutations [128]. In chronic myeloid leukemia in chronic phase (CML-CP) hematopoietic stem cells, the kinase activity associated with the *BCL-ABL1* translocation was found to inhibit UNG2 activity, thereby promoting mutations arising from increased ROS-mediated oxidative base lesions [156].

4.1.3. Guanine Lesions

Two bifunctional double-strand-specific DNA glycosylases, 8-oxoguanine DNA glycosylase 1 (OGG1) and mutY homologue (MUTYH), act upon 8-oxoG, a highly miscoding lesion that instructs A incorporation (8-oxoG:A base-pairs) by the replicative DNA Pol δ/ε. Knockout mice for both enzymes are strongly prone to lung, ovarian cancers and lymphomas, and have shortened life spans [157], whereas human germline biallelic *MUTYH* mutations have been implicated in MUTYH-associated polyposis, a condition associated with increased risk of colorectal cancer [158]. OGG1 cleaves 8-oxoG only when paired with C, owing to specific contacts made with both bases, which trigger catalysis [159]. By contrast, MUTYH specifically cleaves the A base in 8-oxoG:A base-pairs by using a central interconnector domain (ICD) to coordinate the action between the N-terminal catalytic domain and the C-terminal 8-oxoG recognition domain. In addition, the ICD serves as a structural scaffold to direct MUTYH activity to replication foci through specific interactions, including PCNA and Rad9-Rad1-Hus1 (the 9-1-1 complex) [160]. After induction of oxidative stress, the co-localization of OGG1-containing BER patches with H3meK4 or acetylated histone H4 in euchromatic regions and the exclusion from heterochromatic regions suggests that chromatin compaction hinders BER [161]. These conclusions are in line with dinucleosome reconstitution experiments *in vitro*, in which 8-oxoG cleavage in the linker region separating the two nucleosomes was unhindered in the absence of H1, but was decreased ~10-fold upon H1 binding to the linker [162].

In vitro, Oz in Oz:C and Gh in Gh:C mispairs in duplex oligonucleotides were cleaved by NEIL1 as efficiently as the well-recognized pyrimidine lesion 5hU:G, implying that guanine oxidation lesions, in addition to pyrimidine lesions, are good substrates for the enzyme. Similar results were obtained with NTHL1, although Gh was cleaved slightly more efficiently than Oz [163]. NEIL1 and NHTL1 also cleaved Oz:G as efficiently as Oz:C, which would give rise to G→C transversions; Oz:G mispairs are predicted to arise from Pol α or ε replication across Oz, as mentioned below.

Although the ability of NEIL1 and NEIL2 to remove Sp and Gh lesions has been well documented [164–166], a critical role for NEIL3 has recently emerged [167]. *Neil3*$^{-/-}$ mice exhibited learning and memory deficits, impaired proliferation of neural stem/progenitor cells [168], and tissue extracts from *Neil3*$^{-/-}$ mice, but not from wild-type littermates, displayed defective nicking activities on hydantoin lesions only when present on single-stranded DNA [169]. The purified human catalytic domain of NEIL3 was also found to display strong preference for Sp and Gh when compared to several other lesions, including 5-OHC and 5-OHU, with the greatest turnover number (0.035 s^{-1}) on Gh in single-stranded DNA, as assessed from single turnover experiments, ~2-fold higher than on double-stranded Gh and single-/double-stranded Sp. Thus, removal of Sp and Gh lesions appears to depend critically on NEIL3. Interestingly, the enzyme elicited uncoordinated cleavage and β-lyase activities, suggesting that it can act both as a monofunctional and as a bifunctional glycosylase [170].

4.2. Lesion Bypass

Base lesions that remain unrepaired can serve as templates during DNA synthesis, and are either copied by the normal replicative DNA polymerases (Pol α, δ and ε) or alternatively may block replication, in which case they are bypassed by one of 10 specialized polymerases lacking 3'→5' proofreading exonuclease activity during TLS, which limits DNA synthesis to a few bases across the blocking lesion [43,142]. Of the modified bases described above, C5 modified cytosines direct mostly guanine incorporation, and hence are not mutagenic; by contrast, their deaminated counterparts (thymine and uracil derivatives) enable the incorporation of adenine by replicative polymerases [171], which yields C→T (G→A) transitions. Tg stereroisomers block replication; investigations *in vitro* with DNA Polymerase gp43 from bacteriophage RB69 (a polymerase of the B-family, which in humans includes Pol α, δ, ε and ζ) revealed that whereas Tg is weakly bypassed and correctly paired with A, extension is inhibited by the enzymatic exonuclease activity [172] and, thus, extended through TLS by Pol α, ζ, ν, η or θ, either alone or in combination [43,173–179]. Thus, TLS across Tg:G base-pairs originating from 5mC oxidation and subsequent deamination are expected to lead almost quantitatively to C→T (G→A) transitions. 8-oxoG is able to functionally mimic thymine in the *syn* conformation, and DNA synthesis by replicative polymerases has been shown to yield both the correct 8-oxoG:C base-pair and the incorrect 8-oxoG(*syn*):A(*anti*) Hoogsteen base-pair [180–182] (Figure 6a–c) in ~3:2 ratios and to be extended, mostly by Pol λ [182–187]. The 8-oxoG:A base-pair would therefore give rise to G→T (C→A) transversions.

The crystal structure of bacteriophage B-family Polymerase RB69 bound to the templating *R*-stereoisomer of Gh revealed that the base was flipped in a non-templating position. However, the results also suggested that either slow rotation by the *R*-isomer or, more effectively, the *S*-isomer would present the pyrimidine-like hydantoin side to the enzyme, thereby instructing incorporation of a purine (A or G) [188]. Indeed, a Y567A mutant of RB69 was found to insert both bases with >100-fold increased efficiency, whereas extension was blocked [189]. These data suggest that, as in the case with Tg stereoisomers, Gh:A and Gh:G are switched to the enzymatic exonucleolytic domain, thereby triggering a futile incorporation/degradation cycle that effectively blocks DNA replication and renders Gh an obligate mutagen. Thus, a TLS extender polymerase may assist in lesion bypass across Gh *in vivo*; in addition, Pol η was found to bypass Gh efficiently and to incorporate either A or G in primer extension assays [190]. Although Sp was also found to yield G→C and G→T transversions [191], no structural data are currently available for polymerease:Sp complexes.

In primer extension assays, Pol η was found to partially extend past the Oz, Iz and Gf lesions (Oz > Iz > Gf), whereas Pol κ was almost completely blocked [135]. Sequence analyses of the extended products indicated that Pol η incorporated either C (55%–65%) or G (35%–45%), whereas Pol κ incorporated C (41%–58%), G (25%–37%) and A (16%–33%), across all lesions. Thus, T incorporation seems to be limited with both TLS polymerases. How these lesions are processed by replicative DNA polymerases and the extent to which each leads to SBSs remains to be determined. The stabilization energies (*ΔE*) for a number of isomers of the oxidative guanine products Gh, Sp, Iz and Oz base-paired with G (Figure 6d–g) have been computed by *ab initio* molecular orbital calculations; using density functional theory, *ΔE* values varied from 28.2 kcal/mol for Sp:G to 20.7 kcal/mol for Oz:G (30.9 kcal/mol for the canonical G:C base-pair) [130,192].

Figure 6. (**a**) Canonical G:C base-pair; (**b**) Canonical A:T base-pair; (**c**) 8-oxoG(*syn*):A(*anti*) base-pair; (**d**) Gh:G base-pair; (**e**) Sp:G base-pair; (**f**) Iz:G base-pair; (**g**) Oz:G base-pair. For noncanonical base-pairs, the templating base is shown in blue.

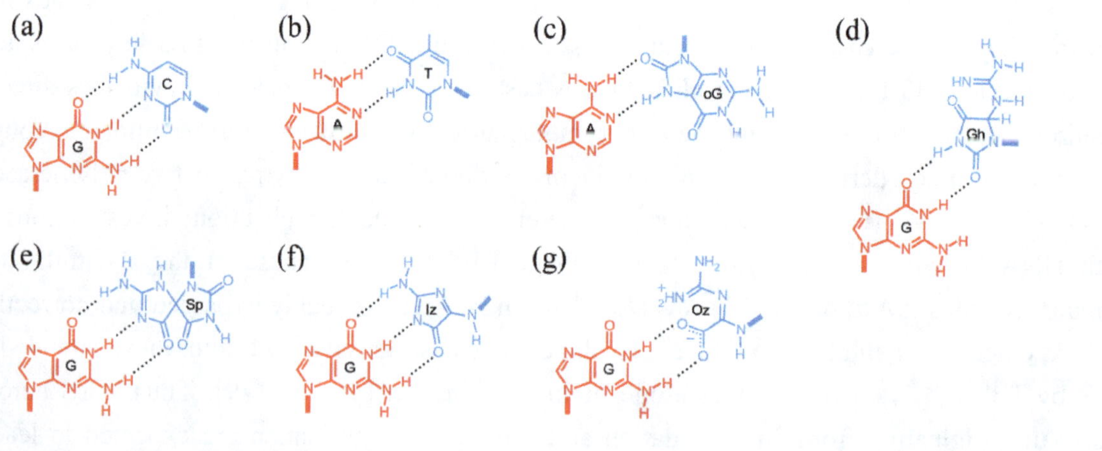

4.3. Transcription-Coupled Repair

Given that cancer genome analyses have indicated that the number of SBSs is often lower at sites of putative base lesions on the transcribed strand than on the non-transcribed strand [13,30,193], we will briefly discuss some key aspects of transcription-coupled repair. Nucleotide excision repair (NER) is considered to be the main pathway for the repair of UV photo-induced DNA lesions; CPDs and 6-4 pyrimidine-pyrimidone photo products (6-4PP) [194], and defects in NER components are associated with inherited DNA repair syndromes such as xeroderma pigmentosum [195] and Cockayne syndrome [194], that display severe hypersensitivity to sunlight. NER also actively repairs many types of DNA adducts that cause helical distortions, including environmental mutagens such as benzo[*a*]pyrene and other PAHs, aromatic amines, oxidative endogenous lesions such as cyclopurines, and adducts caused by cancer chemotherapeutic agents, including cisplatin [196]. NER comprises two distinctive subpathways, global genome NER (GG-NER) and transcription-coupled NER (TC-NER), which acts specifically on the transcribed strand of actively transcribed genes [194]. TC-NER is activated by the physical blockage imposed by bulky DNA adducts and abasic sites on RNA Polymerase II (RNAPII), but not by small lesions such as thymine glycols and 8-oxoG [197,198]. RNAPII is found in transient association with Cockayne syndrome type B protein (CSB, the product of the *ERCC6* gene) and the UV-stimulated scaffold protein A (UVSSA)–ubiquitin-specific peptidase 7 (USP7) complex. RNAPII stalling is believed to stabilize RNAPII/CSB interactions and activate the UVSSA/USP7 complex, thus protecting CSB from degradation through deubiquitination. Stabilized CSB recruits a complex that includes Cockayne syndrome WD repeat protein (CSA, product of the *ERCC8* gene), damage-specific DNA binding protein 1 (DDB1), cullin 4A (CUL4A) and others, which mediates downstream events, including chromatin remodeling, permitting backtracking of the RNAPII complex and exposing the lesion for repair by the GG-NER complex. Specific DNA adducts have been shown to be repaired exclusively by TC-NER but not by GG-NER. For example, exome sequencing of urothelial carcinomas of the upper urinary tract associated with chronic exposure to aristolochic acid, a natural compound from traditional herb medicine, revealed a characteristic A→T (T→A) mutational signature on non-transcribed strands leading to splicing defects [199,200], which was attributed to a failure of GG-NER, but not TC-NER, to recognize aristolactam-DNA adducts [201]. In addition to its role in TC-NER, CSB has been reported to facilitate lesion bypass by the RNAPII complex [202], to associate with components of BER, including OGG1, NEIL1 and AP1, and to elicit critical functions in mitochondria related to DNA repair, ROS homeostasis and others [203–205].

More recently, immunoprecipitation experiments in human gastric epithelial AGS cells revealed a direct association between the DNA glycosylase NEIL2, RNAPII and heterogeneous nuclear ribonucleoprotein U (hnRNP-U) [206]. On a 51-mer oligonucleotide, NEIL2 activity on a single 5-OHU lesion was stimulated 5- to 6-fold by hnRNP-U and likewise, on a plasmid system, a reconstituted transcription-repair complex comprising RNAPII, NEIL2, hnRNP-U and the BER components Pol β, Lig IIIα, PNK was proficient in repairing a 5-OHU lesion on the transcribed, but not on the non-transcribed, strand under conditions of active transcription. Co-localizations of NEIL2 with hnRNP-U with actively transcribed genes were confirmed by pulling down NEIL2-FLAG expressing AGS and neuroblastoma SK-N-BE2-(C) cells, followed by chromatin immunoprecipitation with hnRNP-U antibodies. Finally, a rise in mutations was inferred on selected, transcribed, genes in *NEIL2* knock-down compared to control cells. These composite data extend previous investigations [207], and document the preferential removal of oxidized DNA bases in actively transcribed genes by a system linking BER to the transcriptional apparatus. As already pointed out [206], it remains to be seen whether components of NER, such as CSB, also take part in BER related to transcription. In summary, whereas common oxidative DNA lesions may not be recognized by TC-NER, they are likely to represent a substrate for BER on the transcribed strand in association with transcription.

5. Proposed Mutational Mechanisms

Herein we have provided information on the most prominent SBS mutation patterns found in cancer genomes, *i.e.*, C→T transitions at CG:CG sites and substitutions at C:G base-pairs in the context of YC̲:G̲R dinucleotides, in an attempt to examine the validity of currently proposed models of mutagenesis. The main points may be summarized as follows: (i) higher fractions of SBSs have been found in cancer genomes within gene-poor regions, which are associated with heterochromatin and replicate late during the cell cycle, than in euchromatin, a pattern that mimics the one observed in population analyses; (ii) there are several established mechanisms for base substitutions at C residues, and their relative importance is only now being recognized. These involve deamination, oxidation, BER, TC-NER, TLS, and APOBEC activities; (iii) both guanine and 8-oxoG undergo sequence context- and ROS-dependent oxidation reactions that are consistent with an electron transfer (charge or hole migration) mechanism, whose efficiency is enhanced in the context of nucleosomal DNA; (iv) BER displays widely overlapping substrate specificity; however, NEIL enzymes appear to play a more prominent role in repairing guanine lesions, such as Gh and Sp, which are obligate mutagens during TLS; (v) most endogenous DNA lesions do not activate TC-NER; however, there are hints for a functional link between BER and transcription, leading to preferential repair on the transcribed strand.

C→T transitions at CG:CG sites have generally been attributed to faster spontaneous deamination of 5mC relative to C [13,16,17,19]. Our current analysis suggests that this explanation may be somewhat too restrictive. First, the fact that a family of C5-substituted Cs exists at CG:CG sites implies that C→T transitions are not limited to 5mC but, rather, to any of the C5-substituted species, *i.e.*, 5hmC, 5fC and 5caC. Second, as noted [208], the modest (5-fold at the most) increase in spontaneous deamination rates for C5-substituted Cs relative to Cs contrasts with the larger fractions of mutated Cs at CG:CG sites relative to non-CG:CG sites (up to 10–50 times), both in cancer and the germline [17,209]. Third, there does not seem to be a rational barrier to the possibility that 5mC oxidation and further deamination may yield Tg:G mismatches at CG:CG sites, which would then lead to C→T transitions during TLS. In fact, the observation that such a type of oxidation is facilitated by nucleosome occupancy raises the prospect for thymine glycols in Tg:G mispairs being a more prominent source of mutation at CG:CG sites than T:G mismatches, as previously pointed out [208]. The finding that C:G→T:A substitutions at CG:CG dinucleotides showed a strong positive correlation with the age at cancer diagnosis in ER− cancers, but not in ER+ cancers [193], further supports the conclusion that spontaneous deamination of 5mC may not be the only mechanism leading to mutations at CG:CG.

With respect to mutations at YC̲:G̲R sites, the gene expression data on APOBEC3B enzymes provide strong support for a role in cancer mutagenesis through U:G mismatch intermediates, as in the case of SHM. However, the extent to which the enzyme preference for single-stranded DNA may limit their activity genome-wide, possibly at sites of clustered and strand-coordinated mutations (kataegis) [13,210], remains to be determined. For example, whole-genome sequencing of gastric cancers, in which a prominent C→T signature at GC dinucleotides in coding-regions did not overlap with a preponderance of C→A transversions at CCT or TCA motifs genome-wide, were attributed to AID activation (on single-stranded DNA during transcription) and ROS/NOS, respectively, following *H. pylori* infection [56]. The findings that: (i) oxidative DNA damage occurs at Y<u>C</u>:<u>G</u>R sites, which overlap with APOBEC3B specificity; (ii) numerous oxidation products can form at the target <u>G</u>R; (iii) efficient charge transfer and high mutation rates co-localize with heterochromatin; and (iv) some oxidation products of guanine are obligate mutagens during TLS, make it tempting to attribute a significant role for oxidative damage in mutations at Y<u>C</u>:<u>G</u>R sites genome-wide in cancer mutagenesis. Thus, for SBSs at both CG:CG and Y<u>C</u>:<u>G</u>R sites, we suggest a prominent role for oxidative damage by ROS and other electron-abstracting species.

One severe limitation in elucidating mechanisms of SBS is obviously the lack of information on the steady-state levels of modified bases and mismatches in cancer cells. Current efforts to address this critical issue [211–215] should at least temper these limitations, not only for the limited number of base modifications discussed here, but also for the much larger repertoire that may be formed by both endogenous and environmental agents, and which have not been addressed herein.

Acknowledgments

We thank Richard D. Wood and Robert D. Wells for comments and suggestions. This work was supported by grants from the National Institutes of Health (CA097175 and CA093729) to K.M.V.

Author Contributions

Wrote the paper: AB, DNC and KMV.

References

1. Garraway, L.A.; Lander, E.S. Lessons from the cancer genome. *Cell* **2013**, *153*, 17–37.
2. Vogelstein, B.; Papadopoulos, N.; Velculescu, V.E.; Zhou, S.; Diaz, L.A., Jr.; Kinzler, K.W. Cancer genome landscapes. *Science* **2013**, *339*, 1546–1558.
3. Nik-Zainal, S.; Alexandrov, L.B.; Wedge, D.C.; van Loo, P.; Greenman, C.D.; Raine, K.; Jones, D.; Hinton, J.; Marshall, J.; Stebbings, L.A.; *et al.* Mutational processes molding the genomes of 21 breast cancers. *Cell* **2012**, *149*, 979–993.
4. Forment, J.V.; Kaidi, A.; Jackson, S.P. Chromothripsis and cancer: Causes and consequences of chromosome shattering. *Nat. Rev. Cancer* **2012**, *12*, 663–670.
5. Yang, L.; Luquette, L.J.; Gehlenborg, N.; Xi, R.; Haseley, P.S.; Hsieh, C.H.; Zhang, C.; Ren, X.; Protopopov, A.; Chin, L.; *et al.* Diverse mechanisms of somatic structural variations in human cancer genomes. *Cell* **2013**, *153*, 919–929.
6. Drier, Y.; Lawrence, M.S.; Carter, S.L.; Stewart, C.; Gabriel, S.B.; Lander, E.S.; Meyerson, M.; Beroukhim, R.; Getz, G. Somatic rearrangements across cancer reveal classes of samples with distinct patterns of DNA breakage and rearrangement-induced hypermutability. *Genome Res.* **2013**, *23*, 228–235.
7. Bunting, S.F.; Nussenzweig, A. End-joining, translocations and cancer. *Nat. Rev. Cancer* **2013**, *13*, 443–454.

8. Zack, T.I.; Schumacher, S.E.; Carter, S.L.; Cherniack, A.D.; Saksena, G.; Tabak, B.; Lawrence, M.S.; Zhang, C.Z.; Wala, J.; Mermel, C.H.; *et al.* Pan-cancer patterns of somatic copy number alteration. *Nat. Genet.* **2013**, *45*, 1134–1140.
9. Watson, I.R.; Takahashi, K.; Futreal, P.A.; Chin, L. Emerging patterns of somatic mutations in cancer. *Nat. Rev. Genet.* **2013**, *14*, 703–718.
10. Ciriello, G.; Miller, M.L.; Aksoy, B.A.; Senbabaoglu, Y.; Schultz, N.; Sander, C. Emerging landscape of oncogenic signatures across human cancers. *Nat. Genet.* **2013**, *45*, 1127–1133.
11. Lange, S.S.; Takata, K.; Wood, R.D. DNA polymerases and cancer. *Nat. Rev. Cancer* **2011**, *11*, 96–110.
12. Alexandrov, L.B.; Nik-Zainal, S.; Wedge, D.C.; Campbell, P.J.; Stratton, M.R. Deciphering signatures of mutational processes operative in human cancer. *Cell. Rep.* **2013**, *3*, 246–259.
13. Alexandrov, L.B.; Nik-Zainal, S.; Wedge, D.C.; Aparicio, S.A.; Behjati, S.; Biankin, A.V.; Bignell, G.R.; Bolli, N.; Borg, A.; Borresen-Dale, A.L.; *et al.* Signatures of mutational processes in human cancer. *Nature* **2013**, *500*, 415–421.
14. Fischer, A.; Illingworth, C.J.; Campbell, P.J.; Mustonen, V. EMu: Probabilistic inference of mutational processes and their localization in the cancer genome. *Genome Biol.* **2013**, *14*, R39.
15. Burns, M.B.; Temiz, N.A.; Harris, R.S. Evidence for APOBEC3B mutagenesis in multiple human cancers. *Nat. Genet.* **2013**, *45*, 977–983.
16. Kandoth, C.; McLellan, M.D.; Vandin, F.; Ye, K.; Niu, B.; Lu, C.; Xie, M.; Zhang, Q.; McMichael, J.F.; Wyczalkowski, M.A.; *et al.* Mutational landscape and significance across 12 major cancer types. *Nature* **2013**, *502*, 333–339.
17. Bacolla, A.; Temiz, N.A.; Yi, M.; Ivanic, J.; Cer, R.Z.; Donohue, D.E.; Ball, E.V.; Mudunuri, U.S.; Wang, G.; Jain, A.; *et al.* Guanine holes are prominent targets for mutation in cancer and inherited disease. *PLoS Genet.* **2013**, *9*, e1003816.
18. Lawrence, M.S.; Stojanov, P.; Polak, P.; Kryukov, G.V.; Cibulskis, K.; Sivachenko, A.; Carter, S.L.; Stewart, C.; Mermel, C.H.; Roberts, S.A.; *et al.* Mutational heterogeneity in cancer and the search for new cancer-associated genes. *Nature* **2013**, *499*, 214–218.
19. Ivanov, D.; Hamby, S.E.; Stenson, P.D.; Phillips, A.D.; Kehrer-Sawatzki, H.; Cooper, D.N.; Chuzhanova, N. Comparative analysis of germline and somatic microlesion mutational spectra in 17 human tumor suppressor genes. *Hum. Mutat.* **2011**, *32*, 620–632.
20. Roberts, S.A.; Lawrence, M.S.; Klimczak, L.J.; Grimm, S.A.; Fargo, D.; Stojanov, P.; Kiezun, A.; Kryukov, G.V.; Carter, S.L.; Saksena, G.; *et al.* An APOBEC cytidine deaminase mutagenesis pattern is widespread in human cancers. *Nat. Genet.* **2013**, *45*, 970–976.
21. Greenman, C.; Stephens, P.; Smith, R.; Dalgliesh, G.L.; Hunter, C.; Bignell, G.; Davies, H.; Teague, J.; Butler, A.; Stevens, C.; *et al.* Patterns of somatic mutation in human cancer genomes. *Nature* **2007**, *446*, 153–158.
22. Govindan, R.; Ding, L.; Griffith, M.; Subramanian, J.; Dees, N.D.; Kanchi, K.L.; Maher, C.A.; Fulton, R.; Fulton, L.; Wallis, J.; *et al.* Genomic landscape of non-small cell lung cancer in smokers and never-smokers. *Cell* **2012**, *150*, 1121–1134.
23. Imielinski, M.; Berger, A.H.; Hammerman, P.S.; Hernandez, B.; Pugh, T.J.; Hodis, E.; Cho, J.; Suh, J.; Capelletti, M.; Sivachenko, A.; *et al.* Mapping the hallmarks of lung adenocarcinoma with massively parallel sequencing. *Cell* **2012**, *150*, 1107–1120.

24. Seo, J.S.; Ju, Y.S.; Lee, W.C.; Shin, J.Y.; Lee, J.K.; Bleazard, T.; Lee, J.; Jung, Y.J.; Kim, J.O.; Shin, J.Y.; *et al.* The transcriptional landscape and mutational profile of lung adenocarcinoma. *Genome Res.* **2012**, *22*, 2109–2119.
25. Ding, L.; Getz, G.; Wheeler, D.A.; Mardis, E.R.; McLellan, M.D.; Cibulskis, K.; Sougnez, C.; Greulich, H.; Muzny, D.M.; Morgan, M.B.; *et al.* Somatic mutations affect key pathways in lung adenocarcinoma. *Nature* **2008**, *455*, 1069–1075.
26. TCGARN. Comprehensive genomic characterization of squamous cell lung cancers. *Nature* **2012**, *489*, 519–525.
27. Xiong, D.; Li, G.; Li, K.; Xu, Q.; Pan, Z.; Ding, F.; Vedell, P.; Liu, P.; Cui, P.; Hua, X.; *et al.* Exome sequencing identifies *MXRA5* as a novel cancer gene frequently mutated in non-small cell lung carcinoma from Chinese patients. *Carcinogenesis* **2012**, *33*, 1797–1805.
28. Liu, P.; Morrison, C.; Wang, L.; Xiong, D.; Vedell, P.; Cui, P.; Hua, X.; Ding, F.; Lu, Y.; James, M.; *et al.* Identification of somatic mutations in non-small cell lung carcinomas using whole-exome sequencing. *Carcinogenesis* **2012**, *33*, 1270–1276.
29. Lee, W.; Jiang, Z.; Liu, J.; Haverty, P.M.; Guan, Y.; Stinson, J.; Yue, P.; Zhang, Y.; Pant, K.P.; Bhatt, D.; *et al.* The mutation spectrum revealed by paired genome sequences from a lung cancer patient. *Nature* **2010**, *465*, 473–477.
30. Pleasance, E.D.; Stephens, P.J.; O'Meara, S.; McBride, D.J.; Meynert, A.; Jones, D.; Lin, M.L.; Beare, D.; Lau, K.W.; Greenman, C.; *et al.* A small-cell lung cancer genome with complex signatures of tobacco exposure. *Nature* **2010**, *463*, 184–190.
31. Drablos, F.; Feyzi, E.; Aas, P.A.; Vaagbo, C.B.; Kavli, B.; Bratlie, M.S.; Pena-Diaz, J.; Otterlei, M.; Slupphaug, G.; Krokan, H.E. Alkylation damage in DNA and RNA—Repair mechanisms and medical significance. *DNA Repair* **2004**, *3*, 1389–1407.
32. Matter, B.; Wang, G.; Jones, R.; Tretyakova, N. Formation of diastereomeric benzo[*a*]pyrene diol epoxide-guanine adducts in p53 gene-derived DNA sequences. *Chem. Res. Toxicol.* **2004**, *17*, 731–741.
33. Ziegel, R.; Shallop, A.; Jones, R.; Tretyakova, N. *K-ras* gene sequence effects on the formation of 4-(methylnitrosamino)-1-(3-pyridyl)-1-butanone (NNK)-DNA adducts. *Chem. Res. Toxicol.* **2003**, *16*, 541–550.
34. Anna, L.; Holmila, R.; Kovacs, K.; Gyorffy, E.; Gyori, Z.; Segesdi, J.; Minarovits, J.; Soltesz, I.; Kostic, S.; Csekeo, A.; *et al.* Relationship between *TP53* tumour suppressor gene mutations and smoking-related bulky DNA adducts in a lung cancer study population from Hungary. *Mutagenesis* **2009**, *24*, 475–480.
35. Martinez, V.D.; Thu, K.L.; Vucic, E.A.; Hubaux, R.; Adonis, M.; Gil, L.; MacAulay, C.; Lam, S.; Lam, W.L. Whole-genome sequencing analysis identifies a distinctive mutational spectrum in an arsenic-related lung tumor. *J. Thorac. Oncol.* **2013**, *8*, 1451–1455.
36. Pleasance, E.D.; Cheetham, R.K.; Stephens, P.J.; McBride, D.J.; Humphray, S.J.; Greenman, C.D.; Varela, I.; Lin, M.L.; Ordonez, G.R.; Bignell, G.R.; *et al.* A comprehensive catalogue of somatic mutations from a human cancer genome. *Nature* **2010**, *463*, 191–196.
37. Berger, M.F.; Hodis, E.; Heffernan, T.P.; Deribe, Y.L.; Lawrence, M.S.; Protopopov, A.; Ivanova, E.; Watson, I.R.; Nickerson, E.; Ghosh, P.; *et al.* Melanoma genome sequencing reveals frequent *PREX2* mutations. *Nature* **2012**, *485*, 502–506.

38. Nikolaev, S.I.; Rimoldi, D.; Iseli, C.; Valsesia, A.; Robyr, D.; Gehrig, C.; Harshman, K.; Guipponi, M.; Bukach, O.; Zoete, V.; *et al.* Exome sequencing identifies recurrent somatic *MAP2K1* and *MAP2K2* mutations in melanoma. *Nat. Genet.* **2012**, *44*, 133–139.
39. Stark, M.S.; Woods, S.L.; Gartside, M.G.; Bonazzi, V.F.; Dutton-Regester, K.; Aoude, L.G.; Chow, D.; Sereduk, C.; Niemi, N.M.; Tang, N.; *et al.* Frequent somatic mutations in *MAP3K5* and *MAP3K9* in metastatic melanoma identified by exome sequencing. *Nat. Genet.* **2012**, *44*, 165–169.
40. Krauthammer, M.; Kong, Y.; Ha, B.H.; Evans, P.; Bacchiocchi, A.; McCusker, J.P.; Cheng, E.; Davis, M.J.; Goh, G.; Choi, M.; *et al.* Exome sequencing identifies recurrent somatic *RAC1* mutations in melanoma. *Nat. Genet.* **2012**, *44*, 1006–1014.
41. Wei, X.; Walia, V.; Lin, J.C.; Teer, J.K.; Prickett, T.D.; Gartner, J.; Davis, S.; Program, N.C.S.; Stemke-Hale, K.; Davies, M.A.; *et al.* Exome sequencing identifies *GRIN2A* as frequently mutated in melanoma. *Nat. Genet.* **2011**, *43*, 442–446.
42. Dutton-Regester, K.; Kakavand, H.; Aoude, L.G.; Stark, M.S.; Gartside, M.G.; Johansson, P.; O'Connor, L.; Lanagan, C.; Tembe, V.; Pupo, G.M.; *et al.* Melanomas of unknown primary have a mutation profile consistent with cutaneous sun-exposed melanoma. *Pigment Cell Melanoma Res.* **2013**, *26*, 852–860.
43. Sale, J.E. Translesion DNA synthesis and mutagenesis in eukaryotes. *Cold Spring Harb. Perspect. Biol.* **2013**, *5*, a012708.
44. Pham, P.; Landolph, A.; Mendez, C.; Li, N.; Goodman, M.F. A biochemical analysis linking APOBEC3A to disparate HIV-1 restriction and skin cancer. *J. Biol. Chem.* **2013**, *288*, 29294–29304.
45. Korona, D.A.; Lecompte, K.G.; Pursell, Z.F. The high fidelity and unique error signature of human DNA polymerase epsilon. *Nucleic Acids Res.* **2011**, *39*, 1763–1773.
46. Agbor, A.A.; Goksenin, A.Y.; LeCompte, K.G.; Hans, S.H.; Pursell, Z.F. Human Pol epsilon-dependent replication errors and the influence of mismatch repair on their correction. *DNA Repair* **2013**, *12*, 954–963.
47. Jiricny, J. Postreplicative mismatch repair. *Cold Spring Harb. Perspect. Biol.* **2013**, *5*, a012633.
48. Pluciennik, A.; Dzantiev, L.; Iyer, R.R.; Constantin, N.; Kadyrov, F.A.; Modrich, P. PCNA function in the activation and strand direction of MutLalpha endonuclease in mismatch repair. *Proc. Natl. Acad. Sci. USA* **2010**, *107*, 16066–16071.
49. Arana, M.E.; Kunkel, T.A. Mutator phenotypes due to DNA replication infidelity. *Semin. Cancer Biol.* **2010**, *20*, 304–311.
50. Preston, B.D.; Albertson, T.M.; Herr, A.J. DNA replication fidelity and cancer. *Semin. Cancer Biol.* **2010**, *20*, 281–293.
51. Aquilina, G.; Bignami, M. Mismatch repair in correction of replication errors and processing of DNA damage. *J. Cell. Physiol.* **2001**, *187*, 145–154.
52. Thompson, B.A.; Spurdle, A.B.; Plazzer, J.P.; Greenblatt, M.S.; Akagi, K.; Al-Mulla, F.; Bapat, B.; Bernstein, I.; Capella, G.; den Dunnen, J.T.; *et al.* Application of a 5-tiered scheme for standardized classification of 2,360 unique mismatch repair gene variants in the InSiGHT locus-specific database. *Nat. Genet.* **2014**, *46*, 107–115.
53. Vasen, H.F.; Blanco, I.; Aktan-Collan, K.; Gopie, J.P.; Alonso, A.; Aretz, S.; Bernstein, I.; Bertario, L.; Burn, J.; Capella, G.; *et al.* Revised guidelines for the clinical management of Lynch

syndrome (HNPCC): Recommendations by a group of European experts. *Gut* **2013**, *62*, 812–823.

54. TCGAN. Comprehensive molecular characterization of human colon and rectal cancer. *Nature* **2012**, *487*, 330–337.
55. TCGARN; Kandoth, C.; Schultz, N.; Cherniack, A.D.; Akbani, R.; Liu, Y.; Shen, H.; Robertson, A.G.; Pashtan, I.; Shen, R.; *et al.* Integrated genomic characterization of endometrial carcinoma. *Nature* **2013**, *497*, 67–73.
56. Nagarajan, N.; Bertrand, D.; Hillmer, A.M.; Zang, Z.J.; Yao, F.; Jacques, P.E.; Teo, A.S.; Cutcutache, I.; Zhang, Z.; Lee, W.H.; *et al.* Whole-genome reconstruction and mutational signatures in gastric cancer. *Genome Biol.* **2012**, *13*, R115.
57. Kim, T.M.; Laird, P.W.; Park, P.J. The landscape of microsatellite instability in colorectal and endometrial cancer genomes. *Cell* **2013**, *155*, 858–868.
58. Yoshida, R.; Miyashita, K.; Inoue, M.; Shimamoto, A.; Yan, Z.; Egashira, A.; Oki, E.; Kakeji, Y.; Oda, S.; Maehara, Y. Concurrent genetic alterations in DNA polymerase proofreading and mismatch repair in human colorectal cancer. *Eur. J. Hum. Genet.* **2011**, *19*, 320–325.
59. Palles, C.; Cazier, J.B.; Howarth, K.M.; Domingo, E.; Jones, A.M.; Broderick, P.; Kemp, Z.; Spain, S.L.; Guarino, E.; Salguero, I.; *et al.* Germline mutations affecting the proofreading domains of POLE and POLD1 predispose to colorectal adenomas and carcinomas. *Nat. Genet.* **2013**, *45*, 136–144.
60. Church, D.N.; Briggs, S.E.; Palles, C.; Domingo, E.; Kearsey, S.J.; Grimes, J.M.; Gorman, M.; Martin, L.; Howarth, K.M.; Hodgson, S.V.; *et al.* DNA polymerase epsilon and delta exonuclease domain mutations in endometrial cancer. *Hum. Mol. Genet.* **2013**, *22*, 2820–2828.
61. Briggs, S.; Tomlinson, I. Germline and somatic polymerase epsilon and delta mutations define a new class of hypermutated colorectal and endometrial cancers. *J. Pathol.* **2013**, *230*, 148–153.
62. Dai, H.; Hickey, R.J.; Liu, J.; Bigsby, R.M.; Lanner, C.; Malkas, L.H. Error-promoting DNA synthesis in ovarian cancer cells. *Gynecol. Oncol.* **2013**, *131*, 198–206.
63. Vieira, V.C.; Soares, M.A. The role of cytidine deaminases on innate immune responses against human viral infections. *Biomed Res. Int.* **2013**, *2013*, 683095.
64. Smith, H.C.; Bennett, R.P.; Kizilyer, A.; McDougall, W.M.; Prohaska, K.M. Functions and regulation of the APOBEC family of proteins. *Semin. Cell Dev. Biol.* **2012**, *23*, 258–268.
65. Koito, A.; Ikeda, T. Intrinsic immunity against retrotransposons by APOBEC cytidine deaminases. *Front. Microbiol.* **2013**, *4*, 28.
66. Bohn, M.F.; Shandilya, S.M.; Albin, J.S.; Kouno, T.; Anderson, B.D.; McDougle, R.M.; Carpenter, M.A.; Rathore, A.; Evans, L.; Davis, A.N.; *et al.* Crystal structure of the DNA cytosine deaminase APOBEC3F: The catalytically active and HIV-1 Vif-binding domain. *Structure* **2013**, *21*, 1042–1050.
67. Kitamura, S.; Ode, H.; Nakashima, M.; Imahashi, M.; Naganawa, Y.; Kurosawa, T.; Yokomaku, Y.; Yamane, T.; Watanabe, N.; Suzuki, A.; *et al.* The APOBEC3C crystal structure and the interface for HIV-1 Vif binding. *Nat. Struct. Mol. Biol.* **2012**, *19*, 1005–1010.
68. Burns, M.B.; Lackey, L.; Carpenter, M.A.; Rathore, A.; Land, A.M.; Leonard, B.; Refsland, E.W.; Kotandeniya, D.; Tretyakova, N.; Nikas, J.B.; *et al.* APOBEC3B is an enzymatic source of mutation in breast cancer. *Nature* **2013**, *494*, 366–370.

69. Lewis, P.D.; Harvey, J.S.; Waters, E.M.; Skibinski, D.O.; Parry, J.M. Spontaneous mutation spectra in *supF*: Comparative analysis of mammalian cell line base substitution spectra. *Mutagenesis* **2001**, *16*, 503–515.
70. Bacolla, A.; Wang, G.; Jain, A.; Chuzhanova, N.A.; Cer, R.Z.; Collins, J.R.; Cooper, D.N.; Bohr, V.A.; Vasquez, K.M. Non-B DNA-forming sequences and WRN deficiency independently increase the frequency of base substitution in human cells. *J. Biol. Chem.* **2011**, *286*, 10017–10026.
71. Giese, B. Long-distance electron transfer through DNA. *Annu. Rev. Biochem.* **2002**, *71*, 51–70.
72. Sugiyama, H.; Saito, I. Theoretical studies of GC-specific photocleavage of DNA via electron transfer: Significant lowering of ionization potential and 5'-localization of HOMO of stacked GG bases in B-form DNA. *J. Am. Chem. Soc.* **1996**, *118*, 7063–7068.
73. Saito, I.; Nakamura, T.; Nakatani, K.; Yoshioka, Y.; Yamaguchi, K.; Sugiyama, H. Mapping of the hot spots for DNA damage by one-electron oxidation: Efficacy of GG doublets and GGG triplets as a trap in long-range hole migration. *J. Am. Chem. Soc.* **1998**, *120*, 12686–12687.
74. Senthilkumar, K.; Grozema, F.C.; Guerra, C.F.; Bickelhaupt, F.M.; Siebbeles, L.D. Mapping the sites for selective oxidation of guanines in DNA. *J. Am. Chem. Soc.* **2003**, *125*, 13658–13659.
75. Yamazaki, S.; Hayano, M.; Masai, H. Replication timing regulation of eukaryotic replicons: Rif1 as a global regulator of replication timing. *Trends Genet.* **2013**, *29*, 449–460.
76. Aparicio, O.M. Location, location, location: It's all in the timing for replication origins. *Genes Dev.* **2013**, *27*, 117–128.
77. Liu, L.; De, S.; Michor, F. DNA replication timing and higher-order nuclear organization determine single-nucleotide substitution patterns in cancer genomes. *Nat. Commun.* **2013**, *4*, 1502.
78. Stamatoyannopoulos, J.A.; Adzhubei, I.; Thurman, R.E.; Kryukov, G.V.; Mirkin, S.M.; Sunyaev, S.R. Human mutation rate associated with DNA replication timing. *Nat. Genet.* **2009**, *41*, 393–395.
79. Koren, A.; Polak, P.; Nemesh, J.; Michaelson, J.J.; Sebat, J.; Sunyaev, S.R.; McCarroll, S.A. Differential relationship of DNA replication timing to different forms of human mutation and variation. *Am. J. Hum. Genet.* **2012**, *91*, 1033–1040.
80. Gavin, D.P.; Chase, K.A.; Sharma, R.P. Active DNA demethylation in post-mitotic neurons: A reason for optimism. *Neuropharmacology* **2013**, *75*, 233–245.
81. Schuster-Bockler, B.; Lehner, B. Chromatin organization is a major influence on regional mutation rates in human cancer cells. *Nature* **2012**, *488*, 504–507.
82. Kim, N.; Jinks-Robertson, S. Transcription as a source of genome instability. *Nat. Rev. Genet.* **2012**, *13*, 204–214.
83. Wang, G.; Christensen, L.A.; Vasquez, K.M. Z-DNA-forming sequences generate large-scale deletions in mammalian cells. *Proc. Natl. Acad. Sci. USA* **2006**, *103*, 2677–2682.
84. Kadyrova, L.Y.; Mertz, T.M.; Zhang, Y.; Northam, M.R.; Sheng, Z.; Lobachev, K.S.; Shcherbakova, P.V.; Kadyrov, F.A. A reversible histone H3 acetylation cooperates with mismatch repair and replicative polymerases in maintaining genome stability. *PLOS Genet.* **2013**, *9*, e1003899.
85. Kinney, S.R.; Pradhan, S. Regulation of expression and activity of DNA (cytosine-5) methyltransferases in mammalian cells. *Prog. Mol. Biol. Transl. Sci.* **2011**, *101*, 311–333.
86. Smith, Z.D.; Meissner, A. DNA methylation: Roles in mammalian development. *Nat. Rev. Genet.* **2013**, *14*, 204–220.

87. Thomson, J.P.; Moggs, J.G.; Wolf, C.R.; Meehan, R.R. Epigenetic profiles as defined signatures of xenobiotic exposure. *Mutat. Res.* **2013**, doi:10.1016/j.mrgentox.2013.08.007.
88. Song, C.X.; Szulwach, K.E.; Dai, Q.; Fu, Y.; Mao, S.Q.; Lin, L.; Street, C.; Li, Y.; Poidevin, M.; Wu, H.; *et al.* Genome-wide profiling of 5-formylcytosine reveals its roles in epigenetic priming. *Cell* **2013**, *153*, 678–691.
89. Kohli, R.M.; Zhang, Y. TET enzymes, TDG and the dynamics of DNA demethylation. *Nature* **2013**, *502*, 472–479.
90. Losman, J.A.; Kaelin, W.G., Jr. What a difference a hydroxyl makes: Mutant IDH, (R)-2-hydroxyglutarate, and cancer. *Genes Dev.* **2013**, *27*, 836–852.
91. Guilhamon, P.; Eskandarpour, M.; Halai, D.; Wilson, G.A.; Feber, A.; Teschendorff, A.E.; Gomez, V.; Hergovich, A.; Tirabosco, R.; Fernanda Amary, M.; *et al.* Meta-analysis of IDH-mutant cancers identifies EBF1 as an interaction partner for TET2. *Nat. Commun.* **2013**, *4*, 2166.
92. Rendina, A.R.; Pietrak, B.; Smallwood, A.; Zhao, H.; Qi, H.; Quinn, C.; Adams, N.D.; Concha, N.; Duraiswami, C.; Thrall, S.H.; *et al.* Mutant IDH1 enhances the production of 2-hydroxyglutarate due to its kinetic mechanism. *Biochemistry* **2013**, *52*, 4563–4577.
93. Lian, C.G.; Xu, Y.; Ceol, C.; Wu, F.; Larson, A.; Dresser, K.; Xu, W.; Tan, L.; Hu, Y.; Zhan, Q.; *et al.* Loss of 5-hydroxymethylcytosine is an epigenetic hallmark of melanoma. *Cell* **2012**, *150*, 1135–1146.
94. Sandoval, J.; Mendez-Gonzalez, J.; Nadal, E.; Chen, G.; Carmona, F.J.; Sayols, S.; Moran, S.; Heyn, H.; Vizoso, M.; Gomez, A.; *et al.* A prognostic DNA methylation signature for stage I non-small-cell lung cancer. *J. Clin. Oncol.* **2013**, *31*, 4140–4147.
95. Schiesser, S.; Pfaffeneder, T.; Sadeghian, K.; Hackner, B.; Steigenberger, B.; Schroder, A.S.; Steinbacher, J.; Kashiwazaki, G.; Hofner, G.; Wanner, K.T.; *et al.* Deamination, oxidation, and C-C bond cleavage reactivity of 5-hydroxymethylcytosine, 5-formylcytosine, and 5-carboxycytosine. *J. Am. Chem. Soc.* **2013**, *135*, 14593–14599.
96. Ehrlich, M.; Norris, K.F.; Wang, R.Y.; Kuo, K.C.; Gehrke, C.W. DNA cytosine methylation and heat-induced deamination. *Biosci. Rep.* **1986**, *6*, 387–393.
97. Shen, J.C.; Rideout, W.M., 3rd; Jones, P.A. The rate of hydrolytic deamination of 5-methylcytosine in double-stranded DNA. *Nucleic Acids Res.* **1994**, *22*, 972–976.
98. Frederico, L.A.; Kunkel, T.A.; Shaw, B.R. A sensitive genetic assay for the detection of cytosine deamination: Determination of rate constants and the activation energy. *Biochemistry* **1990**, *29*, 2532–2537.
99. Zhang, X.; Mathews, C.K. Effect of DNA cytosine methylation upon deamination-induced mutagenesis in a natural target sequence in duplex DNA. *J. Biol. Chem.* **1994**, *269*, 7066–7069.
100. Sowers, L.C.; Sedwick, W.D.; Shaw, B.R. Hydrolysis of N3-methyl-2'-deoxycytidine: Model compound for reactivity of protonated cytosine residues in DNA. *Mutat. Res.* **1989**, *215*, 131–138.
101. Nikolova, E.N.; Goh, G.B.; Brooks, C.L., 3rd; Al-Hashimi, H.M. Characterizing the protonation state of cytosine in transient G.C Hoogsteen base pairs in duplex DNA. *J. Am. Chem. Soc.* **2013**, *135*, 6766–6769.
102. Fryxell, K.J.; Zuckerkandl, E. Cytosine deamination plays a primary role in the evolution of mammalian isochores. *Mol. Biol. Evol.* **2000**, *17*, 1371–1383.

103. Chen, X.; Chen, Z.; Chen, H.; Su, Z.; Yang, J.; Lin, F.; Shi, S.; He, X. Nucleosomes suppress spontaneous mutations base-specifically in eukaryotes. *Science* **2012**, *335*, 1235–1238.
104. Labet, V.; Morell, C.; Cadet, J.; Eriksson, L.A.; Grand, A. Hydrolytic deamination of 5-methylcytosine in protic medium—A theoretical study. *J. Phys. Chem. A* **2009**, *113*, 2524–2533.
105. Patel, J.C.; Rice, M.E. Classification of H(2)O(2) as a neuromodulator that regulates striatal dopamine release on a subsecond time scale. *ACS Chem. Neurosci.* **2012**, *3*, 991–1001.
106. Nogueira, V.; Hay, N. Molecular pathways: Reactive oxygen species homeostasis in cancer cells and implications for cancer therapy. *Clin. Cancer Res.* **2013**, *19*, 4309–4314.
107. Gupte, A.; Mumper, R.J. Elevated copper and oxidative stress in cancer cells as a target for cancer treatment. *Cancer Treat. Rev.* **2009**, *35*, 32–46.
108. Nieborowska-Skorska, M.; Kopinski, P.K.; Ray, R.; Hoser, G.; Ngaba, D.; Flis, S.; Cramer, K.; Reddy, M.M.; Koptyra, M.; Penserga, T.; *et al.* Rac2-MRC-cIII-generated ROS cause genomic instability in chronic myeloid leukemia stem cells and primitive progenitors. *Blood* **2012**, *119*, 4253–4263.
109. Cao, H.; Jiang, Y.; Wang, Y. Kinetics of deamination and Cu(II)/H_2O_2/ascorbate-induced formation of 5-methylcytosine glycol at CpG sites in duplex DNA. *Nucleic Acids Res.* **2009**, *37*, 6635–6643.
110. Wagner, J.R.; Cadet, J. Oxidation reactions of cytosine DNA components by hydroxyl radical and one-electron oxidants in aerated aqueous solutions. *Acc. Chem. Res.* **2010**, *43*, 564–571.
111. Liang, Q.; Dedon, P.C. Cu(II)/H_2O_2-induced DNA damage is enhanced by packaging of DNA as a nucleosome. *Chem. Res. Toxicol.* **2001**, *14*, 416–422.
112. Tan, S.; Davey, C.A. Nucleosome structural studies. *Curr. Opin. Struct. Biol.* **2011**, *21*, 128–136.
113. Zavitsanos, K.; Nunes, A.M.; Malandrinos, G.; Hadjiliadis, N. Copper effective binding with 32–62 and 94–125 peptide fragments of histone H2B. *J. Inorg. Biochem.* **2011**, *105*, 102–110.
114. Ishida, S.; Andreux, P.; Poitry-Yamate, C.; Auwerx, J.; Hanahan, D. Bioavailable copper modulates oxidative phosphorylation and growth of tumors. *Proc. Natl. Acad. Sci. USA* **2013**, *110*, 19507–19512.
115. Bienvenu, C.; Cadet, J. Synthesis and kinetic study of the deamination of the cis diastereomers of 5,6-dihydroxy-5,6-dihydro-5-methyl-2'-deoxycytidine. *J. Org. Chem.* **1996**, *61*, 2632–2637.
116. Bransteitter, R.; Prochnow, C.; Chen, X.S. The current structural and functional understanding of APOBEC deaminases. *Cell. Mol. Life Sci.* **2009**, *66*, 3137–3147.
117. Larijani, M.; Martin, A. The biochemistry of activation-induced deaminase and its physiological functions. *Semin. Immunol.* **2012**, *24*, 255–263.
118. Iwatani, Y.; Takeuchi, H.; Strebel, K.; Levin, J.G. Biochemical activities of highly purified, catalytically active human APOBEC3G: Correlation with antiviral effect. *J. Virol.* **2006**, *80*, 5992–6002.
119. Nabel, C.S.; Jia, H.; Ye, Y.; Shen, L.; Goldschmidt, H.L.; Stivers, J.T.; Zhang, Y.; Kohli, R.M. AID/APOBEC deaminases disfavor modified cytosines implicated in DNA demethylation. *Nat. Chem. Biol.* **2012**, *8*, 751–758.
120. Wijesinghe, P.; Bhagwat, A.S. Efficient deamination of 5-methylcytosines in DNA by human APOBEC3A, but not by AID or APOBEC3G. *Nucleic Acids Res.* **2012**, *40*, 9206–9217.

121. Morgan, H.D.; Dean, W.; Coker, H.A.; Reik, W.; Petersen-Mahrt, S.K. Activation-induced cytidine deaminase deaminates 5-methylcytosine in DNA and is expressed in pluripotent tissues: Implications for epigenetic reprogramming. *J. Biol. Chem.* **2004**, *279*, 52353–52360.
122. Bishop, K.N.; Holmes, R.K.; Sheehy, A.M.; Davidson, N.O.; Cho, S.J.; Malim, M.H. Cytidine deamination of retroviral DNA by diverse APOBEC proteins. *Curr. Biol.* **2004**, *14*, 1392–1396.
123. Leonard, B.; Hart, S.N.; Burns, M.B.; Carpenter, M.A.; Temiz, N.A.; Rathore, A.; Isaksson Vogel, R.; Nikas, J.B.; Law, E.K.; Brown, W.L.; *et al.* APOBEC3B upregulation and genomic mutation patterns in serous ovarian carcinoma. *Cancer Res.* **2013**, *73*, 7222–7231.
124. Langlois, M.A.; Beale, R.C.; Conticello, S.G.; Neuberger, M.S. Mutational comparison of the single-domained APOBEC3C and double-domained APOBEC3F/G anti-retroviral cytidine deaminases provides insight into their DNA target site specificities. *Nucleic Acids Res.* **2005**, *33*, 1913–1923.
125. Wiegand, H.L.; Doehle, B.P.; Bogerd, H.P.; Cullen, B.R. A second human antiretroviral factor, APOBEC3F, is suppressed by the HIV-1 and HIV-2 Vif proteins. *EMBO J.* **2004**, *23*, 2451–2458.
126. Liddament, M.T.; Brown, W.L.; Schumacher, A.J.; Harris, R.S. APOBEC3F properties and hypermutation preferences indicate activity against HIV-1 *in vivo*. *Curr. Biol.* **2004**, *14*, 1385–1391.
127. Kohli, R.M.; Maul, R.W.; Guminski, A.F.; McClure, R.L.; Gajula, K.S.; Saribasak, H.; McMahon, M.A.; Siliciano, R.F.; Gearhart, P.J.; Stivers, J.T. Local sequence targeting in the AID/APOBEC family differentially impacts retroviral restriction and antibody diversification. *J. Biol. Chem.* **2010**, *285*, 40956–40964.
128. Krijger, P.H.; Tsaalbi-Shtylik, A.; Wit, N.; van den Berk, P.C.; de Wind, N.; Jacobs, H. Rev1 is essential in generating G to C transversions downstream of the Ung2 pathway but not the Msh2+Ung2 hybrid pathway. *Eur. J. Immunol.* **2013**, *43*, 2765–2770.
129. Cui, L.; Ye, W.; Prestwich, E.G.; Wishnok, J.S.; Taghizadeh, K.; Dedon, P.C.; Tannenbaum, S.R. Comparative analysis of four oxidized guanine lesions from reactions of DNA with peroxynitrite, singlet oxygen, and gamma-radiation. *Chem. Res. Toxicol.* **2013**, *26*, 195–202.
130. Kino, K.; Sugiyama, H. UVR-induced G-C to C-G transversions from oxidative DNA damage. *Mutat. Res.* **2005**, *571*, 33–42.
131. Douki, T.; Angelov, D.; Cadet, J. UV laser photolysis of DNA: Effect of duplex stability on charge-transfer efficiency. *J. Am. Chem. Soc.* **2001**, *123*, 11360–11366.
132. Giese, B.; Amaudrut, J.; Kohler, A.K.; Spormann, M.; Wessely, S. Direct observation of hole transfer through DNA by hopping between adenine bases and by tunnelling. *Nature* **2001**, *412*, 318–320.
133. Prat, F.; Houk, K.N.; Foote, C.S. Effect of guanine stacking on the oxidation of 8-oxoguanine in B-DNA. *J. Am. Chem. Soc.* **1998**, *120*, 845–846.
134. Cadet, J.; Wagner, J.R. DNA base damage by reactive oxygen species, oxidizing agents, and UV radiation. *Cold Spring Harb. Perspect. Biol.* **2013**, *5*, a012559.
135. Stathis, D.; Lischke, U.; Koch, S.C.; Deiml, C.A.; Carell, T. Discovery and mutagenicity of a guanidinoformimine lesion as a new intermediate of the oxidative deoxyguanosine degradation pathway. *J. Am. Chem. Soc.* **2012**, *134*, 4925–4930.

136. Hickerson, R.P.; Prat, F.; Muller, J.G.; Foote, C.S.; Burrows, C.J. Sequence and stacking dependence of 8-oxoguanine oxidation: Comparison of one-electron *vs.* singlet oxygen mechanisms. *J. Am. Chem. Soc.* **1999**, *121*, 9423–9428.
137. Lim, K.S.; Taghizadeh, K.; Wishnok, J.S.; Babu, I.R.; Shafirovich, V.; Geacintov, N.E.; Dedon, P.C. Sequence-dependent variation in the reactivity of 8-Oxo-7,8-dihydro-2'-deoxyguanosine toward oxidation. *Chem. Res. Toxicol.* **2012**, *25*, 366–373.
138. Lim, K.S.; Cui, L.; Taghizadeh, K.; Wishnok, J.S.; Chan, W.; DeMott, M.S.; Babu, I.R.; Tannenbaum, S.R.; Dedon, P.C. *In situ* analysis of 8-oxo-7,8-dihydro-2'-deoxyguanosine oxidation reveals sequence- and agent-specific damage spectra. *J. Am. Chem. Soc.* **2012**, *134*, 18053–18064.
139. McKibbin, P.L.; Fleming, A.M.; Towheed, M.A.; Van Houten, B.; Burrows, C.J.; David, S.S. Repair of hydantoin lesions and their amine adducts in DNA by base and nucleotide excision repair. *J. Am. Chem. Soc.* **2013**, *135*, 13851–13861.
140. Davis, W.B.; Bjorklund, C.C.; Deline, M. Probing the effects of DNA-protein interactions on DNA hole transport: The N-terminal histone tails modulate the distribution of oxidative damage and chemical lesions in the nucleosome core particle. *Biochemistry* **2012**, *51*, 3129–3142.
141. Krokan, H.E.; Bjoras, M. Base excision repair. *Cold Spring Harb. Perspect. Biol.* **2013**, *5*, a012583.
142. Sarasin, A. An overview of the mechanisms of mutagenesis and carcinogenesis. *Mutat. Res.* **2003**, *544*, 99–106.
143. Ray, S.; Menezes, M.R.; Senejani, A.; Sweasy, J.B. Cellular roles of DNA polymerase beta. *Yale J. Biol. Med.* **2013**, *86*, 463–469.
144. Hashimoto, H.; Zhang, X.; Cheng, X. Excision of thymine and 5-hydroxymethyluracil by the MBD4 DNA glycosylase domain: Structural basis and implications for active DNA demethylation. *Nucleic Acids Res.* **2012**, *40*, 8276–8284.
145. Sjolund, A.B.; Senejani, A.G.; Sweasy, J.B. MBD4 and TDG: Multifaceted DNA glycosylases with ever expanding biological roles. *Mutat. Res.* **2013**, *743–744*, 12–25.
146. Millar, C.B.; Guy, J.; Sansom, O.J.; Selfridge, J.; MacDougall, E.; Hendrich, B.; Keightley, P.D.; Bishop, S.M.; Clarke, A.R.; Bird, A. Enhanced CpG mutability and tumorigenesis in MBD4-deficient mice. *Science* **2002**, *297*, 403–405.
147. Wong, E.; Yang, K.; Kuraguchi, M.; Werling, U.; Avdievich, E.; Fan, K.; Fazzari, M.; Jin, B.; Brown, A.M.; Lipkin, M.; *et al. Mbd4* inactivation increases C→T transition mutations and promotes gastrointestinal tumor formation. *Proc. Natl. Acad. Sci. USA* **2002**, *99*, 14937–14942.
148. Sansom, O.J.; Bishop, S.M.; Bird, A.; Clarke, A.R. MBD4 deficiency does not increase mutation or accelerate tumorigenesis in mice lacking MMR. *Oncogene* **2004**, *23*, 5693–5696.
149. Ishibashi, T.; So, K.; Cupples, C.G.; Ausio, J. MBD4-mediated glycosylase activity on a chromatin template is enhanced by acetylation. *Mol. Cell. Biol.* **2008**, *28*, 4734–4744.
150. Vasovcak, P.; Krepelova, A.; Menigatti, M.; Puchmajerova, A.; Skapa, P.; Augustinakova, A.; Amann, G.; Wernstedt, A.; Jiricny, J.; Marra, G.; *et al.* Unique mutational profile associated with a loss of TDG expression in the rectal cancer of a patient with a constitutional PMS2 deficiency. *DNA Repair* **2012**, *11*, 616–623.

151. Ocampo-Hafalla, M.T.; Altamirano, A.; Basu, A.K.; Chan, M.K.; Ocampo, J.E.; Cummings, A., Jr.; Boorstein, R.J.; Cunningham, R.P.; Teebor, G.W. Repair of thymine glycol by hNth1 and hNeil1 is modulated by base pairing and *cis-trans* epimerization. *DNA Repair* **2006**, *5*, 444–454.
152. Chan, M.K.; Ocampo-Hafalla, M.T.; Vartanian, V.; Jaruga, P.; Kirkali, G.; Koenig, K.L.; Brown, S.; Lloyd, R.S.; Dizdaroglu, M.; Teebor, G.W. Targeted deletion of the genes encoding NTH1 and NEIL1 DNA N-glycosylases reveals the existence of novel carcinogenic oxidative damage to DNA. *DNA Repair* **2009**, *8*, 786–794.
153. Hegde, M.L.; Hegde, P.M.; Bellot, L.J.; Mandal, S.M.; Hazra, T.K.; Li, G.M.; Boldogh, I.; Tomkinson, A.E.; Mitra, S. Prereplicative repair of oxidized bases in the human genome is mediated by NEIL1 DNA glycosylase together with replication proteins. *Proc. Natl. Acad. Sci. USA* **2013**, *110*, E3090–E3099.
154. Chaisaingmongkol, J.; Popanda, O.; Warta, R.; Dyckhoff, G.; Herpel, E.; Geiselhart, L.; Claus, R.; Lasitschka, F.; Campos, B.; Oakes, C.C.; *et al.* Epigenetic screen of human DNA repair genes identifies aberrant promoter methylation of *NEIL1* in head and neck squamous cell carcinoma. *Oncogene* **2012**, *31*, 5108–5116.
155. Doseth, B.; Ekre, C.; Slupphaug, G.; Krokan, H.E.; Kavli, B. Strikingly different properties of uracil-DNA glycosylases UNG2 and SMUG1 may explain divergent roles in processing of genomic uracil. *DNA Repair* **2012**, *11*, 587–593.
156. Slupianek, A.; Falinski, R.; Znojek, P.; Stoklosa, T.; Flis, S.; Doneddu, V.; Pytel, D.; Synowiec, E.; Blasiak, J.; Bellacosa, A.; *et al.* *BCR-ABL1* kinase inhibits uracil DNA glycosylase UNG2 to enhance oxidative DNA damage and stimulate genomic instability. *Leukemia* **2013**, *27*, 629–634.
157. Xie, Y.; Yang, H.; Cunanan, C.; Okamoto, K.; Shibata, D.; Pan, J.; Barnes, D.E.; Lindahl, T.; McIlhatton, M.; Fishel, R.; *et al.* Deficiencies in mouse *Myh* and *Ogg1* result in tumor predisposition and G to T mutations in codon 12 of the *K-ras* oncogene in lung tumors. *Cancer Res.* **2004**, *64*, 3096–3102.
158. Ruggieri, V.; Pin, E.; Russo, M.T.; Barone, F.; Degan, P.; Sanchez, M.; Quaia, M.; Minoprio, A.; Turco, E.; Mazzei, F.; *et al.* Loss of MUTYH function in human cells leads to accumulation of oxidative damage and genetic instability. *Oncogene* **2013**, *32*, 4500–4508.
159. Bjoras, M.; Seeberg, E.; Luna, L.; Pearl, L.H.; Barrett, T.E. Reciprocal "flipping" underlies substrate recognition and catalytic activation by the human 8-oxo-guanine DNA glycosylase. *J. Mol. Biol.* **2002**, *317*, 171–177.
160. Luncsford, P.J.; Chang, D.Y.; Shi, G.; Bernstein, J.; Madabushi, A.; Patterson, D.N.; Lu, A.L.; Toth, E.A. A structural hinge in eukaryotic MutY homologues mediates catalytic activity and Rad9-Rad1-Hus1 checkpoint complex interactions. *J. Mol. Biol.* **2010**, *403*, 351–370.
161. Amouroux, R.; Campalans, A.; Epe, B.; Radicella, J.P. Oxidative stress triggers the preferential assembly of base excision repair complexes on open chromatin regions. *Nucleic Acids Res.* **2010**, *38*, 2878–2890.
162. Menoni, H.; Shukla, M.S.; Gerson, V.; Dimitrov, S.; Angelov, D. Base excision repair of 8-oxoG in dinucleosomes. *Nucleic Acids Res.* **2012**, *40*, 692–700.
163. Kino, K.; Takao, M.; Miyazawa, H.; Hanaoka, F. A DNA oligomer containing 2,2,4-triamino-5(2H)-oxazolone is incised by human NEIL1 and NTH1. *Mutat Res.* **2012**, *734*, 73–77.

164. Dou, H.; Mitra, S.; Hazra, T.K. Repair of oxidized bases in DNA bubble structures by human DNA glycosylases NEIL1 and NEIL2. *J. Biol. Chem.* **2003**, *278*, 49679–49684.
165. Hailer, M.K.; Slade, P.G.; Martin, B.D.; Rosenquist, T.A.; Sugden, K.D. Recognition of the oxidized lesions spiroiminodihydantoin and guanidinohydantoin in DNA by the mammalian base excision repair glycosylases NEIL1 and NEIL2. *DNA Repair* **2005**, *4*, 41–50.
166. Krishnamurthy, N.; Zhao, X.; Burrows, C.J.; David, S.S. Superior removal of hydantoin lesions relative to other oxidized bases by the human DNA glycosylase hNEIL1. *Biochemistry* **2008**, *47*, 7137–7146.
167. Liu, M.; Doublie, S.; Wallace, S.S. Neil3, the final frontier for the DNA glycosylases that recognize oxidative damage. *Mutat. Res.* **2013**, *743–744*, 4–11.
168. Regnell, C.E.; Hildrestrand, G.A.; Sejersted, Y.; Medin, T.; Moldestad, O.; Rolseth, V.; Krokeide, S.Z.; Suganthan, R.; Luna, L.; Bjoras, M.; *et al.* Hippocampal adult neurogenesis is maintained by Neil3-dependent repair of oxidative DNA lesions in neural progenitor cells. *Cell Rep.* **2012**, *2*, 503–510.
169. Rolseth, V.; Krokeide, S.Z.; Kunke, D.; Neurauter, C.G.; Suganthan, R.; Sejersted, Y.; Hildrestrand, G.A.; Bjoras, M.; Luna, L. Loss of Neil3, the major DNA glycosylase activity for removal of hydantoins in single stranded DNA, reduces cellular proliferation and sensitizes cells to genotoxic stress. *Biochim. Biophys. Acta* **2013**, *1833*, 1157–1164.
170. Krokeide, S.Z.; Laerdahl, J.K.; Salah, M.; Luna, L.; Cederkvist, F.H.; Fleming, A.M.; Burrows, C.J.; Dalhus, B.; Bjoras, M. Human NEIL3 is mainly a monofunctional DNA glycosylase removing spiroimindiohydantoin and guanidinohydantoin. *DNA Repair* **2013**, *12*, 1159–1164.
171. Wardle, J.; Burgers, P.M.; Cann, I.K.; Darley, K.; Heslop, P.; Johansson, E.; Lin, L.J.; McGlynn, P.; Sanvoisin, J.; Stith, C.M.; *et al.* Uracil recognition by replicative DNA polymerases is limited to the archaea, not occurring with bacteria and eukarya. *Nucleic Acids Res.* **2008**, *36*, 705–711.
172. Aller, P.; Duclos, S.; Wallace, S.S.; Doublie, S. A crystallographic study of the role of sequence context in thymine glycol bypass by a replicative DNA polymerase serendipitously sheds light on the exonuclease complex. *J. Mol. Biol.* **2011**, *412*, 22–34.
173. Fischhaber, P.L.; Gerlach, V.L.; Feaver, W.J.; Hatahet, Z.; Wallace, S.S.; Friedberg, E.C. Human DNA polymerase kappa bypasses and extends beyond thymine glycols during translesion synthesis *in vitro*, preferentially incorporating correct nucleotides. *J. Biol. Chem.* **2002**, *277*, 37604–37611.
174. Yoon, J.H.; Bhatia, G.; Prakash, S.; Prakash, L. Error-free replicative bypass of thymine glycol by the combined action of DNA polymerases kappa and zeta in human cells. *Proc. Natl. Acad. Sci. USA* **2010**, *107*, 14116–14121.
175. Takata, K.; Shimizu, T.; Iwai, S.; Wood, R.D. Human DNA polymerase N (POLN) is a low fidelity enzyme capable of error-free bypass of 5S-thymine glycol. *J. Biol. Chem.* **2006**, *281*, 23445–23455.
176. Takata, K.; Arana, M.E.; Seki, M.; Kunkel, T.A.; Wood, R.D. Evolutionary conservation of residues in vertebrate DNA polymerase N conferring low fidelity and bypass activity. *Nucleic Acids Res.* **2010**, *38*, 3233–3244.
177. Kusumoto, R.; Masutani, C.; Iwai, S.; Hanaoka, F. Translesion synthesis by human DNA polymerase eta across thymine glycol lesions. *Biochemistry* **2002**, *41*, 6090–6099.

178. Hogg, M.; Seki, M.; Wood, R.D.; Doublie, S.; Wallace, S.S. Lesion bypass activity of DNA polymerase theta (POLQ) is an intrinsic property of the pol domain and depends on unique sequence inserts. *J. Mol. Biol.* **2011**, *405*, 642–652.
179. Seki, M.; Masutani, C.; Yang, L.W.; Schuffert, A.; Iwai, S.; Bahar, I.; Wood, R.D. High-efficiency bypass of DNA damage by human DNA polymerase Q. *EMBO J.* **2004**, *23*, 4484–4494.
180. Shibutani, S.; Takeshita, M.; Grollman, A.P. Insertion of specific bases during DNA synthesis past the oxidation-damaged base 8-oxodG. *Nature* **1991**, *349*, 431–434.
181. Fazlieva, R.; Spittle, C.S.; Morrissey, D.; Hayashi, H.; Yan, H.; Matsumoto, Y. Proofreading exonuclease activity of human DNA polymerase delta and its effects on lesion-bypass DNA synthesis. *Nucleic Acids Res.* **2009**, *37*, 2854–2866.
182. Markkanen, E.; Castrec, B.; Villani, G.; Hubscher, U. A switch between DNA polymerases delta and lambda promotes error-free bypass of 8-oxo-G lesions. *Proc. Natl. Acad. Sci. USA* **2012**, *109*, 20401–20406.
183. Markkanen, E.; Hubscher, U.; van Loon, B. Regulation of oxidative DNA damage repair: The adenine:8-oxo-guanine problem. *Cell Cycle* **2012**, *11*, 1070–1075.
184. Brown, J.A.; Duym, W.W.; Fowler, J.D.; Suo, Z. Single-turnover kinetic analysis of the mutagenic potential of 8-oxo-7,8-dihydro-2'-deoxyguanosine during gap-filling synthesis catalyzed by human DNA polymerases lambda and beta. *J. Mol. Biol.* **2007**, *367*, 1258–1269.
185. Freudenthal, B.D.; Beard, W.A.; Wilson, S.H. DNA polymerase minor groove interactions modulate mutagenic bypass of a templating 8-oxoguanine lesion. *Nucleic Acids Res.* **2013**, *41*, 1848–1858.
186. Avkin, S.; Livneh, Z. Efficiency, specificity and DNA polymerase-dependence of translesion replication across the oxidative DNA lesion 8-oxoguanine in human cells. *Mutat. Res.* **2002**, *510*, 81–90.
187. Maga, G.; Crespan, E.; Markkanen, E.; Imhof, R.; Furrer, A.; Villani, G.; Hubscher, U.; van Loon, B. DNA polymerase delta-interacting protein 2 is a processivity factor for DNA polymerase lambda during 8-oxo-7,8-dihydroguanine bypass. *Proc. Natl. Acad. Sci. USA* **2013**, *110*, 18850–18855.
188. Aller, P.; Ye, Y.; Wallace, S.S.; Burrows, C.J.; Doublie, S. Crystal structure of a replicative DNA polymerase bound to the oxidized guanine lesion guanidinohydantoin. *Biochemistry* **2010**, *49*, 2502–2509.
189. Beckman, J.; Wang, M.; Blaha, G.; Wang, J.; Konigsberg, W.H. Substitution of Ala for Tyr567 in RB69 DNA polymerase allows dAMP and dGMP to be inserted opposite Guanidinohydantoin. *Biochemistry* **2010**, *49*, 8554–8563.
190. Kino, K.; Ito, N.; Sugasawa, K.; Sugiyama, H.; Hanaoka, F. Translesion synthesis by human DNA polymerase eta across oxidative products of guanine. *Nucleic Acids Symp. Ser. (Oxf.)* **2004**, *48*, 171–172.
191. Henderson, P.T.; Delaney, J.C.; Muller, J.G.; Neeley, W.L.; Tannenbaum, S.R.; Burrows, C.J.; Essigmann, J.M. The hydantoin lesions formed from oxidation of 7,8-dihydro-8-oxoguanine are potent sources of replication errors *in vivo*. *Biochemistry* **2003**, *42*, 9257–9262.

192. Suzuki, M.; Kino, K.; Morikawa, M.; Kobayashi, T.; Komori, R.; Miyazawa, H. Calculation of the stabilization energies of oxidatively damaged guanine base pairs with guanine. *Molecules* **2012**, *17*, 6705–6715.
193. Stephens, P.J.; Tarpey, P.S.; Davies, H.; van Loo, P.; Greenman, C.; Wedge, D.C.; Nik-Zainal, S.; Martin, S.; Varela, I.; Bignell, G.R.; *et al.* The landscape of cancer genes and mutational processes in breast cancer. *Nature* **2012**, *486*, 400–404.
194. Vermeulen, W.; Fousteri, M. Mammalian transcription-coupled excision repair. *Cold Spring Harb. Perspect. Biol.* **2013**, *5*, a012625.
195. Lehmann, A.R.; McGibbon, D.; Stefanini, M. Xeroderma pigmentosum. *Orphanet J. Rare Dis.* **2011**, *6*, 70.
196. Scharer, O.D. Nucleotide excision repair in eukaryotes. *Cold Spring Harb. Perspect. Biol.* **2013**, *5*, a012609.
197. Tornaletti, S.; Maeda, L.S.; Lloyd, D.R.; Reines, D.; Hanawalt, P.C. Effect of thymine glycol on transcription elongation by T7 RNA polymerase and mammalian RNA polymerase II. *J. Biol. Chem.* **2001**, *276*, 45367–45371.
198. Tornaletti, S.; Maeda, L.S.; Kolodner, R.D.; Hanawalt, P.C. Effect of 8-oxoguanine on transcription elongation by T7 RNA polymerase and mammalian RNA polymerase II. *DNA Repair* **2004**, *3*, 483–494.
199. Hoang, M.L.; Chen, C.H.; Sidorenko, V.S.; He, J.; Dickman, K.G.; Yun, B.H.; Moriya, M.; Niknafs, N.; Douville, C.; Karchin, R.; *et al.* Mutational signature of aristolochic acid exposure as revealed by whole-exome sequencing. *Sci. Transl. Med.* **2013**, *5*, 197ra102.
200. Poon, S.L.; Pang, S.T.; McPherson, J.R.; Yu, W.; Huang, K.K.; Guan, P.; Weng, W.H.; Siew, E.Y.; Liu, Y.; Heng, H.L.; *et al.* Genome-wide mutational signatures of aristolochic acid and its application as a screening tool. *Sci. Transl. Med.* **2013**, *5*, 197ra101.
201. Sidorenko, V.S.; Yeo, J.E.; Bonala, R.R.; Johnson, F.; Scharer, O.D.; Grollman, A.P. Lack of recognition by global-genome nucleotide excision repair accounts for the high mutagenicity and persistence of aristolactam-DNA adducts. *Nucleic Acids Res.* **2012**, *40*, 2494–2505.
202. Charlet-Berguerand, N.; Feuerhahn, S.; Kong, S.E.; Ziserman, H.; Conaway, J.W.; Conaway, R.; Egly, J.M. RNA polymerase II bypass of oxidative DNA damage is regulated by transcription elongation factors. *EMBO J.* **2006**, *25*, 5481–5491.
203. Khobta, A.; Epe, B. Repair of oxidatively generated DNA damage in Cockayne syndrome. *Mech. Ageing Dev.* **2013**, *134*, 253–260.
204. D'Errico, M.; Pascucci, B.; Iorio, E.; van Houten, B.; Dogliotti, E. The role of CSA and CSB protein in the oxidative stress response. *Mech. Ageing Dev.* **2013**, *134*, 261–269.
205. Aamann, M.D.; Muftuoglu, M.; Bohr, V.A.; Stevnsner, T. Multiple interaction partners for Cockayne syndrome proteins: Implications for genome and transcriptome maintenance. *Mech. Ageing Dev.* **2013**, *134*, 212–224.
206. Banerjee, D.; Mandal, S.M.; Das, A.; Hegde, M.L.; Das, S.; Bhakat, K.K.; Boldogh, I.; Sarkar, P.S.; Mitra, S.; Hazra, T.K. Preferential repair of oxidized base damage in the transcribed genes of mammalian cells. *J. Biol. Chem.* **2011**, *286*, 6006–6016.

207. Inukai, N.; Yamaguchi, Y.; Kuraoka, I.; Yamada, T.; Kamijo, S.; Kato, J.; Tanaka, K.; Handa, H. A novel hydrogen peroxide-induced phosphorylation and ubiquitination pathway leading to RNA polymerase II proteolysis. *J. Biol. Chem.* **2004**, *279*, 8190–8195.
208. Pfeifer, G.P.; Besaratinia, A. Mutational spectra of human cancer. *Hum. Genet.* **2009**, *125*, 493–506.
209. Walser, J.C.; Furano, A.V. The mutational spectrum of non-CpG DNA varies with CpG content. *Genome Res.* **2010**, *20*, 875–882.
210. Roberts, S.A.; Sterling, J.; Thompson, C.; Harris, S.; Mav, D.; Shah, R.; Klimczak, L.J.; Kryukov, G.V.; Malc, E.; Mieczkowski, P.A.; *et al.* Clustered mutations in yeast and in human cancers can arise from damaged long single-strand DNA regions. *Mol. Cell* **2012**, *46*, 424–435.
211. Galashevskaya, A.; Sarno, A.; Vagbo, C.B.; Aas, P.A.; Hagen, L.; Slupphaug, G.; Krokan, H.E. A robust, sensitive assay for genomic uracil determination by LC/MS/MS reveals lower levels than previously reported. *DNA Repair* **2013**, *12*, 699–706.
212. Mangerich, A.; Knutson, C.G.; Parry, N.M.; Muthupalani, S.; Ye, W.; Prestwich, E.; Cui, L.; McFaline, J.L.; Mobley, M.; Ge, Z.; *et al.* Infection-induced colitis in mice causes dynamic and tissue-specific changes in stress response and DNA damage leading to colon cancer. *Proc. Natl. Acad. Sci. USA* **2012**, *109*, E1820–E1829.
213. Chan, S.W.; Dedon, P.C. The biological and metabolic fates of endogenous DNA damage products. *J. Nucleic Acids* **2010**, *2010*, 929047.
214. Gaikwad, N.W. Metabolomic profiling unravels DNA adducts in human breast that are formed from peroxidase mediated activation of estrogens to quinone methides. *PLoS One* **2013**, *8*, e65826.
215. Tretyakova, N.; Goggin, M.; Sangaraju, D.; Janis, G. Quantitation of DNA adducts by stable isotope dilution mass spectrometry. *Chem. Res. Toxicol.* **2012**, *25*, 2007–2035.

11

Reading and Language Disorders: The Importance of Both Quantity and Quality

Dianne F. Newbury [1,*], Anthony P. Monaco [2] and Silvia Paracchini [3]

[1] Wellcome Trust Centre for Human Genetics, University of Oxford, Oxford OX3 7BN, UK
[2] Tufts University, Ballou Hall, Medford, MA 02155, USA; E-Mail: Anthony.Monaco@tufts.edu
[3] School of Medicine, University of St. Andrews, St. Andrews, KY16 9TF, UK;
E-Mail: sp58@st-andrews.ac.uk

* Author to whom correspondence should be addressed; E-Mail: dianne@well.ox.ac.uk;

Abstract: Reading and language disorders are common childhood conditions that often co-occur with each other and with other neurodevelopmental impairments. There is strong evidence that disorders, such as dyslexia and Specific Language Impairment (SLI), have a genetic basis, but we expect the contributing genetic factors to be complex in nature. To date, only a few genes have been implicated in these traits. Their functional characterization has provided novel insight into the biology of neurodevelopmental disorders. However, the lack of biological markers and clear diagnostic criteria have prevented the collection of the large sample sizes required for well-powered genome-wide screens. One of the main challenges of the field will be to combine careful clinical assessment with high throughput genetic technologies within multidisciplinary collaborations.

Keywords: Specific Language Impairment (SLI); dyslexia; genetics

---

1. Introduction

The disturbance of speech and language development is a common feature of many neurodevelopmental disorders [1]. Language impairment is often secondary to more pressing clinical features (e.g., in autistic disorders, epilepsy or periventricular heterotopia), but in some cases may represent the primary clinical concern ("specific language disorders", e.g., in Specific Language Impairment (SLI) and dyslexia) [1]. Specific language disorders typically occur in the absence of any gross neurodevelopmental difficulties, neurological or sensory impairments and with normal non-verbal intelligence. SLI is defined as a disturbance of oral language skills, whereas dyslexia is an impairment in the use and/or understanding of written language [2]. Both show a strong familial bias, and heritability estimates indicate that a high proportion of the phenotypic variation in each of these disorders can be attributed to genetic variation [3–5].

2. Complex Traits, Complex Definitions

Whilst the terms "dyslexia" and "SLI" are widely used in the clinical and research literature, both disorders lack clear diagnostic guidelines and are often defined chiefly in terms of exclusionary criteria [6]. DSM-5 (Diagnostic and Statistical Manual) classifies SLI as a language disorder, while dyslexia is categorized as a specific learning disorder. However, both diagnoses require that "the individual's difficulties must not be better explained by developmental, neurological, sensory (vision or hearing), or motor disorders and must significantly interfere with academic achievement, occupational performance, or activities of daily living" [7]. To complicate things further, deficits of language may vary considerably both between individuals and over the developmental trajectory in addition to the modality of language affected [6,8]. It is generally accepted that dyslexia primarily reflects a difficulty in the domain of phonological decoding (translating written letters into speech sounds) [9]. Nonetheless, it is very common to observe a phonological deficit in combination with other manifestations, like sensory or fine motor control problems [10,11]. SLI has been proposed to reflect a deficit in phonological short-term memory (the retention of verbal information for short periods of time) [12], auditory perception (the processing of brief and/or rapid auditory stimuli) [13] and/or the development and application of grammatical rules [14]. However, all of these theories are supported by varying amounts of evidence in the primary research literature.

3. Changing and Heterogeneous Phenotypes

In addition, the difficulties experienced by any given individual may cross linguistic and cognitive domains and often change as the child develops [15,16]. Such observations perhaps suggest that SLI and dyslexia cannot be treated as discrete clinical conditions. Instead, it is possible that these language disorders might represent complex end effects of the disruption of multiple cognitive development processes that overlap with, and are related to, the secondary language disorders mentioned above [17]. Under this model, the investigation of dimensions of impairment may be more relevant than the ascertainment of clinical cohorts. The observed co-occurrence of SLI and dyslexia (~50%, [18]) support such a hypothesis and has led researchers to suggest that both disorders may result from an impairment in phonological representation [17]. The clinical presentation of the deficit may represent the severity of the underlying impairment or the presence of additional language- or cognitive-related difficulties [17,19]. Thus, "specific" language disorders may be the exception, rather than the rule, since co-morbidities often extend outside of the linguistic domains. For example, weaknesses in motor skills and executive function and reduced functional brain laterality are commonly described in children with dyslexia or SLI [20–26]. Nonetheless, the causal relationships between symptoms and cognitive, linguistic and developmental markers have yet to be elucidated. Others maintain that SLI and dyslexia may still have separate etiologies, and the co-incidence may simply represent comorbidity, as seen with other neurodevelopmental disorders [19]. Identifying the genetic underpinning of these disorders is required to inform this debate and reach more definitive conclusions about the diagnosis. For example, are there shared or partially overlapping genetic factors that contribute to separate disorders? Can we talk about common etiologies? Or do the DSM-5 definitions correlate with distinct biological pathways? While it would be tempting to simplistically ask if candidate genes for a single disorder can also influence a disorder with a different diagnosis, we might be asking the wrong questions if the initial diagnoses are artificial clinical constructs and misleading with regards to etiology.

4. Monogenic Conditions Back in the Picture

For many language disorders that are associated with known syndromes, the genetic cause of the syndrome itself is known. These involve a wide range of genetic mutations, from point mutations (for example, *MECP2* mutations in Rett syndrome [27]), to nucleotide expansions (for example, the expansion of the *FMR1* gene in Fragile X [16]) and deletion syndromes, such as velo-cardio-facial syndrome, which results from a 3-Mb deletion on chromosome 22q11.2 [28], or the duplication or deletion of an entire chromosome, such as Down syndrome and Turner syndrome [29–31]. Thus, it is likely that any population selected to have language impairment will harbor a subset of children with these recognized syndromes [32]. Single gene mutations have also been described for some specific language disorders. Mutations and disruptions of the *FOXP2* gene lead to childhood apraxia of speech [33,34], and point mutations in genes in the lysosomal pathway (*GNPTAB*, *GNPTG* and *NAGPA*) have been associated with persistent stuttering [35]. However, even in these severe and exceptional cases, there is often a high degree of heterogeneity between individuals. While some generalizations can be made, there is still considerable inter-individual variation. For example, individuals with *FOXP2* disruptions invariably present with dyspraxic speech (the inability to make fine-tuned oromotor movements necessary for coherent speech), while others can present with both receptive and expressive language difficulties, only an expressive deficit, only with intellectual deficits or good performance on non-verbal intelligence tasks [36].

5. Genetic Windows into Development

One argument against the utility of understanding the genetic underpinnings of these rare syndromic language disorders has been to question their relevance to the biology of common forms of SLI and dyslexia. However, the identification of a specific candidate gene and mutations thereof can allow the development of targeted investigations in cellular or animal models, which, in turn, can point to mechanisms that might be relevant to more common forms of language-related conditions affecting thousands of children. An example of this is how the FOXP2 transcription factor regulates the expression of target genes, such as *CNTNAP2*, which has been shown to play a role in more common forms of language impairment [37], as well as other neurodevelopmental disorders [38–41] and normal language variation [42]. The increased resolution and power of genetic screening demonstrates that the boundary between monogenic and common traits may actually be less defined than that predicted previously. Recent large-scale studies have clearly shown that genes disrupted by highly penetrant mutations and leading to well-defined diseases can play a role in complex disorders, although this may only be relevant in a subset of cases [43]. Nonetheless, genetic contributions to the majority of specific language disorders are expected to be complex in nature and involve genetic variation in many genes, which combine to determine an overall risk of disorder [44].

6. Classical Approaches

Genetic contributions to neurodevelopmental disorders (both syndromic and specific forms) can be traced by linkage analyses. This approach can be applied to extended multi-generational families or large collections of small nuclear families and can involve the investigation of linkage-disequilibrium patterns [45]. Linkage analysis allows the identification of broad chromosome regions that co-segregate with a disorder in a given family unit. Since such studies consider segregation patterns rather than specific genetic variants, they can enable the identification of shared chromosome regions even when the pathological variants in each region differ between family units. This methodology is particularly powerful for the identification of genetic variants with high penetrance and expressivity and can be applied to more complex situations, such as that expected for both SLI and dyslexia.

7. SLI and Dyslexia Linkage Loci

Genome-wide screens or targeted analyses of dyslexic families have allowed the identification of linkage loci on chromosomes 15q21 (DYX1—OMIM#127700) [46,47], 6p22.3-p21.3 (DYX2—OMIM#600202) [46,48–51], 2p16-p15 (DYX3—OMIM#604254) [52,53], 3p12-q13 (DYX5—OMIM#606896) [54,55], 18p11.2 (DYX6—OMIM#606616) [56], 11 (DYX7—OMIM#127700) [57], 1p36-p34 (DYX8—OMIM#608995) [58,59] and Xq27.2-q28 (DYX9—OMIM#300509) [60]. Subsequent fine-mapping efforts across linked loci, through the investigation of specific genetic variations or the characterization of individuals with chromosome imbalances, have led to the identification of putative risk variants in the *DCDC2* [61], *KIAA0319* [62–64], *DYX1C1* [65,66], *C2orf3*/*MRPL19* [67], *CYP19A1* [68] and *ROBO1* [54] genes. Each of these candidate genes has a variable amount of support, ranging from observations limited to a single family to replication across multiple cohorts [69]. Nonetheless, functional analyses have led to an intriguing conversion upon pathways involving neuronal migration [70–76] and cilia motility [72,77–79], as discussed in later sections. Investigations of SLI are less advanced, but linkage studies have identified four loci of interest on chromosomes 7q35-q36 (SLI4—OMIM#612514) [80], 13q21 (SLI3—OMIM#607134) [81,82], 16q (SLI1—OMIM#606711) [83–86] and 19q (SLI2—OMIM#606712) [83–86], and subsequent studies have identified two candidate genes; *CMIP* and *ATP2C2*, both in SLI1 [87]. Microdeletions in the *ZNF277* gene on chromosome 7 have also been implicated in SLI [88], as has the Human Leukocyte Antigen (HLA) locus [89].

Although the linkage loci described for SLI and dyslexia do not overlap, studies of other complex genetic disorders indicate that there may be hundreds of genetic variants contributing to any one phenotypic status. Since genetic analyses are likely to detect only the major gene effects within any given cohort (the so-called winners curse [90]), this observation may simply be a result of the number of studies performed rather than the reflection of separate pathologies *per se*.

8. GWA Studies

Advances in genetic technologies over the last decade have allowed enormous leaps in our characterization and understanding of both rare and common genetic variations at the sequence level. Projects, such as the HapMap ([91]) and 1000 Genomes ([92]), have provided us with catalogues of expected variations across multiple populations. Methodological advances, such as microarrays and high throughput genotyping and sequencing platforms, have allowed us to characterize known variants efficiently. Accordingly, gene identification shifted from linkage analysis to genome-wide association (GWA) studies [93]. GWA studies usually interrogate large cohorts of cases and controls (typical sample sizes range from 1000 up to 1,000,000), but can be extended to a regression analysis of variant frequency upon performance in phenotype-related quantitative tasks (quantitative GWA study). Since GWA studies consider the frequency of each variant independently, there is an underlying assumption that the causal variation (or a Single Nucleotide Polymorphism (SNP) in linkage disequilibrium with the causal variant) will be common across cases. As such, association studies do not allow for a high level of genetic heterogeneity between cases.

9. GWA Studies of Speech and Language-Related Traits

GWA studies of speech and language cohorts to date have not yielded consistent findings. Meaburn *et al.* (2008) applied a pooled genotyping method across two extreme samples selected from a twin cohort on the basis of reading ability (755 low reading ability cases and 747 high reading ability controls) using 100,000 SNPs, but did not identify any significant associations [94]. This may reflect the sparse density of the genotyping arrays employed at that time. Roeske *et al.* analyzed a discovery (N = 200) and replication (N = 186) cohort both selected for dyslexia and found association for a specific electrophysiological endophenotype of dyslexia ("mismatch negativity component" or MMN) ($p = 5.14 \times e^{-8}$ in combined dataset) pointing to the *SLC2A3* gene, which is implicated in glucose transport in the brain [95]. Field *et al.* 2013, performed a joint linkage and association study on 718 individuals from 101 dyslexia families with 100,000 SNPs. Again, they did not find any associations that met the threshold of genome-wide significance (1×10^{-8} [96]) [97]. They did however, find suggestive association ($p = 6.2 \times 10^{-7}$) with an SNP 77 Kb downstream of the *FGF18* gene, which has been implicated in lateralization [97]. Luciano *et al.* reported a meta-analysis of quantitative reading and language measures across two relatively large population-based samples (1177 individuals from 538 families and approximately 5000 cases) [98]. They found a suggestive level of association ($p = 7.34 \times 10^{-8}$) between variants in the *ABCC13* gene on chromosome 21 and non-word repetition (a marker of phonological short-term memory). However, they did not replicate the association to *FGF18* [98]. Eicher *et al.* also used a population-based sample, but in their study, they selected low language and reading performers for their GWA study [99]. Their sample included 163 language impaired probands, 353 dyslexic probands and 174 comorbid probands (*i.e.*, those with both language and reading impairment). They compared variant frequencies between these proband groups and the remainder of the population. They observed suggestive association with SNPs in

ZNF385D ($p = 5.45 \times 10^{-7}$) and *COL4A2* ($p = 7.59 \times 10^{-7}$) in the cases with the comorbid phenotype, and to SNPs in the *NDST4* ($p = 1.4 \times 10^{-7}$) gene in language impaired probands [99].

Recently, it has been proposed that parent-of-origin effects could explain part of the missing heritability of complex traits, suggesting that the addition of these effects within GWA studies may be fruitful [100]. A recent study of 278 families with a language-impaired child, investigated child genotype and parent-of-origin effects [101]. They identified significant evidence for paternal parent-of-origin effects on chromosome 14q12 ($p = 3.74 \times 10^{-8}$) and suggestive evidence for maternal parent-of-origin effects on chromosome 5p13 ($p = 1.16 \times 10^{-7}$) [101]. The paternally-associated SNP on chromosome 14 yields a non-synonymous coding change within the *NOP9* gene, which has been reported to be significantly dysregulated in individuals with schizophrenia.

10. GWA Study Design Factors

The ability of GWA studies to identify risk variants depends upon several factors. These include the effect size of the variants, the frequency of the variants in the population under study and the population as a whole and, of course, sample size and study design [102]. The lack of replication described in the previous section is therefore perhaps not surprising: each started with a different hypothesis and definition of disorder and applied different selection procedures and association methodologies. Inconsistency across studies makes it hard to assess whether non-replication indicates the presence of false positives or is simply a function of study design or power to detect an association. In addition, the variations between independent studies means that it is not possible to simply combine existing cohorts to generate adequately powered meta-analyses.

As discussed above, some of the language-related GWA studies used population-based samples and chose to select low language-performing subjects as cases (*i.e.*, the lower tail end of the distribution, marked as "cases" Figure 1a). Others studied speech and language-related traits across the entire distribution in a quantitative GWA approach (*i.e.*, the entire normal distribution in Figure 1a). Such population-based approaches assume that genetic contributions to disorders of speech and language will be the same as those that contribute to speech and language-related traits across the entire distribution. This assumption is dependent upon the way in which disorders, such as SLI and dyslexia, are conceptualized: do they just represent the lower tail of the normal distribution with respect to speech and language (dimensional model) or is there a qualitative difference between dyslexic individuals and poor readers (categorical model)? (*i.e.*, the difference between the "cases" and the lower normal distribution in Figure 1a).

Figure 1. Study design for quantitative phenotypes (**a**) genome-wide association (GWA) studies for speech and language-related traits typically use phenotypes across the entire distribution (population-based quantitative GWA studies). Others might apply a binary affection status under which low language-performing individuals are defined as "cases" and individuals within the "normal" language range (usually performance above the mean) as "controls". Under certain conditions, "super-controls" can provide more power, as they are selected to fall at the upper extreme of the distribution. If controls with phenotype data are not available, they may be derived from standard control populations under the knowledge that they might include a small proportion of cases. Quantitative GWA studies restricted to cases may be based on a phenotypic distribution restricted to the lower tail of the entire distribution or may be based on a phenotypic curve derived across cases samples, as denoted by the two normal distributions in (**a**) (note that in (**a**), the phenotype distribution may not necessarily be expected to be normal, although it is shown as such in the figure). (**b**) The pegboard test generates a quantitative measure for handedness (PegQ) that is normally distributed around a positive mean. PegQ strongly correlates with hand preference, so that individuals with positive scores are very likely to be right-handed (roughly 90% of the population), and individuals with negative scores are likely to be left handed. Typically genetic studies for handedness have used the categorical measures of hand-preference using a case-control (left *vs.* right) study design.

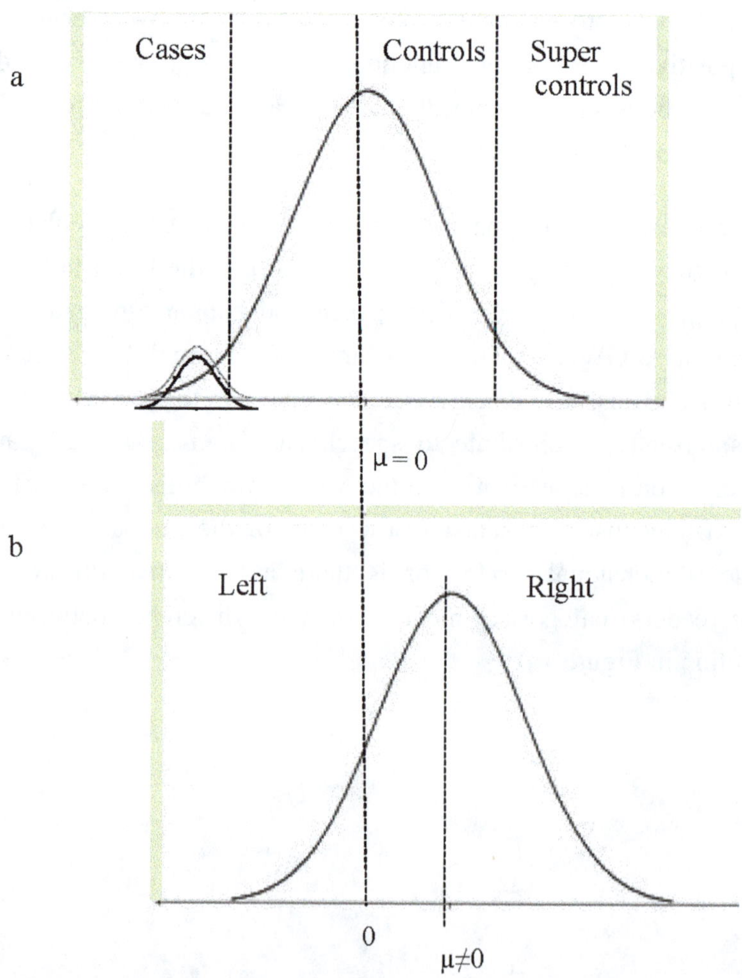

11. Dimensional and Categorical Models of Language Disorder

The distinction between dimensional and categorical models of language disorders is still a matter of debate. Leonard argues that perhaps there is no tangible cause for SLI, and the mental representations of children with the label SLI are not distinct from other children [103]. Taxonomic and principal component analyses support this view, demonstrating that relationships between language-related measures do not differ between children affected by SLI and those with normal language development [104–106]. In addition, the ease of the acquisition of specific language features has also been shown to be consistent between children affected by SLI and those with normal language development [107]. Under this dimensional model, one can consider variation across the range when attempting to identify underlying genetic effects. Twin studies of SLI instead support a categorical distinction [4,108–111]. Such studies indicate that children who have speech and language difficulties that are severe enough to warrant clinical referral have a qualitatively different profile of impairment, which shows increased levels of heritability [110]. Thus, it is possible that some forms of speech and language impairment, at least, have different underlying pathology from those mechanisms that are important in normal language variation (*i.e.*, the distinct lower normal distribution in Figure 1a). Under this hypothesis, one needs to specifically select children with SLI to identify these distinct underlying genetic effects. It is, of course, possible that both models contribute to some level: studies of the effects of specific genetic risk variants upon language development indicate that some genetic risk variants play a role across the entire distribution of ability [87,112–114], while others appear to play a role that is specific to impairment [87,113]. These data suggest that a variety of approaches will be required to delineate genetic effects underlying language impairment: GWA studies of individuals across a range of abilities will identify genes that contribute across the entire distribution, while studies of specific disorder subsets will be required to identify genes with more specific effects.

12. Cross-Linguistic Difficulties of Speech and Language Disorders

Given all these complexities, we extol the value of combining quantity with quality in the collection of larger cohorts carefully characterized at the phenotypic level. This is a major investment both in time and resources, but clearly represents the most promising way forward. The success of genetic mapping for complex traits has been largely facilitated by the collaboration of scientists and clinicians in large consortia, which have facilitated the collection of large sample sizes for genome mapping. This normally requires international collaborations, which pose additional complications for reading and language disorders. The psychometric tests used to assess reading and language skills cannot be separated from the language spoken in different countries and are not always directly comparable. In this context, the European NeuroDys project is working to define common guidelines for the collection and assessment of a large dyslexia cohort throughout research groups spread across Europe [115]. NeuroDys aims to exclude comorbidity and select severe dyslexic cases by selecting those more than 1.25 SD below grade level on a standardized word-reading test. Their strategy also includes the screening of control samples. Since dyslexia has a relatively high frequency, 5%–10% of population samples, routinely employed in GWA studies, will be expected to consist of individuals with dyslexia. GWA analysis of the NeuroDys cohort is currently underway. Analysis of dyslexia candidate genes in this cohort did not reveal any statistically significant association, highlighting many of the challenges covered in this review [116]. As discussed previously, universal inclusion/exclusion criteria are a good strategy to facilitate meta-analyses or cross-linguistic categorical GWA studies. Nonetheless, they do not entirely solve the challenge of obtaining large sample sizes for quantitative analyses. For example, single word reading tests tend to assess accuracy in English *versus* speed in transparent languages, like Italian or German. A composite score, including both accuracy and speed, as used by the NeuroDys study, can address this, but does not necessarily reflect the real reading difficulty in different populations. It is interesting to note that despite these differences, associations with candidate genes have been replicated in different languages; association with the *DCDC2* gene has been reported in English-speaking (Meng *et al.* 2005 [61], Scerri *et al.* 2011 [113]) and German-speaking (Schumacher *et al.* 2006 [117]) cohorts. Similarly, cross-linguistic studies find that features of the native language can act to modulate given aspects of the SLI phenotype. In general, those linguistic features that are considered "hardest" for a normally developing child to understand will represent areas of particular problem for children with SLI [118,119]. Within-child differences indicate that this generalization extends to bilingual children. Bilingual children with SLI encounter similar problems to monolingual children with SLI, and these problems appear to be language-specific [120]. These cross-linguistic variations represent an extra complication for meta-analyses and collaborative studies of speech and language impairments. If the features of the disorder vary across populations, the measurement of disability will be language-specific. Thus, multi-site efforts will not only need to consider the best way to overcome cross-ethnicity genetic differences, but also cross-ethnicity language differences.

13. Mega-, Meta- and Mixed-GWA Studies

Many successful GWA studies in the literature now include hundreds of thousands of individuals (diabetes, body mass index, height) [121–125] compared to the hundreds employed in the studies described above. Such "mega-GWA studies" have been relatively successful in identifying significant and consistent loci that contribute to either disease status or continuous traits across the general population. However, it is clear from these large studies that each genetic risk variant is likely to explain only a small amount of the heritability and that several hundred risk loci are likely to underlie any given trait or disorder [126]. Besides an increased sample sizes, a possible route of future investigation may be provided by a model adopted in the study of psychiatric disorders. Much like SLI and dyslexia, there is much evidence for the existence of shared genetic effects between psychiatric disorders, such as bipolar disorder, depressive disorder and schizophrenia. "Mixed-GWA study" investigations of mixed cohorts across these disorders have recently identified risk factors that span these clinical boundaries, suggesting that there may be some common pathophysiologies across related disorders [127,128]. The above studies suggest that meta-analyses across the existing SLI and dyslexia cohorts would be a worthwhile effort, despite the challenges involved in the amalgamation of these highly heterogeneous samples.

14. Quantitative *vs.* Qualitative, or Both

Throughout this manuscript, we have clearly advocated the combined analyses of existing cohorts or the collection of new large homogeneous samples relevant to developmental disorders. However, is sample size all that matters? If that were the case, the field should probably shift towards questionnaire-based phenotypic assessments, which would be an efficient way to boost numbers. Nonetheless, there is an intrinsic value in investing in phenotypic assessments that are detailed and quantitative. Firstly, this strategy allows cohorts to be stratified in distinct subgroups according to distinct criteria, which might be required for different hypotheses. For example, it makes it possible to select for the severity of disorders or to change features along with revised diagnostic criteria (as in Figure 1b). In addition, association analyses that use quantitative measures are potentially more powerful, as they exploit the full range of variability of a given phenotype, allowing the direct investigation of dimensional and categorical models. A recent story describing a GWA study for handedness is emblematic of how small, detailed cohorts can provide power to detect relevant biological effects [129,130]. This GWA study was conducted in a small sample (N = 728, well below the recommended standard) originally selected for a dyslexia diagnosis. In addition to language-related measures, this cohort was characterized with the pegboard test, which allows the derivation of a quantitative measure of relative hand skills (PegQ). This measure is normally distributed with a right-shifted mean and strongly correlates with hand preference (Figure 1b). Despite the small sample size, the screening led to a genome-wide significant result within the *PCSK6* gene, known to regulate the NODAL protein pathway, essential for the left/right patterning in early embryonic development [131], and therefore represents an extremely interesting candidate for a lateralized phenotype. Although this association was replicated in two independent cohorts with developmental dyslexia, the risk variant appeared to confer an opposite effect (reduced relative laterality) in a general population cohort (N = 2666). In support of these findings, subsequent gene enrichment analysis showed that genes controlling structural asymmetries were associated with the handedness measure in both cohorts regardless of a dyslexia diagnosis [130]. These findings demonstrate the importance of studying phenotypes within specific selected cohorts, as well as across the general population, as discussed above. It is possible that these effects are the result of risk variant interactions with different genotypic/biological backgrounds, as discussed below. A dyslexia-specific effect has also been observed for other traits. A variant in the *MYO18B* gene has also been implicated in mathematical ability specifically in children with dyslexia [132]. Hand preference (left *vs.* right) can readily be collected as a simple add-on question in any questionnaire battery. Thus, it would be very straightforward to re-analyze or meta-analyze existing GWA cohorts (even those collected for the investigation of different traits) using a case-control definition based on hand preference. A recent conference abstract describing a genome-wide analysis of hand preference data from more than 20,000 individuals did not find any genome-wide significant loci (Medland *et al.* ASHG, 2009 [133]). Similarly, a study of 4268 subjects from a population-based cohort, which included broad measures of laterality (hand or foot preference, ocular dominance or hand clasp), did not yield genome-wide significant associations [134]. These data suggest that the genetic effects at the population level are extremely low or, alternatively, that a categorical approach is not suitable for dissecting the genetics of this trait. It is likely that the biological regulation of hand preference involves complex and integrated processes that are not efficiently represented by the reductive phenotypes of left- *versus* right-handedness. If this can be said for a relatively transparent trait, like handedness, then we might expect these correlations to be even lower when considering a phenotype as complex as language. In addition, we must consider that the genetic effects might depend on sample stratification for a neurodevelopmental disorder definition.

15. Filling the Gap

Collecting large-scale GWA cohorts takes time, effort, funding and usually a concerted and collaborative effort between multiple research and clinical teams. Does this mean that we cannot make progress in this field until such statistical criteria are met? Perhaps the application of next-generation sequencing technologies to existing sample sets with detailed phenotypic information has the potential to fill this gap. While small individual family-based units do not provide enough power to map variants through linkage analyses alone, the application of this technique in combination with high-throughput sequencing can provide a powerful paradigm. The whole-exome or whole-genome sequencing of sporadic cases and their parents under the assumption of causative *de novo* mutations has proven successful in disorders, such as autism [135–138] and intellectual disability [139]. The sequencing of larger family units with multiple recurrent cases have allowed the identification of "ultra-rare" (often defined as <0.01%) or private (unique to the given family) mutations with high functional impact (*i.e.*, the gain of a stop codon and frame shift mutations) that cosegregate with disorder [140,141]. A recent exome sequencing study of children with childhood apraxia of speech investigated known candidate genes for language development [142]. They reported potentially deleterious variants in *FOXP1*, *CNTNAP2*, *ATP13A4*, *CNTNAP1*, *KIAA0319* and *SETX*, providing support for the role of these candidate genes [142].

16. Difficulties of High-Throughput Sequencing Studies

Although relatively simplistic in design, high throughput sequencing studies are far from straightforward. Rigorous quality control procedures and expert bioinformatics are needed, and the variation between platforms, capture assays and algorithms can be problematic. The main problem, though, is perhaps the proof of causality. The number of *de novo* mutations can be affected by environmental factors (e.g., paternal age [143]), but all being equal, the expected number of *de novo* mutations is approximately one per exome per generation. In contrast, the number of private mutations identified in an exome can be substantial, particularly as sequencing sensitivity and coverage increase. In addition, by design, these methods focus upon mutations that are likely to be private to the given individual and will not generalize between families. Studies of autistic disorder indicate that recurrent mutations in specific genes will be rare [135–137]. Furthermore, while the role of rare disruptive mutations is perhaps more tangible than that of common variation, these effects are still likely to function as part of a complex genetic model and represent risk variants rather than a causative mutation. Incomplete segregation, even within family units, is often observed for both rare mutations and copy number events, which can also prove a fruitful avenue of research in relatively small sample sizes. While large, disruptive events of genes that are known to be important can clearly be assigned some functionality, when a mutation or copy number variant is private, it can be very hard to distinguish between a functional effect that is subject to incomplete penetrance or modifier gene effects and a non-functional change. Such observations have led to the double-hit hypothesis in developmental disorders in which mutations and/or copy number events combine in an additive or epistatic manner to modulate the exact clinical presentation [136,144–148].

17. Translational Relevance

The pathway from association to functional evidence is a long, but necessary road if we are to truly elucidate the biological mechanisms underlying speech and language disorders. While a p-value $< 0.5 \times 10^{-8}$ or the observation or private mutations in conserved motifs are certainly robust indicators of a genuine genetic susceptibility, these findings remain statistical predictions if they are not coupled by functional data. It is important to keep in mind that the risk variants identified by GWA studies are not necessarily functional and often represent proxies for the functional variant [102]. Functional investigations are being further progressed by projects, such as the Encyclopedia Of DNA Elements (ENCODE) and the Genetic European Variation in Health and Disease (GEUVADIS) project, which apply high-throughput methods to the study of gene expression and regulation to facilitate our understanding of the findings of GWA studies [149,150]. The functionality of rare coding mutations or copy number events is more tractable, but still not straightforward, especially when they are private. Changes to the coding sequence do not always result in protein dysfunction, and the interpretation of the severity of a given mutation often relies on *in silico* predictions [151]. Once the identity of the functional variant is established, its biological contribution to disorder needs to be clarified, a step which is usually achieved in animal models. The development of translational targets is not always clear, even for fully penetrant monogenic mutations with clear functional effects [152,153]. These limitations reflect the importance of considering genetic variants within relevant pathways and networks and are likely to be exacerbated for complex disorders.

18. Biology beyond the *p*-Values

The association of variants in the *PCSK6* gene and relative hand skill introduced earlier acts to illustrate the importance of taking statistical associations forward with molecular investigations of the mechanism. As described above, the *PCSK6* association with relative hand skill appears to confer specific effects in individuals with dyslexia. Given that the prevalence of left-handers is not higher among individuals with dyslexia compared to controls, at first glance, it appears difficult to explain this specificity. The very fact that this handedness measure was collected in individuals with dyslexia stems from a long-sought link between laterality and neurodevelopmental disorders [154]. Language is a lateralized behavior under the control of one specialized cerebral hemisphere (the left one, in most cases), and it has been suggested that language-related disorders might therefore be linked to handedness, which is the most obvious lateralized behavior. For many decades, researchers have looked with little success for a link between handedness and different psychiatric disorders, mainly by assessing the frequency of left-handers in patient cohorts [155]. Whilst largely inconclusive, an increased frequency of left-handers has been reported for schizophrenia [156]. Furthermore, imaging studies have shown that disorders, such as dyslexia and SLI, present with atypical and weaker cerebral lateralization [25]. Although, there is some controversy over whether these effects are causative or consequential [157].

19. Converging Evidence towards Biological Pathways

More direct evidence in support of the dyslexia-specific *PCSK6* association comes from very recent functional studies. We have already eluded to the apparently connected functional role of dyslexia candidate genes. Functional studies and *in vivo* techniques have demonstrated that several of the dyslexia candidates described above (*KIAA0319*, *DCDC2* and *DYX1C1*) are involved in early phases of fetal brain development and, in particular, in neuronal migration processes. In addition, *ROBO1* is a neuronal axon guidance receptor that is also important for cortex development. The importance of neuronal migration processes in dyslexia is thus now generally accepted in the literature [158]. However, it is not clear which exact mechanisms these candidate genes mediate. More recently, a role for cilia function and development has also been suggested as a common biological pathway between dyslexia candidate genes [159]. Cilia are essential in the establishment of left/right axis determination in the first weeks of embryogenesis by the activation of the NODAL pathway on the left side of the embryo. Mutations in genes controlling cilia function lead to a class of conditions often characterized by laterality defects (ciliopathies). *DCDC2*, *DYX1C1* and *KIAA0319* have recently been reported to form a novel co-expressed module in ciliated cells [79]. *Dcdc2* has been implicated in the control of cilia length [72], and disruption of *Dyx1c1* in mice and zebrafish [77,78] leads to laterality defects and impairments in cilia function, resembling that observed in primary ciliary dyskinesia (PCD). In humans, *DYX1C1* mutations have been identified in 12 families with PCD [77]. *KIAA0319* is characterized by five polycystic kidney disease (PKD) domains [160] typically found in proteins (e.g., *PKD1* and *PKD2*) playing key roles in cilia function [161].

Cilial structures control many processes. In particular, they are important in neuronal migration, where they play a guiding role during cortical development. It has therefore been suggested that the roles of cilia in neuronal migration may be directly implicated in leading to cortical defects, which are at the basis of cognitive deficits in neurodevelopmental disorders [162]. Taken together, these observations support a possible interaction between biological pathways controlling the establishment of left/right structural asymmetries and neuronal migration early in development with genes implicated in dyslexia [159]. These interactions could be mediated by concomitant actions in controlling cilia function and could explain the *PCSK6* association observed specifically in individuals with dyslexia. Furthermore, these data suggest for the first time that the mechanisms that control left/right body asymmetries maybe also be relevant in establishing functional brain asymmetries, contrary to previous evidence [163]. Incorrect reference order

20. Conclusions

We have discussed the various factors that can impact upon the relative success of genetic investigations in general and highlighted those factors that may be particularly pertinent to the investigation of speech and language disorders. While GWA studies have often been criticized for their high economic cost and little clinical benefits, they have contributed enormously to our understanding of human genome variation and appreciation of the complexity behind human biology and genetic disorders. Nonetheless, the application of these technologies to the study of genetic contributions to language and reading has not progressed as fast as other complex traits. One obvious limitation to the study of speech and language is the challenge of defining cases and ascertaining homogeneous cohorts with phenotypic measures that are universally relevant at the biological level. The translational application of genetic research to complex disorders will require the integration of large genomic datasets, functional genomic screenings and basic research projects aimed at studying the human brain. We know that genes and proteins do not act in isolation, and their functions differ between individuals, tissues, environments and over time; yet, we still consider risk variants as independent entities with fixed effects. Although we must accept that it is not currently possible to simultaneously consider all these effects within a single model, perhaps it is time to question the adherence to clinically defined strata in the genetic study of developmental disorders and, instead, promote the acceptance of cross-disorder studies that fully consider comorbidities across clinical symptoms and environmental factors, as well as regulatory effects and epigenetics. This reiterates the need for multidisciplinary collaborations to enable an increase in sample sizes, while maintaining detailed phenotypic assessments that ultimately will inform the definition of diagnostic criteria. In terms of future directions, it is difficult to establish whether resources would be better spent collecting large cohorts that meet a superficial categorical cutoff or, alternatively, in assembling smaller cohorts characterized across a range of detailed developmental phenotypes. What is clear is that the definition of universal guidelines will facilitate the coordinated collection of uniform and international cohorts, easing the burden of downstream analyses. Although larger collaborations are needed, a meta-analysis of existing resources would be a good starting point to establish such efforts. The reporting of complete data sets, even when events are considered non-significant or uninteresting in isolation, will also be crucial to the planning of future collections. Ultimately, functional evidence is necessary to definitely prove the downstream effects of genetic variants and to understand the biology of underlying disorders and neurodevelopment. Despite the lack of advances from genome-wide screening in speech and language disorders, the functional assessment of candidate genes has allowed considerable progress in the identification of specific biological processes that may be important in these phenotypes. In particular, neuronal migration and ciliogenesis have been highlighted as two, perhaps related, processes that may play a role in developmental dyslexia. These encouraging findings demonstrate the importance of the systematic integration of functional studies and genetic association or sequencing studies.

Acknowledgments

We would like to thank all the families, professionals and individuals who participated in our research. Dianne Newbury is a Medical Research Council (MRC) Career Development Fellow and a Junior Research Fellow at St. John's College, University of Oxford. The work of the Newbury lab is funded by the Medical Research Council (G1000569/1 and MR/J003719/1). Anthony P. Monaco is the President of Tufts University and a Visiting Professor at the Wellcome Trust Centre for Human Genetics. The work of the Monaco lab is funded by a program grant from the Wellcome Trust (092071). Silvia Paracchini is a Royal Society University Research Fellow. The work of the Wellcome Trust Centre in Oxford is supported by the Wellcome Trust (090532/Z/09/Z).

Author Contributions

All authors conceptualized and wrote this review.

References

1. Harel, S.; Greenstein, Y.; Kramer, U.; Yifat, R.; Samuel, E.; Nevo, Y.; Leitner, Y.; Kutai, M.; Fattal, A.; Shinnar, S. Clinical characteristics of children referred to a child development center for evaluation of speech, language, and communication disorders. *Pediatr. Neurol.* **1996**, *15*, 305–311.
2. American-Psychiatric-Association. *Diagnostic and Statistical Manual of Mental Disorders*, 4th ed. (dsm-iv); American Psychiatric Association: Washington, DC, USA, 1994.
3. Bishop, D.V. The role of genes in the etiology of specific language impairment. *J. Commun. Disord.* **2002**, *35*, 311–328.
4. Hayiou-Thomas, M.E. Genetic and environmental influences on early speech, language and literacy development. *J. Commun. Disord.* **2008**, *41*, 397–408.
5. DeFries, J.C.; Fulker, D.W.; LaBuda, M.C. Evidence for a genetic aetiology in reading disability of twins. *Nature* **1987**, *329*, 537–539.
6. Bishop, D.V.M. *Uncommon Understanding: Development and Disorders of Language Comprehension in Children*; Psychology Press: Hove, UK, 1997.
7. American Psychiatric Association. *Diagnostic and Statistical Manual of Mental Disorders: DSM-5*, 5th ed.; American Psychiatric Publishing: Arlington, VA, USA, 2013.
8. Conti-Ramsden, G.; Botting, N. Characteristics of children attending language units in england: A national study of 7-year-olds. *Int. J. Lang. Commun. Disord.* **1999**, *34*, 359–366.
9. Snowling, M.J. Phonemic deficits in developmental dyslexia. *Psychol. Res.* **1981**, *43*, 219–234.
10. Ramus, F. Neurobiology of dyslexia: A reinterpretation of the data. *Trends Neurosci.* **2004**, *27*, 720–726.
11. Peterson, R.L.; Pennington, B.F. Developmental dyslexia. *Lancet* **2012**, *379*, 1997–2007.
12. Gathercole, S.E.; Baddeley, A.D. Phonological memory deficits in language disordered children: Is there a causal connection? *J. Mem. Lang.* **1990**, *29*, 336–360.
13. Tallal, P.; Piercy, M. Defects of non-verbal auditory perception in children with developmental aphasia. *Nature* **1973**, *241*, 468–469.
14. Gopnik, M. Feature-blind grammar and dysphagia. *Nature* **1990**, *344*, 715.
15. Bishop, D.V. The underlying nature of specific language impairment. *J. Child Psychol. Psychiatry Allied Discip.* **1992**, *33*, 3–66.
16. Abbeduto, L.; Brady, N.; Kover, S.T. Language development and fragile x syndrome: Profiles, syndrome-specificity, and within-syndrome differences. *Ment. Retard. Dev. Disabil. Res. Rev.* **2007**, *13*, 36–46.
17. Bishop, D.V.; Snowling, M.J. Developmental dyslexia and specific language impairment: Same or different? *Psychol. Bull.* **2004**, *130*, 858–886.
18. McArthur, G.M.; Hogben, J.H.; Edwards, V.T.; Heath, S.M.; Mengler, E.D. On the "specifics" of

specific reading disability and specific language impairment. *J. Child Psychol. Psychiatry Allied Discip.* **2000**, *41*, 869–874.

19. Catts, H.W.; Adlof, S.M.; Hogan, T.P.; Weismer, S.E. Are specific language impairment and dyslexia distinct disorders? *J. Speech Lang. Hear. Res.* **2005**, *48*, 1378–1396.
20. Pieters, S.; de Block, K.; Scheiris, J.; Eyssen, M.; Desoete, A.; Deboutte, D.; van Waelvelde, H.; Roeyers, H. How common are motor problems in children with a developmental disorder: Rule or exception? *Child Care Health Dev.* **2012**, *38*, 139–145.
21. Flapper, B.C.; Schoemaker, M.M. Developmental coordination disorder in children with specific language impairment: Co-morbidity and impact on quality of life. *Res. Dev. Disabil.* **2013**, *34*, 756–763.
22. Gooch, D.; Hulme, C.; Nash, H.M.; Snowling, M.J. Comorbidities in preschool children at family risk of dyslexia. *J. Child Psychol. Psychiatry Allied Discip.* **2013**, *55*, 237–246.
23. Bloom, J.S.; Garcia-Barrera, M.A.; Miller, C.J.; Miller, S.R.; Hynd, G.W. Planum temporale morphology in children with developmental dyslexia. *Neuropsychologia* **2013**, *51*, 1684–1692.
24. Johnson, B.W.; McArthur, G.; Hautus, M.; Reid, M.; Brock, J.; Castles, A.; Crain, S. Lateralized auditory brain function in children with normal reading ability and in children with dyslexia. *Neuropsychologia* **2013**, *51*, 633–641.
25. Badcock, N.A.; Bishop, D.V.; Hardiman, M.J.; Barry, J.G.; Watkins, K.E. Co-localisation of abnormal brain structure and function in specific language impairment. *Brain Lang.* **2012**, *120*, 310–320.
26. Stoodley, C.J.; Fawcett, A.J.; Nicolson, R.I.; Stein, J.F. Impaired balancing ability in dyslexic children. *Exp. Brain Res.* **2005**, *167*, 370–380.
27. Budden, S. Clinical variability in early speech-language development in females with rett syndrome. *Dev. Med. Child Neurol.* **2012**, *54*, 392–393.
28. Roizen, N.J.; Antshel, K.M.; Fremont, W.; AbdulSabur, N.; Higgins, A.M.; Shprintzen, R.J.; Kates, W.R. 22q11.2ds deletion syndrome: Developmental milestones in infants and toddlers. *J. Dev. Behav. Pediatr.* **2007**, *28*, 119–124.
29. Starke, M.; Albertsson Wikland, K.; Moller, A. Parents' descriptions of development and problems associated with infants with turner syndrome: A retrospective study. *J. Paediatr. Child Health* **2003**, *39*, 293–298.
30. Laws, G.; Bishop, D.V. Verbal deficits in down's syndrome and specific language impairment: A comparison. *Int. J. Lang. Commun. Disord.* **2004**, *39*, 423–451.
31. Laws, G.; Bishop, D.V. A comparison of language abilities in adolescents with down syndrome and children with specific language impairment. *J. Speech Lang. Hear. Res.* **2003**, *46*, 1324–1339.
32. Simpson, N.H.; Addis, L.; Brandler, W.M.; Slonims, V.; Clark, A.; Watson, J.; Scerri, T.S.; Hennessy, E.R.; Bolton, P.F.; Conti-Ramsden, G.; *et al.* Increased prevalence of sex chromosome aneuploidies in specific language impairment and dyslexia. *Dev. Med. Child Neurol.* **2013**, doi:10.1111/dmcn.12294.
33. Lai, C.S.; Fisher, S.E.; Hurst, J.A.; Vargha-Khadem, F.; Monaco, A.P. A forkhead-domain gene is mutated in a severe speech and language disorder. *Nature* **2001**, *413*, 519–523.

34. Fisher, S.E.; Scharff, C. Foxp2 as a molecular window into speech and language. *Trends Genet.* **2009**, *25*, 166–177.
35. Kang, C.; Riazuddin, S.; Mundorff, J.; Krasnewich, D.; Friedman, P.; Mullikin, J.C.; Drayna, D. Mutations in the lysosomal enzyme-targeting pathway and persistent stuttering. *N. Engl. J. Med.* **2010**, *362*, 677–685.
36. Vernes, S.C.; Nicod, J.; Elahi, F.M.; Coventry, J.A.; Kenny, N.; Coupe, A.M.; Bird, L.E.; Davies, K.E.; Fisher, S.E. Functional genetic analysis of mutations implicated in a human speech and language disorder. *Hum. Mol. Genet.* **2006**, *15*, 3154–3167.
37. Vernes, S.C.; Newbury, D.F.; Abrahams, B.S.; Winchester, L.; Nicod, J.; Groszer, M.; Alarcon, M.; Oliver, P.L.; Davies, K.E.; Geschwind, D.H.; *et al.* A functional genetic link between distinct developmental language disorders. *N. Engl. J. Med.* **2008**, *359*, 2337–2345.
38. Alarcon, M.; Abrahams, B.S.; Stone, J.L.; Duvall, J.A.; Perederiy, J.V.; Bomar, J.M.; Sebat, J.; Wigler, M.; Martin, C.L.; Ledbetter, D.H.; *et al.* Linkage, association, and gene-expression analyses identify cntnap2 as an autism-susceptibility gene. *Am. J. Hum. Genet.* **2008**, *82*, 150–159.
39. Arking, D.E.; Cutler, D.J.; Brune, C.W.; Teslovich, T.M.; West, K.; Ikeda, M.; Rea, A.; Guy, M.; Lin, S.; Cook, E.H.; *et al.* A common genetic variant in the neurexin superfamily member cntnap2 increases familial risk of autism. *Am. J. Hum. Genet.* **2008**, *82*, 160–164.
40. Friedman, J.I.; Vrijenhoek, T.; Markx, S.; Janssen, I.M.; van der Vliet, W.A.; Faas, B.H.; Knoers, N.V.; Cahn, W.; Kahn, R.S.; Edelmann, L.; *et al.* Cntnap2 gene dosage variation is associated with schizophrenia and epilepsy. *Mol. Psychiatry* **2008**, *13*, 261–266.
41. Allison, D.B.; Thiel, B.; St Jean, P.; Elston, R.C.; Infante, M.C.; Schork, N.J. Multiple phenotype modeling in gene-mapping studies of quantitative traits: Power advantages. *Am. J. Hum. Genet.* **1998**, *63*, 1190–1201.
42. Whitehouse, A.J.; Bishop, D.V.; Ang, Q.W.; Pennell, C.E.; Fisher, S.E. Cntnap2 variants affect early language development in the general population. *Genes Brain Behav.* **2011**, *10*, 451–456.
43. Blair, D.R.; Lyttle, C.S.; Mortensen, J.M.; Bearden, C.F.; Jensen, A.B.; Khiabanian, H.; Melamed, R.; Rabadan, R.; Bernstam, E.V.; Brunak, S.; *et al.* A nondegenerate code of deleterious variants in mendelian loci contributes to complex disease risk. *Cell* **2013**, *155*, 70–80.
44. Tallal, P.; Ross, R.; Curtiss, S. Familial aggregation in specific language impairment. *J. Speech Hear. Disord.* **1989**, *54*, 167–173.
45. Almasy, L.; Blangero, J. Human qtl linkage mapping. *Genetica* **2009**, *136*, 333–340.
46. Grigorenko, E.L.; Wood, F.B.; Meyer, M.S.; Hart, L.A.; Speed, W.C.; Shuster, A.; Pauls, D.L. Susceptibility loci for distinct components of developmental dyslexia on chromosomes 6 and 15. *Am. J. Hum. Genet.* **1997**, *60*, 27–39.
47. Schulte-Korne, G.; Grimm, T.; Nothen, M.M.; Muller-Myhsok, B.; Cichon, S.; Vogt, I.R.; Propping, P.; Remschmidt, H. Evidence for linkage of spelling disability to chromosome 15. *Am. J. Hum. Genet.* **1998**, *63*, 279–282.
48. Cardon, L.R.; Smith, S.D.; Fulker, D.W.; Kimberling, W.J.; Pennington, B.F.; DeFries, J.C. Quantitative trait locus for reading disability on chromosome 6. *Science* **1994**, *266*, 276–279.
49. Fisher, S.E.; Marlow, A.J.; Lamb, J.; Maestrini, E.; Williams, D.F.; Richardson, A.J.; Weeks, D.E.; Stein, J.F.; Monaco, A.P. A quantitative-trait locus on chromosome 6p influences different aspects of developmental dyslexia. *Am. J. Hum. Genet.* **1999**, *64*, 146–156.

50. Gayan, J.; Smith, S.D.; Cherny, S.S.; Cardon, L.R.; Fulker, D.W.; Brower, A.M.; Olson, R.K.; Pennington, B.F.; DeFries, J.C. Quantitative-trait locus for specific language and reading deficits on chromosome 6p. *Am. J. Hum. Genet.* **1999**, *64*, 157–164.
51. Kaplan, D.E.; Gayan, J.; Ahn, J.; Won, T.W.; Pauls, D.; Olson, R.K.; DeFries, J.C.; Wood, F.; Pennington, B.F.; Page, G.P.; *et al.* Evidence for linkage and association with reading disability on 6p21.3–22. *Am. J. Hum. Genet.* **2002**, *70*, 1287–1298.
52. Fagerheim, T.; Raeymaekers, P.; Tonnessen, F.E.; Pedersen, M.; Tranebjaerg, L.; Lubs, H.A. A new gene (dyx3) for dyslexia is located on chromosome 2. *J. Med. Genet.* **1999**, *36*, 664–669.
53. Francks, C.; Fisher, S.E.; Olson, R.K.; Pennington, B.F.; Smith, S.D.; DeFries, J.C.; Monaco, A.P. Fine mapping of the chromosome 2p12–16 dyslexia susceptibility locus: Quantitative association analysis and positional candidate genes sema4f and otx1. *Psychiatr. Genet.* **2002**, *12*, 35–41.
54. Hannula-Jouppi, K.; Kaminen-Ahola, N.; Taipale, M.; Eklund, R.; Nopola-Hemmi, J.; Kaariainen, H.; Kere, J. The axon guidance receptor gene robo1 is a candidate gene for developmental dyslexia. *PLoS Genet.* **2005**, *1*, e50.
55. Nopola-Hemmi, J.; Myllyluoma, B.; Haltia, T.; Taipale, M.; Ollikainen, V.; Ahonen, T.; Voutilainen, A.; Kere, J.; Widen, E. A dominant gene for developmental dyslexia on chromosome 3. *J. Med. Genet.* **2001**, *38*, 658–664.
56. Fisher, S.E.; Francks, C.; Marlow, A.J.; MacPhie, I.L.; Newbury, D.F.; Cardon, L.R.; Ishikawa-Brush, Y.; Richardson, A.J.; Talcott, J.B.; Gayan, J.; *et al.* Independent genome-wide scans identify a chromosome 18 quantitative-trait locus influencing dyslexia. *Nat. Genet.* **2002**, *30*, 86–91.
57. Hsiung, G.Y.; Kaplan, B.J.; Petryshen, T.L.; Lu, S.; Field, L.L. A dyslexia susceptibility locus (dyx7) linked to dopamine d4 receptor (drd4) region on chromosome 11p15.5. *Am. J. Med. Genet. Part B Neuropsychiatr. Genet.* **2004**, *125B*, 112–119.
58. Grigorenko, E.L.; Wood, F.B.; Meyer, M.S.; Pauls, J.E.; Hart, L.A.; Pauls, D.L. Linkage studies suggest a possible locus for developmental dyslexia on chromosome 1p. *Am. J. Med. Genet.* **2001**, *105*, 120–129.
59. Rabin, M.; Wen, X.L.; Hepburn, M.; Lubs, H.A.; Feldman, E.; Duara, R. Suggestive linkage of developmental dyslexia to chromosome 1p34-p36. *Lancet* **1993**, *342*, 178.
60. De Kovel, C.G.; Franke, B.; Hol, F.A.; Lebrec, J.J.; Maassen, B.; Brunner, H.; Padberg, G.W.; Platko, J.; Pauls, D. Confirmation of dyslexia susceptibility loci on chromosomes 1p and 2p, but not 6p in a dutch sib-pair collection. *Am. J. Med. Genet. Part B Neuropsychiatr. Genet.* **2008**, *147*, 294–300.
61. Meng, H.; Smith, S.D.; Hager, K.; Held, M.; Liu, J.; Olson, R.K.; Pennington, B.F.; DeFries, J.C.; Gelernter, J.; O'Reilly-Pol, T.; *et al.* Dcdc2 is associated with reading disability and modulates neuronal development in the brain. *Proc. Natl. Acad. Sci. USA* **2005**, *102*, 17053–17058.
62. Francks, C.; Paracchini, S.; Smith, S.D.; Richardson, A.J.; Scerri, T.S.; Cardon, L.R.; Marlow, A.J.; MacPhie, I.L.; Walter, J.; Pennington, B.F.; *et al.* A 77-kilobase region of chromosome 6p22.2 is associated with dyslexia in families from the united kingdom and from the united states. *Am. J. Hum. Genet.* **2004**, *75*, 1046–1058.
63. Cope, N.; Harold, D.; Hill, G.; Moskvina, V.; Stevenson, J.; Holmans, P.; Owen, M.J.; O'Donovan, M.C.; Williams, J. Strong evidence that kiaa0319 on chromosome 6p is a susceptibility gene for developmental dyslexia. *Am. J. Hum. Genet.* **2005**, *76*, 581–591.

64. Paracchini, S.; Thomas, A.; Castro, S.; Lai, C.; Paramasivam, M.; Wang, Y.; Keating, B.J.; Taylor, J.M.; Hacking, D.F.; Scerri, T.; *et al.* The chromosome 6p22 haplotype associated with dyslexia reduces the expression of kiaa0319, a novel gene involved in neuronal migration. *Hum. Mol. Genet.* **2006**, *15*, 1659–1666.
65. Taipale, M.; Kaminen, N.; Nopola-Hemmi, J.; Haltia, T.; Myllyluoma, B.; Lyytinen, H.; Muller, K.; Kaaranen, M.; Lindsberg, P.J.; Hannula-Jouppi, K.; *et al.* A candidate gene for developmental dyslexia encodes a nuclear tetratricopeptide repeat domain protein dynamically regulated in brain. *Proc. Natl. Acad. Sci. USA* **2003**, *100*, 11553–11558.
66. Bates, T.C.; Lind, P.A.; Luciano, M.; Montgomery, G.W.; Martin, N.G.; Wright, M.J. Dyslexia and dyx1c1: Deficits in reading and spelling associated with a missense mutation. *Mol. Psychiatry* **2009**, *15*, 1190–1196.
67. Anthoni, H.; Zucchelli, M.; Matsson, H.; Muller-Myhsok, B.; Fransson, I.; Schumacher, J.; Massinen, S.; Onkamo, P.; Warnke, A.; Griesemann, H.; *et al.* A locus on 2p12 containing the co-regulated mrpl19 and c2orf3 genes is associated to dyslexia. *Hum. Mol. Genet.* **2007**, *16*, 667–677.
68. Anthoni, H.; Sucheston, L.E.; Lewis, B.A.; Tapia-Paez, I.; Fan, X.; Zucchelli, M.; Taipale, M.; Stein, C.M.; Hokkanen, M.E.; Castren, E.; *et al.* The aromatase gene cyp19a1: Several genetic and functional lines of evidence supporting a role in reading, speech and language. *Behav. Genet.* **2012**, *42*, 509–527.
69. Scerri, T.S.; Schulte-Korne, G. Genetics of developmental dyslexia. *Eur. Child Adolesc. Psychiatry* **2010**, *19*, 179–197.
70. Currier, T.A.; Etchegaray, M.A.; Haight, J.L.; Galaburda, A.M.; Rosen, G.D. The effects of embryonic knockdown of the candidate dyslexia susceptibility gene homologue dyx1c1 on the distribution of gabaergic neurons in the cerebral cortex. *Neuroscience* **2011**, *172*, 535–546.
71. Tammimies, K.; Vitezic, M.; Matsson, H.; Le Guyader, S.; Burglin, T.R.; Ohman, T.; Stromblad, S.; Daub, C.O.; Nyman, T.A.; Kere, J.; *et al.* Molecular networks of dyx1c1 gene show connection to neuronal migration genes and cytoskeletal proteins. *Biol. Psychiatry* **2013**, *73*, 583–590.
72. Massinen, S.; Hokkanen, M.E.; Matsson, H.; Tammimies, K.; Tapia-Paez, I.; Dahlstrom-Heuser, V.; Kuja-Panula, J.; Burghoorn, J.; Jeppsson, K.E.; Swoboda, P.; *et al.* Increased expression of the dyslexia candidate gene dcdc2 affects length and signaling of primary cilia in neurons. *PLoS One* **2011**, *6*, e20580.
73. Adler, W.T.; Platt, M.P.; Mehlhorn, A.J.; Haight, J.L.; Currier, T.A.; Etchegaray, M.A.; Galaburda, A.M.; Rosen, G.D. Position of neocortical neurons transfected at different gestational ages with shrna targeted against candidate dyslexia susceptibility genes. *PLoS One* **2013**, *8*, e65179.
74. Platt, M.P.; Adler, W.T.; Mehlhorn, A.J.; Johnson, G.C.; Wright, K.A.; Choi, R.T.; Tsang, W.H.; Poon, M.W.; Yeung, S.Y.; Waye, M.M.; *et al.* Embryonic disruption of the candidate dyslexia susceptibility gene homolog kiaa0319-like results in neuronal migration disorders. *Neuroscience* **2013**, *248C*, 585–593.
75. Wang, Y.; Yin, X.; Rosen, G.; Gabel, L.; Guadiana, S.M.; Sarkisian, M.R.; Galaburda, A.M.; Loturco, J.J. Dcdc2 knockout mice display exacerbated developmental disruptions following knockdown of doublecortin. *Neuroscience* **2011**, *190*, 398–408.

76. Szalkowski, C.E.; Fiondella, C.G.; Galaburda, A.M.; Rosen, G.D.; Loturco, J.J.; Fitch, R.H. Neocortical disruption and behavioral impairments in rats following in utero rnai of candidate dyslexia risk gene kiaa0319. *Int. J. Dev. Neurosci.* **2012**, *30*, 293–302.
77. Tarkar, A.; Loges, N.T.; Slagle, C.E.; Francis, R.; Dougherty, G.W.; Tamayo, J.V.; Shook, B.; Cantino, M.; Schwartz, D.; Jahnke, C.; *et al.* Dyx1c1 is required for axonemal dynein assembly and ciliary motility. *Nat. Genet.* **2013**, *45*, 995–1003.
78. Chandrasekar, G.; Vesterlund, L.; Hultenby, K.; Tapia-Paez, I.; Kere, J. The zebrafish orthologue of the dyslexia candidate gene dyx1c1 is essential for cilia growth and function. *PLoS One* **2013**, *8*, e63123.
79. Ivliev, A.E.; 't Hoen, P.A.C.; van Roon-Mom, W.M.; Peters, D.J.M.; Sergeeva, M.G. Exploring the transcriptome of ciliated cells using in silico dissection of human tissues. *PLoS One* **2012**, *7*, e35618.
80. Villanueva, P.; Newbury, D.F.; Jara, L.; de Barbieri, Z.; Mirza, G.; Palomino, H.M.; Fernandez, M.A.; Cazier, J.B.; Monaco, A.P.; Palomino, H. Genome-wide analysis of genetic susceptibility to language impairment in an isolated chilean population. *Eur. J. Hum. Genet.* **2011**, *19*, 687–695.
81. Bartlett, C.W.; Flax, J.F.; Logue, M.W.; Smith, B.J.; Vieland, V.J.; Tallal, P.; Brzustowicz, L.M. Examination of potential overlap in autism and language loci on chromosomes 2, 7, and 13 in two independent samples ascertained for specific language impairment. *Hum. Hered.* **2004**, *57*, 10–20.
82. Bartlett, C.W.; Flax, J.F.; Logue, M.W.; Vieland, V.J.; Bassett, A.S.; Tallal, P.; Brzustowicz, L.M. A major susceptibility locus for specific language impairment is located on 13q21. *Am. J. Hum. Genet.* **2002**, *71*, 45–55.
83. SLI Consortium (SLIC). Highly significant linkage to the sli1 locus in an expanded sample of individuals affected by specific language impairment. *Am. J. Hum. Genet.* **2004**, *74*, 1225–1238.
84. SLI Consortium. A genomewide scan identifies two novel loci involved in specific language impairment. *Am. J. Hum. Genet.* **2002**, *70*, 384–398.
85. Monaco, A.P. Multivariate linkage analysis of specific language impairment (sli). *Ann. Hum. Genet.* **2007**, *71*, 660–673.
86. Falcaro, M.; Pickles, A.; Newbury, D.F.; Addis, L.; Banfield, E.; Fisher, S.E.; Monaco, A.P.; Simkin, Z.; Conti-Ramsden, G. Genetic and phenotypic effects of phonological short-term memory and grammatical morphology in specific language impairment. *Genes Brain Behav.* **2008**, *7*, 393–402.
87. Newbury, D.F.; Winchester, L.; Addis, L.; Paracchini, S.; Buckingham, L.L.; Clark, A.; Cohen, W.; Cowie, H.; Dworzynski, K.; Everitt, A.; *et al.* Cmip and atp2c2 modulate phonological short-term memory in language impairment. *Am. J. Hum. Genet.* **2009**, *85*, 264–272.
88. Ceroni, F.; Simpson, N.H.; Francks, C.; Baird, G.; Conti-Ramsden, G.; Clark, A.; Bolton, P.F.; Hennessy, E.R.; Donnelly, P.; Bentley, D.R.; *et al.* Homozygous microdeletion of exon 5 in znf277 in a girl with specific language impairment. *Eur. J. Hum. Genet.* **2014**, doi:10.1038/ejhg.2014.4.
89. Nudel, R.; Simpson, N.H.; Baird, G.; O'Hare, A.; Conti-Ramsden, G.; Bolton, P.F.; Hennessy, E.R.; Consortium, S.L.I.; Monaco, A.P.; Knight, J.C.; *et al.* Associations of hla alleles with specific language impairment. *J. Neurodev. Disord.* **2014**, doi:10.1186/1866-1955-6-1.
90. Zollner, S.; Pritchard, J.K. Overcoming the winner's curse: Estimating penetrance parameters from case-control data. *Am. J. Hum. Genet.* **2007**, *80*, 605–615.

91. The International Hapmap Project. Available online: http://www.hapmap.org/ (accessed on 31 March 2014).
92. 1000 Genomes: A Deep Catalogue of Human Genetic Variation. Available online: http://www.1000genomes.org/ (accessed on 31 March 2014).
93. Visscher, P.M.; Brown, M.A.; McCarthy, M.I.; Yang, J. Five years of GWAS discovery. *Am. J. Hum. Genet.* **2012**, *90*, 7–24.
94. Meaburn, E.L.; Harlaar, N.; Craig, I.W.; Schalkwyk, L.C.; Plomin, R. Quantitative trait locus association scan of early reading disability and ability using pooled DNA and 100k snp microarrays in a sample of 5760 children. *Mol. Psychiatry* **2008**, *13*, 729–740.
95. Roeske, D.; Ludwig, K.U.; Neuhoff, N.; Becker, J.; Bartling, J.; Bruder, J.; Brockschmidt, F.F.; Warnke, A.; Remschmidt, H.; Hoffmann, P.; *et al.* First genome-wide association scan on neurophysiological endophenotypes points to trans-regulation effects on slc2a3 in dyslexic children. *Mol. Psychiatry* **2011**, *16*, 97–107.
96. Risch, N.; Merikangas, K. The future of genetic studies of complex human diseases. *Science* **1996**, *273*, 1516–1517.
97. Field, L.L.; Shumansky, K.; Ryan, J.; Truong, D.; Swiergala, E.; Kaplan, B.J. Dense-map genome scan for dyslexia supports loci at 4q13, 16p12, 17q22; suggests novel locus at 7q36. *Genes Brain Behav.* **2013**, *12*, 56–69.
98. Luciano, M.; Evans, D.M.; Hansell, N.K.; Medland, S.E.; Montgomery, G.W.; Martin, N.G.; Wright, M.J.; Bates, T.C. A genome-wide association study for reading and language abilities in two population cohorts. *Genes Brain Behav.* **2013**, *12*, 645–652.
99. Eicher, J.D.; Powers, N.R.; Miller, L.L.; Akshoomoff, N.; Amaral, D.G.; Bloss, C.S.; Libiger, O.; Schork, N.J.; Darst, B.F.; Casey, B.J.; *et al.* Genome-wide association study of shared components of reading disability and language impairment. *Genes Brain Behav.* **2013**, *12*, 792–801.
100. Mott, R.; Yuan, W.; Kaisaki, P.; Gan, X.; Cleak, J.; Edwards, A.; Baud, A.; Flint, J. The architecture of parent-of-origin effects in mice. *Cell* **2014**, *156*, 332–342.
101. Nudel, R.; Simpson, N.H.; Baird, G.; O'Hare, A.; Conti-Ramsden, G.; Bolton, P.F.; Hennessy, E.R.; SLIC; Ring, S.M.; Davey Smith, G.; *et al.* Genome-wide association analyses of child genotype effects and parent-of-origin effects in specific language impairment (sli). *Genes Brain Behav.* **2014**, doi:10.1111/gbb.12127.
102. McCarthy, M.I.; Hirschhorn, J.N. Genome-wide association studies: Potential next steps on a genetic journey. *Hum. Mol. Genet.* **2008**, *17*, R156–R165.
103. Leonard, L. Is specific language impairment a useful construct? In *Cambridge Monographs and Texts in Applied Psycholinguistics*; Rosenberg, S., Ed.; Cambridge University Press: Cambridge, UK, 1987; pp. 1–39.
104. Rescorla, L.; Ratner, N.B. Phonetic profiles of toddlers with specific expressive language impairment (sli-e). *J. Speech Hear. Res.* **1996**, *39*, 153–165.
105. Dollaghan, C.A. Taxometric analyses of specific language impairment in 3- and 4-year-old children. *J. Speech Lang. Hear. Res.* **2004**, *47*, 464–475.
106. Dollaghan, C.A. Taxometric analyses of specific language impairment in 6-year-old children. *J. Speech Lang. Hear. Res.* **2011**, *54*, 1361–1371.
107. Leonard, L.B. Specific langauage impairment across languages. *Child Dev. Perspect.* **2014**, *8*, 1–5.

108. Bishop, D.V. Is specific language impairment a valid diagnostic category? Genetic and psycholinguistic evidence. *Philos. Trans. R. Soc. Lond. Ser. B Biol. Sci.* **1994**, *346*, 105–111.

109. Hayiou-Thomas, M.E.; Oliver, B.; Plomin, R. Genetic influences on specific *versus* nonspecific language impairment in 4-year-old twins. *J. Learn. Disabil.* **2005**, *38*, 222–232.

110. Bishop, D.V.; Hayiou-Thomas, M.E. Heritability of specific language impairment depends on diagnostic criteria. *Genes Brain Behav.* **2008**, *7*, 365–372.

111. Eley, T.C.; Bishop, D.V.; Dale, P.S.; Oliver, B.; Petrill, S.A.; Price, T.S.; Purcell, S.; Saudino, K.J.; Simonoff, E.; Stevenson, J.; *et al.* Genetic and environmental origins of verbal and performance components of cognitive delay in 2-year-olds. *Dev. Psychol.* **1999**, *35*, 1122–1131.

112. Paracchini, S.; Steer, C.D.; Buckingham, L.L.; Morris, A.P.; Ring, S.; Scerri, T.; Stein, J.; Pembrey, M.E.; Ragoussis, J.; Golding, J.; *et al.* Association of the kiaa0319 dyslexia susceptibility gene with reading skills in the general population. *Am. J. Psychiatry* **2008**, *165*, 1576–1584.

113. Scerri, T.S.; Morris, A.P.; Buckingham, L.L.; Newbury, D.F.; Miller, L.L.; Monaco, A.P.; Bishop, D.V.; Paracchini, S. Dcdc2, kiaa0319 and cmip are associated with reading-related traits. *Biol. Psychiatry* **2011**, *70*, 237–245.

114. Newbury, D.F.; Paracchini, S.; Scerri, T.S.; Winchester, L.; Addis, L.; Richardson, A.J.; Walter, J.; Stein, J.F.; Talcott, J.B.; Monaco, A.P. Investigation of dyslexia and sli risk variants in reading- and language-impaired subjects. *Behav. Genet.* **2011**, *41*, 90–104.

115. Landerl, K.; Ramus, F.; Moll, K.; Lyytinen, H.; Leppanen, P.H.; Lohvansuu, K.; O'Donovan, M.; Williams, J.; Bartling, J.; Bruder, J.; *et al.* Predictors of developmental dyslexia in european orthographies with varying complexity. *J. Child Psychol. Psychiatry Allied Discip.* **2013**, *54*, 686–694.

116. Becker, J.; Czamara, D.; Scerri, T.S.; Ramus, F.; Csepe, V.; Talcott, J.B.; Stein, J.; Morris, A.; Ludwig, K.U.; Hoffmann, P.; *et al.* Genetic analysis of dyslexia candidate genes in the european cross-linguistic neurodys cohort. *Eur. J. Hum. Genet.* **2013**, doi:10.1038/ejhg.2013.199.

117. Schumacher, J.; Anthoni, H.; Dahdouh, F.; Konig, I.R.; Hillmer, A.M.; Kluck, N.; Manthey, M.; Plume, E.; Warnke, A.; Remschmidt, H.; *et al.* Strong genetic evidence of dcdc2 as a susceptibility gene for dyslexia. *Am. J. Hum. Genet.* **2006**, *78*, 52–62.

118. Bedore, L.M.; Leonard, L.B. Grammatical morphology deficits in spanish-speaking children with specific language impairment. *J. Speech Lang. Hear. Res.* **2001**, *44*, 905–924.

119. Dispaldro, M.; Leonard, L.B.; Deevy, P. Clinical markers in italian-speaking children with and without specific language impairment: A study of non-word and real word repetition as predictors of grammatical ability. *Int. J. Lang. Commun. Disord.* **2013**, *48*, 554–564.

120. Paradis, J.; Crago, M.; Genesee, F.; Rice, M. French-english bilingual children with sli: How do they compare with their monolingual peers? *J. Speech Lang. Hear. Res.* **2003**, *46*, 113–127.

121. Do, R.; Willer, C.J.; Schmidt, E.M.; Sengupta, S.; Gao, C.; Peloso, G.M.; Gustafsson, S.; Kanoni, S.; Ganna, A.; Chen, J.; *et al.* Common variants associated with plasma triglycerides and risk for coronary artery disease. *Nat. Genet.* **2013**, *45*, 1345–1352.

122. Global Lipids Genetics, C.; Willer, C.J.; Schmidt, E.M.; Sengupta, S.; Peloso, G.M.; Gustafsson, S.; Kanoni, S.; Ganna, A.; Chen, J.; Buchkovich, M.L.; *et al.* Discovery and refinement of loci associated with lipid levels. *Nat. Genet.* **2013**, *45*, 1274–1283.

123. Magi, R.; Manning, S.; Yousseif, A.; Pucci, A.; Santini, F.; Karra, E.; Querci, G.; Pelosini, C.; McCarthy, M.I.; Lindgren, C.M.; *et al.* Contribution of 32 GWAS-identified common variants to severe obesity in european adults referred for bariatric surgery. *PLoS One* **2013**, *8*, e70735.
124. Berndt, S.I.; Gustafsson, S.; Magi, R.; Ganna, A.; Wheeler, E.; Feitosa, M.F.; Justice, A.E.; Monda, K.L.; Croteau-Chonka, D.C.; Day, F.R.; *et al.* Genome-wide meta-analysis identifies 11 new loci for anthropometric traits and provides insights into genetic architecture. *Nat. Genet.* **2013**, *45*, 501–512.
125. Yang, J.; Loos, R.J.; Powell, J.E.; Medland, S.E.; Speliotes, E.K.; Chasman, D.I.; Rose, L.M.; Thorleifsson, G.; Steinthorsdottir, V.; Magi, R.; *et al.* Fto genotype is associated with phenotypic variability of body mass index. *Nature* **2012**, *490*, 267–272.
126. Lango, H.; Consortium, U.K.T.D.G.; Palmer, C.N.; Morris, A.D.; Zeggini, E.; Hattersley, A.T.; McCarthy, M.I.; Frayling, T.M.; Weedon, M.N. Assessing the combined impact of 18 common genetic variants of modest effect sizes on type 2 diabetes risk. *Diabetes* **2008**, *57*, 3129–3135.
127. Cross-Disorder Group of the Psychiatric Genomics Consortium; Genetic Risk Outcome of Psychosis (GROUP) Consortium. Identification of risk loci with shared effects on five major psychiatric disorders: A genome-wide analysis. *Lancet* **2013**, *381*, 1371–1379.
128. Cross-Disorder Group of the Psychiatric Genomics Consortium; Lee, S.H.; Ripke, S.; Neale, B.M.; Faraone, S.V.; Purcell, S.M.; Perlis, R.H.; Mowry, B.J.; Thapar, A.; Goddard, M.E.; *et al.* Genetic relationship between five psychiatric disorders estimated from genome-wide snps. *Nat. Genet.* **2013**, *45*, 984–994.
129. Scerri, T.S.; Brandler, W.M.; Paracchini, S.; Morris, A.P.; Ring, S.M.; Talcott, J.B.; Stein, J.; Monaco, A.P. *Pcsk6* is associated with handedness in individuals with dyslexia. *Hum. Mol. Genet.* **2011**, *20*, 608–614.
130. Brandler, W.M.; Morris, A.P.; Evans, D.M.; Scerri, T.S.; Kemp, J.P.; Timpson, N.J.; St Pourcain, B.; Smith, G.D.; Ring, S.M.; Stein, J.; *et al.* Common variants in left/right asymmetry genes and pathways are associated with relative hand skill. *PLoS Genet.* **2013**, *9*, e1003751.
131. Hamada, H.; Meno, C.; Watanabe, D.; Saijoh, Y. Establishment of vertebrate left-right asymmetry. *Nat. Rev. Genet.* **2002**, *3*, 103–113.
132. Ludwig, K.U.; Samann, P.; Alexander, M.; Becker, J.; Bruder, J.; Moll, K.; Spieler, D.; Czisch, M.; Warnke, A.; Docherty, S.J.; *et al.* A common variant in myosin-18b contributes to mathematical abilities in children with dyslexia and intraparietal sulcus variability in adults. *Transl. Psychiatry* **2013**, *3*, e229.
133. Medland, S.E.; Lindgren, C.M.; Magi, R.; Neale, B.M.; Albrecht, E.; Esko, T.; Evans, D.M.; Hottenga, J.J.; Ikram, M.A.; Mangino, M.; *et al.* Meta-Analysis of GWAS for Handedness: Results from the ENGAGE consortium. Available online: http://www.ashg.org/2009meeting/abstracts/fulltext/f21141.htm/ (accessed on 31 March 2014).
134. Eriksson, N.; Macpherson, J.M.; Tung, J.Y.; Hon, L.S.; Naughton, B.; Saxonov, S.; Avey, L.; Wojcicki, A.; Pe'er, I.; Mountain, J. Web-based, participant-driven studies yield novel genetic associations for common traits. *PLoS Genet.* **2010**, *6*, e1000993.
135. Iossifov, I.; Ronemus, M.; Levy, D.; Wang, Z.; Hakker, I.; Rosenbaum, J.; Yamrom, B.; Lee, Y.H.; Narzisi, G.; Leotta, A.; *et al.* *De novo* gene disruptions in children on the autistic spectrum. *Neuron* **2012**, *74*, 285–299.

136. O'Roak, B.J.; Deriziotis, P.; Lee, C.; Vives, L.; Schwartz, J.J.; Girirajan, S.; Karakoc, E.; Mackenzie, A.P.; Ng, S.B.; Baker, C.; *et al.* Exome sequencing in sporadic autism spectrum disorders identifies severe *de novo* mutations. *Nat. Genet.* **2011**, *43*, 585–589.

137. Neale, B.M.; Kou, Y.; Liu, L.; Ma'ayan, A.; Samocha, K.E.; Sabo, A.; Lin, C.F.; Stevens, C.; Wang, L.S.; Makarov, V.; *et al.* Patterns and rates of exonic *de novo* mutations in asd. *Nature* **2012**, *485*, 242–245.

138. O'Roak, B.J.; Vives, L.; Girirajan, S.; Karakoc, E.; Krumm, N.; Coe, B.P.; Levy, R.; Ko, A.; Lee, C.; Smith, J.D.; *et al.* Sporadic autism exomes reveal a highly interconnected protein network of *de novo* mutations. *Nature* **2012**, *485*, 246–250.

139. Vissers, L.E.; de Ligt, J.; Gilissen, C.; Janssen, I.; Steehouwer, M.; de Vries, P.; van Lier, B.; Arts, P.; Wieskamp, N.; del Rosario, M.; *et al.* A *de novo* paradigm for mental retardation. *Nat. Genet.* **2011**, *42*, 1109–1112.

140. Toma, C.; Torrico, B.; Hervas, A.; Valdes-Mas, R.; Tristan-Noguero, A.; Padillo, V.; Maristany, M.; Salgado, M.; Arenas, C.; Puente, X.S.; *et al.* Exome sequencing in multiplex autism families suggests a major role for heterozygous truncating mutations. *Mol. Psychiatry* **2013**, doi:10.1038/mp.2013.106.

141. Yu, T.W.; Chahrour, M.H.; Coulter, M.E.; Jiralerspong, S.; Okamura-Ikeda, K.; Ataman, B.; Schmitz-Abe, K.; Harmin, D.A.; Adli, M.; Malik, A.N.; *et al.* Using whole-exome sequencing to identify inherited causes of autism. *Neuron* **2013**, *77*, 259–273.

142. Worthey, E.A.; Raca, G.; Laffin, J.J.; Wilk, B.M.; Harris, J.M.; Jakielski, K.J.; Dimmock, D.P.; Strand, E.A.; Shriberg, L.D. Whole-exome sequencing supports genetic heterogeneity in childhood apraxia of speech. *J. Neurodev. Disord.* **2013**, *5*, 29.

143. Kong, A.; Frigge, M.L.; Masson, G.; Besenbacher, S.; Sulem, P.; Magnusson, G.; Gudjonsson, S.A.; Sigurdsson, A.; Jonasdottir, A.; Jonasdottir, A.; *et al.* Rate of *de novo* mutations and the importance of father's age to disease risk. *Nature* **2012**, *488*, 471–475.

144. Girirajan, S.; Rosenfeld, J.A.; Coe, B.P.; Parikh, S.; Friedman, N.; Goldstein, A.; Filipink, R.A.; McConnell, J.S.; Angle, B.; Meschino, W.S.; *et al.* Phenotypic heterogeneity of genomic disorders and rare copy-number variants. *N. Engl. J. Med.* **2012**, *367*, 1321–1331.

145. Girirajan, S.; Rosenfeld, J.A.; Cooper, G.M.; Antonacci, F.; Siswara, P.; Itsara, A.; Vives, L.; Walsh, T.; McCarthy, S.E.; Baker, C.; *et al.* A recurrent 16p12.1 microdeletion supports a two-hit model for severe developmental delay. *Nat. Genet.* **2010**, *42*, 203–209.

146. Cooper, G.M.; Coe, B.P.; Girirajan, S.; Rosenfeld, J.A.; Vu, T.H.; Baker, C.; Williams, C.; Stalker, H.; Hamid, R.; Hannig, V.; *et al.* A copy number variation morbidity map of developmental delay. *Nat. Genet.* **2011**, *43*, 838–846.

147. Leblond, C.S.; Jutta, H.; Delorme, R.; Proepper, C.; Betancur, C.; Huguet, G.; Konyukh, M.; Chaste, P.; Ey, E.; Rastam, M.; *et al.* Genetic and functional analyses of shank2 mutations suggest a multiple hit model of autism spectrum disorders. *PLoS Genet.* **2012**, *8*, e1002521.

148. Newbury, D.F.; Mari, F.; Sadighi Akha, E.; Macdermot, K.D.; Canitano, R.; Monaco, A.P.; Taylor, J.C.; Renieri, A.; Fisher, S.E.; Knight, S.J. Dual copy number variants involving 16p11 and 6q22 in a case of childhood apraxia of speech and pervasive developmental disorder. *Eur. J. Hum. Genet.* **2013**, *21*, 361–365.

149. Consortium, E.P.; Birney, E.; Stamatoyannopoulos, J.A.; Dutta, A.; Guigo, R.; Gingeras, T.R.; Margulies, E.H.; Weng, Z.; Snyder, M.; Dermitzakis, E.T.; *et al*. Identification and analysis of functional elements in 1% of the human genome by the encode pilot project. *Nature* **2007**, *447*, 799–816.
150. Lappalainen, T.; Sammeth, M.; Friedlander, M.R.; 't Hoen, P.A.; Monlong, J.; Rivas, M.A.; Gonzalez-Porta, M.; Kurbatova, N.; Griebel, T.; Ferreira, P.G.; *et al*. Transcriptome and genome sequencing uncovers functional variation in humans. *Nature* **2013**, *501*, 506–511.
151. Gratten, J.; Visscher, P.M.; Mowry, B.J.; Wray, N.R. Interpreting the role of *de novo* protein-coding mutations in neuropsychiatric disease. *Nat. Genet.* **2013**, *45*, 234–238.
152. Laederich, M.B.; Horton, W.A. Fgfr3 targeting strategies for achondroplasia. *Expert Rev. Mol. Med.* **2012**, *14*, e11.
153. Foldynova-Trantirkova, S.; Wilcox, W.R.; Krejci, P. Sixteen years and counting: The current understanding of fibroblast growth factor receptor 3 (fgfr3) signaling in skeletal dysplasias. *Hum. Mutat.* **2012**, *33*, 29–41.
154. Eglinton, E.; Annett, M. Handedness and dyslexia: A meta-analysis. *Percept. Mot. Skills* **1994**, *79*, 1611–1616.
155. Bishop, D.V. How to increase your chances of obtaining a significant association between handedness and disorder. *J. Clin. Exp. Neuropsychol.* **1990**, *12*, 812–816.
156. Dragovic, M.; Hammond, G. Handedness in schizophrenia: A quantitative review of evidence. *Acta Psychiatr. Scand.* **2005**, *111*, 410–419.
157. Bishop, D.V. Cerebral asymmetry and language development: Cause, correlate, or consequence? *Science* **2013**, doi:10.1126/science.1230531.
158. Paracchini, S.; Scerri, T.; Monaco, A.P. The genetic lexicon of dyslexia. *Annu. Rev. Genomics Hum. Genet.* **2007**, *8*, 57–79.
159. Brandler, W.M.; Paracchini, S. The genetic relationship between handedness and neurodevelopmental disorders. *Trends Mol. Med.* **2014**, *20*, 83–90.
160. Velayos-Baeza, A.; Toma, C.; da Roza, S.; Paracchini, S.; Monaco, A.P. Alternative splicing in the dyslexia-associated gene kiaa0319. *Mamm. Genome* **2007**, *18*, 627–634.
161. Field, S.; Riley, K.L.; Grimes, D.T.; Hilton, H.; Simon, M.; Powles-Glover, N.; Siggers, P.; Bogani, D.; Greenfield, A.; Norris, D.P. Pkd1l1 establishes left-right asymmetry and physically interacts with pkd2. *Development* **2011**, *138*, 1131–1142.
162. Lee, J.E.; Gleeson, J.G. Cilia in the nervous system: Linking cilia function and neurodevelopmental disorders. *Curr. Opin. Neurol.* **2011**, *24*, 98–105.
163. Sun, T.; Walsh, C.A. Molecular approaches to brain asymmetry and handedness. *Nat. Rev. Neurosci.* **2006**, *7*, 655–662.

12

Epigenetic Variation in Monozygotic Twins: A Genome-Wide Analysis of DNA Methylation in Buccal Cells

Jenny van Dongen [1,*], Erik A. Ehli [2,3], Roderick C. Slieker [4], Meike Bartels [1], Zachary M. Weber [2], Gareth E. Davies [2,3], P. Eline Slagboom [4], Bastiaan T. Heijmans [4] and Dorret I. Boomsma [1]

[1] Department of Biological Psychology, VU University Amsterdam, Van der Boechorststraat 1, 1081 BT Amsterdam, The Netherlands; E-Mails: m.bartels@vu.nl (M.B.); di.boomsma@vu.nl (D.I.B.)

[2] Avera Institute for Human Genetics, 3720 W. 69th Street, Sioux Falls, SD 57108, USA; E-Mails: Erik.Ehli@avera.org (E.A.E.); Zachary.Weber@avera.org (Z.M.W.); Gareth.Davies@avera.org (G.E.D.)

[3] Department of Psychiatry, University of South Dakota, 4400 W. 69th Street, Sioux Falls, SD 57108, USA

[4] Department of Molecular Epidemiology, Leiden University Medical Center, P.O. Box 9600, 2300 RC Leiden, The Netherlands; E-Mails: R.C.Slieker@lumc.nl (R.C.S.); P.Slagboom@lumc.nl (P.E.S.); B.T.Heijmans@lumc.nl (B.T.H.)

* Author to whom correspondence should be addressed; E-Mail: j.van.dongen@vu.nl;

Abstract: DNA methylation is one of the most extensively studied epigenetic marks in humans. Yet, it is largely unknown what causes variation in DNA methylation between individuals. The comparison of DNA methylation profiles of monozygotic (MZ) twins offers a unique experimental design to examine the extent to which such variation is related to individual-specific environmental influences and stochastic events or to familial factors (DNA sequence and shared environment). We measured genome-wide DNA methylation in buccal samples from ten MZ pairs (age 8–19) using the Illumina 450k array and examined twin correlations for methylation level at 420,921 CpGs after QC. After selecting CpGs showing the most variation in the methylation level between subjects, the mean genome-wide correlation (rho) was 0.54. The correlation was higher, on average, for CpGs within CpG islands (CGIs), compared to CGI shores, shelves and non-CGI regions, particularly at hypomethylated CpGs. This finding suggests that individual-specific environmental and stochastic influences account for more variation in DNA methylation in

CpG-poor regions. Our findings also indicate that it is worthwhile to examine heritable and shared environmental influences on buccal DNA methylation in larger studies that also include dizygotic twins.

Keywords: DNA methylation; cytosine; epigenetics; Illumina 450k; twin study; buccal epithelium; genome-wide methylation; children

---

1. Introduction

To date, hundreds of genetic risk variants for complex traits and diseases have been identified, although for most of these variants, the biological mechanisms remain to be elucidated [1]. Interestingly, the majority of disease-associated genetic variation is located in regulatory regions of the genome [2], including transcription-factor-occupied regions and DNase I hypersensitive sites (which correspond to open chromatin) [3]. This suggests that mechanisms that control the activity of genes, including epigenetic mechanisms, may represent an important link between DNA sequence variation and common disease susceptibility [4]. Trying to unravel the molecular biology underlying complex traits and disease, much attention has been drawn recently to these epigenetic mechanisms; non-DNA sequence-based regulation of gene expression by DNA methylation, histone modification, microRNAs, *etc.* [5]. DNA methylation is one of the most extensively studied epigenetic mechanisms in human populations and tissues and is the focus of this paper.

In humans, DNA methylation occurs almost exclusively at cytosines that are part of CpG dinucleotides. The relationship between DNA methylation and expression varies depending on the genomic context: CpG methylation at promoter regions is generally thought to repress gene expression, while gene body methylation is generally associated with active gene expression and has been suggested to regulate splicing [6–8]. In most cell types, the majority of CpGs in the genome (on average, 70%–80%) is typically methylated [9]. Of the unmethylated CpG sites in the genome, most occur in areas of clustered CpGs, called CpG islands, which are often present in promoter regions. Yet, DNA methylation patterns may vary, and differential methylation has been demonstrated to occur across age [10], cell types, tissues [7,11] and disease states [12,13], and it has become clear that widespread variation in methylation patterns exist between individuals [14]. Accumulating evidence suggests that DNA methylation patterns can be affected by genetic variants (mQTLs) [15], environmental exposures [16] and stochastic factors [17,18], but it is largely unknown how much each of these factors account for overall variation between individuals in DNA methylation across the genome. Twin studies provide insight into the proportion of inter-individual variation in DNA methylation that is due to genetic variation, environmental effects and stochastic variation [19].

Because MZ twins derive from a single zygote and, therefore, have (nearly) identical DNA sequences (see, for example, Ye *et al.*, 2013 [20]), the comparison of DNA methylation patterns of MZ twins allows one to examine the extent to which differences in methylation between human individuals are related to environmental and stochastic events. Previous studies have highlighted that various tissues of MZ twins already show differences in DNA methylation at birth [21,22] and that differences between twins for average genome-wide DNA methylation, total histone acetylation levels and methylation at certain loci increase with age (referred to as "epigenetic drift") [23]. Although a cross-sectional study of DNA methylation discordance in saliva from 34 MZ pairs (age range: 21–55 years) found no evidence for larger differences in DNA methylation in older MZ pairs [24], results from a cross-sectional analysis based on 230 MZ pairs (age range: 18–89 years) suggested a gradual increase of DNA methylation discordance in MZ twins from early adulthood to advanced age at various candidate loci, which was supported by longitudinal data from 19 elderly MZ pairs [25].

In the past few years, various studies have examined DNA methylation at a set of candidate genes or particular genomic regions in MZ and dizygotic (DZ) twins [26–31], usually reporting greater similarity of MZ twins compared to DZ twins, suggesting that heritable influences contribute to DNA methylation variation at specific regions. While CpG sites at some imprinted loci showed evidence for moderate to high heritability in blood samples from adolescent and middle-aged twins [29], other genomic regions, including the major histocompatibility complex (MHC) region, showed little evidence for genetic influences on DNA methylation variation [28]. Twin studies also highlighted variation between tissues in the importance of genetic influences on methylation of candidate loci at birth [30]. A longitudinal classical twin study of three candidate genes (*DRD4*, *SLC6A4/SERT* and *MAOA*) based on buccal cells indicated that changes in the methylation of these genes within individuals between age five and 10 are mostly attributable to non-shared environmental influences and stochastic variation [31]. Clearly, twin studies of candidate regions suggest that there is broad variation in the importance of heritable influences and environmental or stochastic variation to DNA methylation at different regions.

To date, only a few genome-scale analyses of DNA methylation have been performed using the classical twin design, including a study of ~12,000 CpG sites within islands [32], two studies that used a promoter-specific array targeting ~27,000 CpG sites (Illumina 27k) [21,33] and two studies that used the Infinium HumanMethylation450 array (Illumina 450k) [22,34], which assesses ~485,000 CpG sites across a variety of regions in the genome, including gene bodies and intergenic regions [35]. The studies that assessed heritability consistently reported that the average heritability of the methylation level at CpGs across the genome is low to moderate when all sites are considered, although the heritability of individual CpGs ranges between 0% and 100%. The following estimates of average heritability across genome-wide CpGs have been reported to date (based on all analyzed CpGs): 18% in blood from 32- to 80-year-old twins (21 MZ pairs and 31 DZ pairs) [33], 5% in placenta, 7% in human umbilical vascular endothelial cells (HUVEC) and 12% in cord-blood mononuclear cells (CBMC) from neonatal twins (22 MZ and 12 DZ pair [21]) and 19% in adipose tissue from adult female twins (97 MZ pairs and 162 DZ pairs) [34]. In two studies of neonatal twin tissues, methylation discordance in MZ and DZ twins increased with increasing distance from CpG islands (CGIs) for certain probes (Type I), *i.e.*, differences were larger in the shores and shelves that flank CGIs [21,22]. In the study of adipose tissue, it was noted that the average genome-wide heritability of DNA methylation was higher when restricting to the most variable CpG sites (for the top 10% CpGs of which the methylation level varied most between subjects, the average heritability was 37%) [34]. It was also found that gene body and intergenic regions showed higher average methylation levels, more variation between subjects and higher heritability compared to promoter regions in adipose tissue [34].

To summarize, there is great interest in unraveling the factors that contribute to variation in DNA methylation between persons, but most previous twin studies of DNA methylation have been limited to candidate genes or a subset of regulatory regions in the genome (mostly promoter regions and CGIs). Two earlier studies used the Illumina 450k to collect genome-wide data in MZ and DZ twins; one in adipose tissue in adults [34] and one in DNA isolated from buccal cells in infants (10 MZ pairs and five DZ pairs, longitudinal design) [22]. In line with earlier findings suggesting the divergence of DNA methylation profiles with age in MZ twins (mostly based on data from adult twins, cross-sectional comparisons and limited genomic coverage), Martino *et al.* [22] showed that widespread DNA methylation changes occur across the genome in buccal cells between birth and 18 months and that some MZ and DZ pairs already show divergence of DNA methylation profiles, whereas other pairs show stable difference levels or became more similar within the first 18 months after birth. In this paper, we analyzed genome-wide DNA methylation profiles (Illumina 450k) from buccal epithelium. We focused on 10 young and adolescent MZ twin pairs (age 8–19). The aim of our study was to examine how similar the DNA methylation profiles of buccal cells from genetically identical subjects are in childhood and adolescence and whether MZ twin similarity varies between different genomic regions.

Previous studies have highlighted differences in mean methylation level, differences in the effect of methylation level on gene expression and differences in the effect size and direction of the effect on methylation for disease associations across different regions in the genome [6]. These findings indicate that the establishment and maintenance of DNA methylation is differentially regulated in different regions and that a given change in methylation in different areas may have different downstream effects, suggesting that DNA methylation in some regions may be more tightly controlled than in others. We questioned whether these regional differences are also accompanied by differences in the importance of environmental and stochastic influences *versus* familial factors (genetic variation and shared environment) to inter-individual variation in methylation levels. Therefore, we describe the MZ twin correlations of individual CpGs as a function of various genomic classifications, including the position relative to CGIs (CGI regions, shores, shelves and non-CGI regions), genes (distal to promoter, proximal to promoter, gene body and intergenic) and ENCODE regulatory regions (DNaseI hypersensitive sites (DHS) and transcription factor binding sites (TFBS)). Hereby, our study gives valuable insight into the factors influencing inter-individual genome-wide DNA methylation variation in buccal cells in childhood and adolescence and into the degree to which these influences vary across functional regions in the genome.

2. Experimental

2.1. Subjects

Ten monozygotic twin pairs, who take part in longitudinal studies of the Netherlands Twin Register (NTR), were selected for the current study. There were five young twin pairs [36] whose buccal samples were collected when the twins were between ages 8 and 10 years and five adolescent pairs [37] who were aged 18–19 years at the time of sample collection. In the young group, there were three male pairs and two female pairs, and in the adolescent group, there were two male pairs and three female pairs. The twins were unselected with respect to phenotypic characteristics. Informed consent was obtained from the parents (children) or from the twins themselves (adolescents). The study was approved by the Central Ethics Committee on Research Involving Human Subjects of the VU University Medical Centre, Amsterdam, an Institutional Review Board certified by the U.S. Office of Human Research Protections (IRB number IRB-2991 under Federal-wide Assurance-3703; IRB/institute codes, NTR 03-180). Participants could indicate if they wished to be informed of the results of zygosity testing. Zygosity testing, based on a set of SNPs and VNTRs, as described in Van Beijsterveldt *et al.* 2013 [36], confirmed that all pairs were MZ. In addition to the twin samples, a single sample was used as a genomic DNA control. This DNA sample (CEPH) was derived from a stable cell line (female) from the HapMap project and was run in four replicates on the methylation BeadChip arrays.

2.2. Buccal DNA Collection

The procedures of buccal swab collection [38] and genomic DNA extraction [39] have been described previously. In short, 16 cotton mouth swabs were individually rubbed against the inside of the cheek by the participants and placed in four separate 15 mL conical tubes (four swabs in each tube) containing 0.5 mL STE buffer (100 mM sodium chloride, 10 mM Tris hydrochloride (pH 8.0) and 10 mM ethylenediaminetetraacetic acid) with proteinase K (0.1 mg/mL) and sodium dodecyl sulfate (SDS) (0.5%) per swab. Individuals were asked to refrain from eating or drinking 1 hour prior to sampling. High molecular weight genomic DNA was extracted from the swabs using a high salt (KAc) precipitation followed by a standard chloroform/isoamyl alcohol (24:1) extraction. The DNA samples were quantified using absorbance at 260 nm with a Nanodrop ND-1000 (Nanodrop Technologies, Wilmington, DE, USA).

2.3. Infinium HumanMethylation450 BeadChip Data Generation

The epigenome-wide methylation data was generated using the Infinium HumanMethylation450 BeadChip Kit (Illumina Inc., San Diego, CA, USA). The Infinium HumanMethylation450 BeadChip is able to interrogate over 450,000 methylation sites across the entire genome, including 99% of RefSeq genes. Content was selected to include gene regulatory regions, such as the promoter, 5' UTR, first exon, gene body and the 3' UTR. Additionally, bead probes were also designed to cover regions adjacent to the CpG islands, such as the shores and shelves [35].

The Infinium DNA methylation assay was performed at the Avera Institute for Human Genetics. The assay was completed exactly as denoted in the manufacturer's protocol. The concentration of genomic DNA used in the Infinium DNA methylation assay was determined by comparing the binding of PicoGreen to known standards (λ DNA) and to the sample DNA. Briefly, 500 ng of genomic DNA was used for bisulfite conversion using the Zymo EZ DNA methylation kit (Zymo Research). Five microliters of bisulfite-converted DNA were whole genome amplified, which was followed by enzymatic end-point fragmentation. The resulting fragments were purified using an isopropanol precipitation, and the resuspended genomic DNA was denatured and hybridized to the BeadChip arrays for 18 hours. Extension, staining and washing were completed manually in flow cells followed by imaging using the iScan system (Illumina, Inc.). The raw data were extracted as *idat* files and were used in the downstream analysis.

2.4. *Quality Control, Normalization and Data Processing*

The raw intensity files (*idat*) were imported into the R environment [40], where further processing, quality control and normalization took place. The performance of bisulfite control probes confirmed successful bisulfite conversion for all samples. For each sample, we compared the overall (median) methylated signal intensity to the overall unmethylated signal intensity across all probes and compared the overall signal intensity from all CpG probes to the overall background signal ("noise"), as assessed using negative control probes. The overall signal from CpG probes was good and well-separated from the background signal for all samples. As a final quality check of the samples, cluster analysis was performed (cluster method = complete linkage) based on the Euclidean distance between samples, which was calculated from the pair-wise correlations between samples using the most variable probes (probes with an SD of the β-value across all 24 samples >0.10, with probes on the X and Y chromosomes and probes containing SNPs, as described in the next paragraph excluded; Nprobes = 38,359). The results of the cluster analysis were visualized in a dendrogram (see the "Results" section), which showed no outlier samples and illustrated tight clustering of the four replicate measures of control DNA.

Several probe-level QC steps were performed to filter out probes with low performance. For all samples, ambiguously-mapped probes were excluded, based on the definition of an overlap of at least 47 bases per probe from Chen *et al.* [41], and all probes containing an SNP, identified in the Dutch population [42], within the CpG site (at the C or G position) were excluded, irrespective of minor allele frequency. For each sample individually, probes with an intensity value of zero (not present on the array of a particular sample), probes with a detection *p*-value > 0.01 (calculated using the function *detectionP* from the *minfi* package) and probes with a bead count <3 were excluded. After these steps, probes with a success rate <0.95 across samples were removed from all samples, and the success rate across probes for each sample was computed (range of per sample success rate: 0.9990–0.9998).

After QC, background and red/green color adjustment were applied to the raw probe intensity values using quantile normalization. Normalized intensity values were converted into beta-values (β). The β-value, which represents the methylation level at a CpG for an individual and ranges from 0 to 1, is calculated as:

$$\beta = \frac{M}{M + U + 100}$$

where M = methylated signal, U = unmethylated signal and 100 represents a correction term to control the β-value of probes with a very low overall signal intensity (*i.e.*, probes for which M + U ~ 0 after background subtraction).

Finally, in anticipation of our categorization of CpGs based on the mean β-value across samples, β-values were adjusted to account for (intra-sample) differences in the distributions of methylation values derived from Type I probes (two bead types per CpG site) *versus* Type II probes (one bead type per CpG site) using the beta-mixture quantile normalization method (BMIQ) [43].

2.5. *Genomic Annotations*

CpGs that passed QC criteria (N = 420,921) were mapped to genomic features, DNase I hypersensitive sites (DHS) and transcription factor binding sites (TFBS), as described by Slieker *et al.* [7]. The genomic feature annotation is based on first assigning CpGs to one of five gene-centric regions:

intergenic region (>10 kb from the nearest transcription start site (TSS)), distal promoter (−10 kb to −1.5 kb from the nearest TSS), proximal promoter (−1.5 kb to +500 bp from the nearest TSS), gene body (+500 bp to 3' end of the gene) and downstream region (3' end to +5 kb from 3' end). Next, CpGs were mapped to CGIs (CG content >50%, length >200 bp and observed/expected ratio of CpGs >0.6; locations were obtained from the UCSC genome browser [44]), CGI shore (2-kb region flanking CGI), CGI shelf (2-kb region flanking CGI shore) or non-CGI regions (Figure 1). According to the gene-annotations, 14.4% of all CpGs were located in intergenic regions, 4.7% mapped to the distal promoter, 40.4% to the proximal promoter, 38.6% to the gene body and 1.9% to the downstream region. Thirty three percent of CpGs were located within CGIs, 23.8% in shores, 9.2% in shelves and 34.0% outside CGIs. The locations of DHS and TFBS, which were described by the ENCODE project [3], were downloaded from the UCSC genome browser. Finally, CpGs were mapped to imprinted genes that were described by Yuen *et al.* [45].

Figure 1. Illustration of a CpG island (CGI) with surrounding CGI shores, CGI-shelves and non-CGI regions.

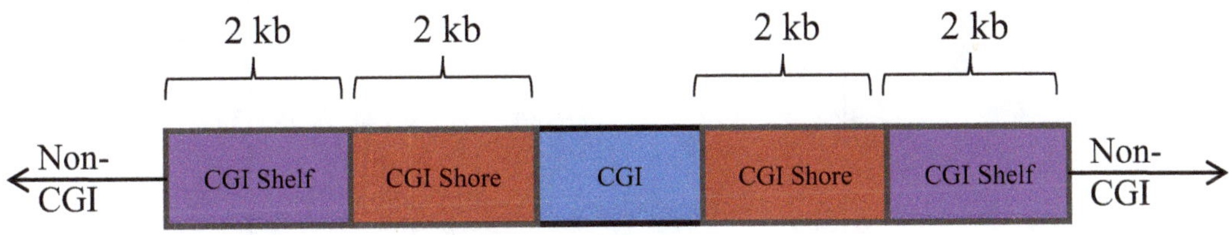

2.6. Statistical Analysis of Twin Data

To examine the s imilarity of DNA methylation profiles of MZ twins, we computed correlations between the normalized β-values of MZ co-twins using the following two approaches: (1) for each MZ twin pair, the Spearman correlation (rho) was computed between the β-values of Twin 1 and the β-values of Twin 2 (across all CpGs, *i.e.*, CpGs are cases), as a measure of the overall similarity of the methylation profiles of each twin pair; (2) for each CpG, the Spearman correlation (rho) was computed between the β-value of Twin 1 and the β-value of Twin 2 (across all 10 MZ twin pairs, *i.e.*, MZ twin pairs are cases), as a measure of the similarity of the methylation level of a CpG in MZ twins. For Scenario 2, we describe the range of correlations for the most variable CpGs. The most variable CpGs were additionally grouped by genomic annotations and average methylation level. For each CpG, the average methylation level (β-value) and the standard deviation (SD) were computed across subjects (20 MZ twins). Based on the average β, CpGs were classified as hypomethylated (mean $\beta < 0.3$), intermediately methylated (mean $\beta \geq 0.3$–0.7), or hypermethylated (mean $\beta \geq 0.7$). Based on the SD, CpGs were classified as "most variable CpGs" if they had an $SD \geq 0.05$.

3. Results and Discussion

3.1. DNA Methylation Level across the Genome

After QC of the methylation data, 420,921 CpGs from 10 monozygotic twin pairs were analyzed. The methylation level across genome-wide CpGs showed the typical bimodal distribution for each subject (Figure 2). Based on our β-value cut-offs (see the "Experimental" section); 184,765 CpGs (43.9%) were classified as hypomethylated, 64,829 CpGs (15.4%) were intermediately methylated and 171,327 CpGs (40.7%) were hypermethylated. CGIs were on average hypomethylated, with CGIs in proximal promoter regions showing a narrow range of average methylation levels across individual CpGs and CGIs in gene bodies, downstream regions and intergenic regions showing a broader range of methylation levels across individual CpGs (see Figure 3). Compared to CGIs, the shores, shelves and non-CGI regions on average had a higher methylation level, except for proximal promoter shores. Shores generally showed the widest range of average methylation levels across individual CpGs, when compared to CGIs, shelves and non-CGI regions (Figure 3).

Figure 2. Density of β-values after normalization for all twin samples.

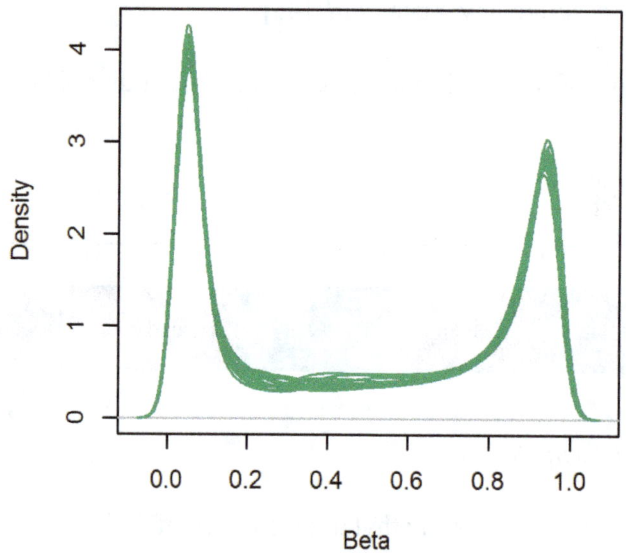

3.2. Similarity of Genome-Wide Methylation Profiles of MZ Twins

A cluster analysis of the methylation data revealed that all but one MZ twin clustered closely together with their co-twin (Figure 4), which could be related to differences in the cellular composition of the samples of this twin pair. Buccal swab samples are mainly composed of buccal epithelial cells with a small proportion of leukocytes, but the exact proportions may vary between persons. Using information from a reference 450k methylation dataset [7], we examined potential variation between twin samples in the proportion of buccal *versus* blood cells, by clustering the twin data based on methylation values at CpGs that showed a large difference in methylation between blood and buccal samples in the reference dataset (see the Supplementary Methods). Although some variation was indicated by this approach, the exclusion of twin samples with putatively deviant cellular proportions yielded similar results for the correlation analyses (see Table S1 and Figure S1), and we therefore decided to keep all samples in the analyses reported in this paper.

Figure 5 shows a typical scatterplot of genome-wide CpG methylation levels in buccal cells from an MZ twin pair. It illustrates that overall, the buccal DNA methylation profiles of MZ pairs are highly concordant when all CpGs are considered (rho = 0.981–0.994 for different MZ pairs, mean rho = 0.991); however, these correlations are to a large extent driven by invariable CpGs that are hypomethylated or hypermethylated in both twins. For pairs of unrelated subjects, the mean correlation was 0.983 (range: 0.970–0.992). When comparing only the most variable CpGs (SD of $\beta \geq 0.05$), the correlations ranged from 0.869 to 0.989 (mean rho = 0.966) in MZ twins (and mean rho = 0.859, range: 0.608–0.963 for unrelated subjects). Thus, when looking only at CpGs that may vary between individuals, the overall pattern of methylation across CpGs is still highly similar within MZ pairs on average, but more variation between individual pairs becomes visible, as the methylation level at variable CpGs overall was more strongly correlated for some MZ pairs than for others. This finding is in line with the results from Martino *et al.* based on buccal cells from twins at birth and at the age of 18 months [22], which also indicated that some MZ pairs are more similar than other pairs with respect to their DNA methylation profiles.

Figure 3. Average methylation level of individual CpGs across gene regions (**a**), CpG islands (CGI) and non-CGI regions (**b**) and for each genomic feature separately (**c**).

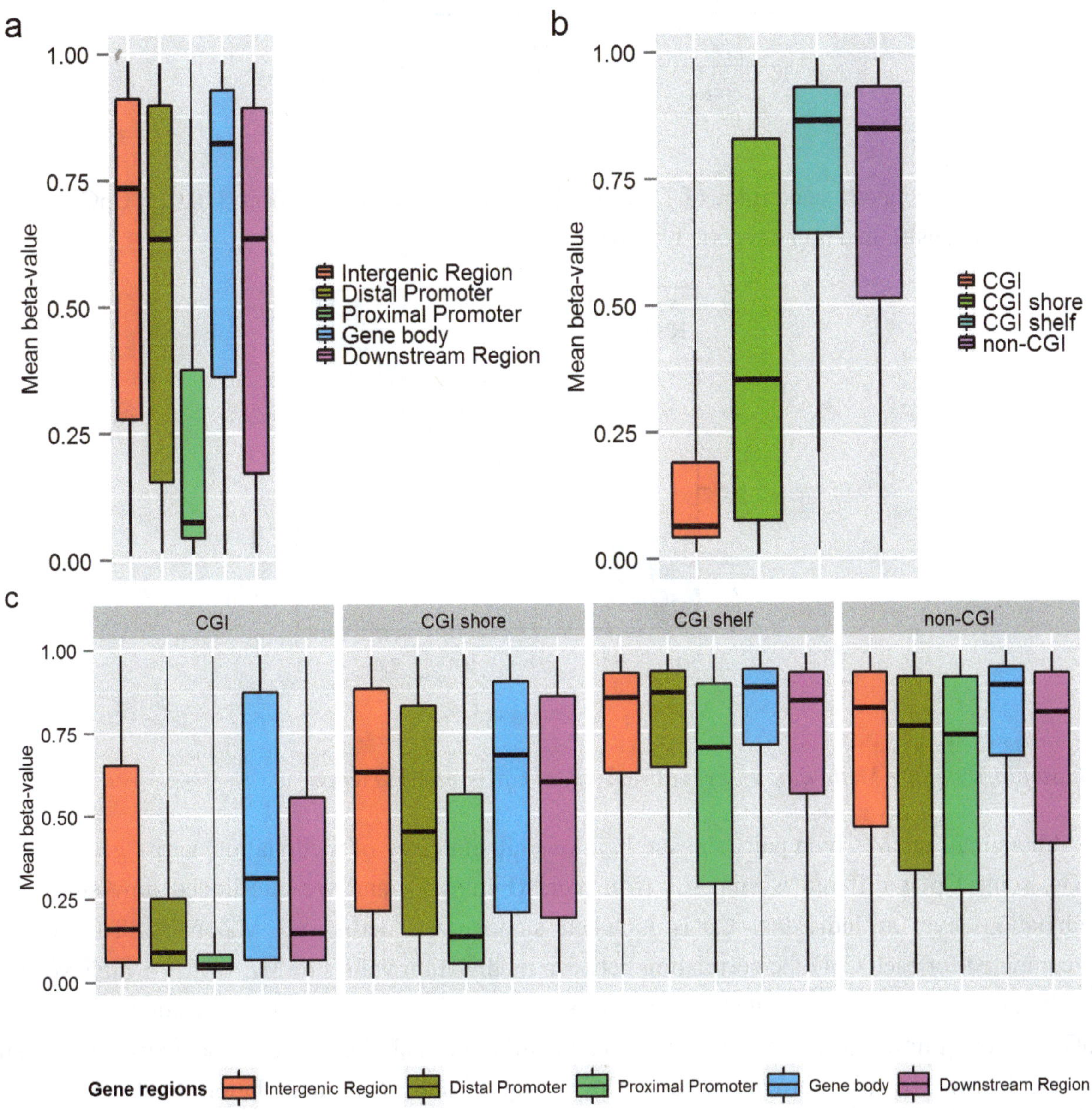

Figure 4. Cluster dendrogram of all twin and control samples. From left to right, the first two branches separate the control samples (HapMap cell line DNA) from the buccal samples from twins.

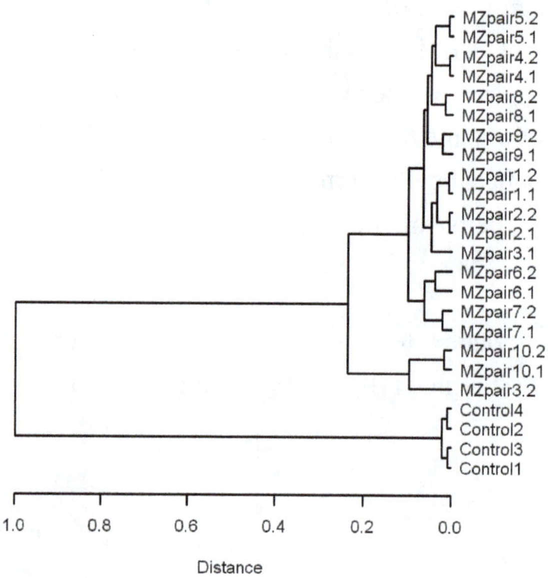

Figure 5. Smooth scatterplot of DNA methylation levels (β-values) at 420,921 CpGs in buccal cells from a monozygotic twin pair.

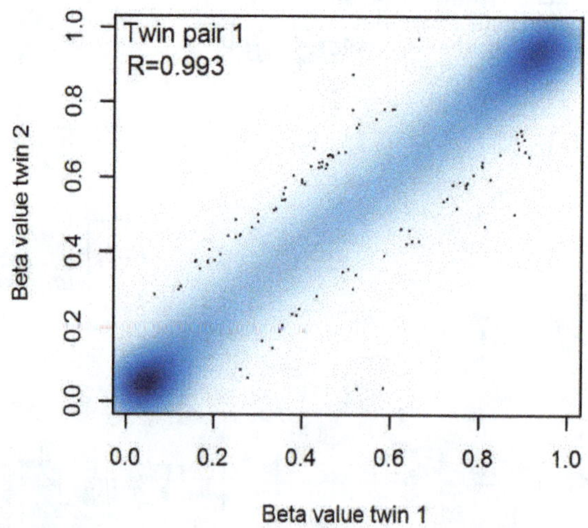

3.3. Similarity of the Methylation Level at Individual CpGs in MZ Twins

Although all ten MZ twin pairs showed high overall similarity of methylation across genome-wide CpGs, some CpGs differed within MZ twin pairs (Figure 5), and we questioned how similar the methylation level at individual CpGs is when summarized across all MZ pairs. To this end, we computed for each CpG the correlation between methylation values of MZ twins. A high MZ twin correlation for a CpG suggests that MZ co-twins consistently show similar methylation levels at this CpG, indicating little stochastic and environmental variation (including measurement error) at this site, whereas a low MZ twin correlation for a CpG suggests dissimilar methylation levels in co-twins, which is indicative of a large degree of stochastic and environmental influences.

Summarizing the individual CpG correlations over all 420,921 CpGs, the average MZ twin correlation was 0.31 (median = 0.35, range: −0.963–1), which is in line with the low heritability across genome-wide CpGs reported by previous studies [21,33,34]. However, as the majority of CpGs showed very little variation in the methylation level between subjects, all subsequent analyses were conducted using only the most variable sites (N = 59,041), which showed an average genome-wide correlation of 0.54 (median = 0.54, range: −0.661–1) in MZ twins. These findings suggest that while the large majority of CpGs are either hyper- or hypo-methylated and show little between-individual variation in DNA methylation in buccal samples, a small portion does vary markedly, and these CpGs are on average moderately to strongly correlated in MZ twins.

Table 1 describes the MZ twin correlations separately for various genomic regions and separately for hypomethylated, intermediately methylated and hypermethylated CpGs. Comparing the different gene-centric classifications, the average MZ twin correlation was highest for CpGs in proximal promoter areas (mean rho = 0.57) and lowest for gene body CpGs (mean rho = 0.51). The MZ twin correlation of methylation values was also lower on average in CGI shores (mean rho = 0.54), shelves (mean rho = 0.50) and non-CGI regions (mean rho = 0.49) compared to CGIs (mean rho = 0.66). Looking at the MZ twin correlations across genome annotations separately for hypomethylated (29.8% of variable CpGs), intermediately methylated (50.0% of variable CpGs) and hypermethylated CpGs (20.2% of variable CpGs), the median MZ twin correlation was consistently lower in the shelves, shores and non-CGI regions compared to CGIs, for all genic and intergenic regions, and this difference was most pronounced for hypomethylated CpGs (Figure 6). This observation suggests that the relative influence of familial *versus* individual-specific influences differs between these regions, with regions of low CpG density showing more variation due to individual-specific environmental and stochastic factors compared to CpG dense regions. Larger methylation discordance of MZ twins in CGI shores and shelves was also previously indicated by studies of neonatal twins [21,22]. Our results thus replicate previous findings and add to these findings that the pattern previously observed in MZ twins at birth is also visible in childhood and adolescence.

Figure 6. MZ twin correlations for individual CpGs grouped by genomic region and average methylation level. Hypo = Hypomethylated. Inter = intermediate methylation. Hyper = Hypermethylated. Results are based on the most variable CpGs (N = 59,041).

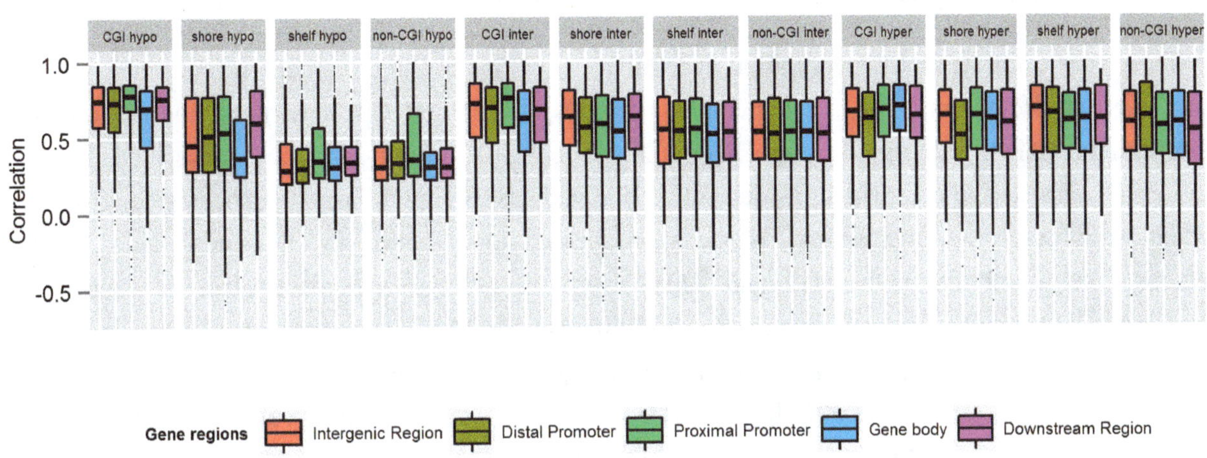

Table 1. Spearman correlation between the methylation values of monozygotic (MZ) twins for individual CpGs. Results are based on the most variable CpGs (N = 59,041).

Category	N CpGs	Mean rho	Median rho	Min rho	Max rho
All CpGs	59,041	0.54	0.54	−0.661	1
Gene-centric annotations	N CpGs (%)	Mean rho	Median rho	Min rho	Max rho
Intergenic (>10 kb from TSS)	11,430 (19.4%)	0.52	0.53	−0.56	1
Distal Promoter (−10 kb to −1.5 kb from TSS)	3193 (5.4%)	0.53	0.53	−0.54	1
Proximal Promoter (−1.5 kb to +500 bp from TSS)	17,880 (30.3%)	0.57	0.62	−0.66	1
Gene Body (+500 bp to 3' end)	25,163 (42.6%)	0.51	0.50	−0.59	1
Downstream region (3' end to +5 kb from 3' end)	1375 (2.3%)	0.55	0.55	−0.66	1
CGI annotations	N CpGs (%)	Mean rho	Median rho	Min rho	Max rho
CGI	10,576 (17.9%)	0.66	0.73	−0.49	1
CGI shore	14,803 (25.1%)	0.54	0.55	−0.59	1
CGI shelf	6001 (10.2%)	0.50	0.49	−0.54	1
Non-CGI	27,661 (46.9%)	0.49	0.47	−0.66	1
Methylation level	N CpGs (%)	Mean rho	Median rho	Min rho	Max rho
Hypomethylated (average beta <0.3)	17,581 (29.8)	0.48	0.42	−0.59	1
Intermediately methylated (average beta ≥0.3–0.7)	29,519 (50.0)	0.55	0.56	−0.66	1
Hypermethylated (average beta ≥0.7)	11,941 (20.2)	0.58	0.61	−0.59	1

The most strongly correlated CpGs in MZ twins (mean rho = 0.73) were hypomethylated CpGs located in proximal promoter CGIs (N = 2547 CpGs, constituting 4.3% of the most variable CpGs, and 2.9% of all CpGs located in proximal promoter CGIs), while MZ twin correlations on average were lowest in hypomethylated non-CGI gene body CpGs (N = 2972 CpGs constituting 5.0% of the most variable CpGs, Mean rho = 0.34). In combination with our observation that most proximal promoter CpGs are on average hypomethylated (Figure 3), these findings indicate that DNA methylation variation is generally depleted in proximal promoter CGIs. Yet, a small proportion of CpGs in proximal promoter CGIs does show marked variation in young and adolescent individuals, and the high average MZ twin correlations at these sites suggest that this variation may be to a large extent under genetic control.

3.4. MZ Twin Resemblance at CpGs in ENCODE Regulatory Regions

To further examine DNA methylation at regulatory regions in the genome, we focused specifically on CpGs located within DNase I hypersensitive sites (DHS) and CpGs within transcription factor binding sites (TFBS) identified by the ENCODE project. It has previously been described that these regions are enriched among disease-associated genetic variants [3], but it has not yet been studied to which extent heritable *versus* other sources of variation account for variation in DNA methylation in these regions. We found that both DHS and TFBS were on average hypomethylated, as expected for transcriptionally active DNA (DHS: mean $\beta = 0.27$, median = 0.09; TFBS: mean $\beta = 0.24$, median = 0.08). The most variable CpGs in these areas (representing 16.2% of all CpGs in DHS and 13.7% of CpGs in TFBS) showed a mean correlation of 0.52 (DHS) and 0.53 (TFBS), respectively, in MZ twins. These results suggest that buccal cells overall show little variation in the methylation level at the majority of CpGs within DHS and TFBS. A small proportion of CpGs in DHS and TFBS, however, does show variation between individuals, and these sites were moderately to strongly correlated in MZ twins, suggesting that these sites may be of particular interest for follow-up in future studies of heritability.

3.5. MZ Twin Resemblance at CpGs in Imprinted Genes

At imprinted genes, one of the alleles is typically methylated to repress expression, while the other allele is unmethylated, depending on the parent from whom the allele was inherited. This results in a methylation level of around 50% at imprinted CpGs when the two alleles are measured simultaneously. A previous twin study demonstrated moderate to high heritability at CpGs at two imprinted loci [29], suggesting that CpGs within imprinted genes may on average show more heritable variation compared to most other genome-wide CpGs. In our dataset, 346 CpGs were located in DMRs (differentially methylated regions) of 59 imprinted genes, described by Yuen *et al.* [45]. These genes were identified as imprinted in human placental tissue, and although some of these genes showed similar methylation patterns in one or multiple fetal tissues, including muscle, brain and kidney, it is unknown whether these genes are also imprinted in buccal cells. From the Yuen *et al.* set, 144 CpGs in 46 genes (see Table S2) showed a methylation level indicative of imprinting in our data (intermediate methylation; mean $\beta \geq 0.3$–0.7). The average MZ twin correlation for this set of CpGs was 0.47 (median rho = 0.50), suggesting that MZ twin correlations at imprinted gene CpGs on average are comparable to the MZ twin correlation at intermediately methylated CpGs in general.

3.6. Interpretation and Future Directions

The average twin correlation of methylation values for MZ twins at individual CpGs was low across all measured genome-wide CpGs, but it was moderate to large on average when focusing only on variably methylated CpGs. This is in line with results from a heritability analysis of DNA methylation in adipose tissue, which showed that the average heritability across all CpGs was higher for the top 10% of CpGs with the largest standard deviation of methylation level across subjects [34].

Importantly, in addition to the effects of environmental and stochastic influences, differences in DNA methylation within MZ twin pairs may result from variation in the cellular composition of samples and from technical variation (including measurement error). Buccal swab samples are mainly composed of buccal epithelial cells with a small proportion of leukocytes, but the exact proportions may vary between persons, which could lead to methylation variation within MZ twin pairs that mainly tag differences in cell type composition. We examined the impact of variation in the proportion of buccal *versus* leukocytes on our data by studying the methylation patterns of all twin samples at CpGs with a large methylation level difference between buccal and blood samples (see Supplementary Methods). Exclusion of four twin pairs, for which this approach indicated a more deviant cellular composition in one or both twins (lower proportion of buccal epithelial cells; see Figure S2 and Table S3), however, had very little impact on the average MZ twin correlations reported in this paper and led to the same conclusions (See Table S1 and Figure S1 for the results based on the exclusion of the putatively more heterogeneous samples).

With respect to technical variation, it is important to note that if the actual methylation status at a particular site is either completely unmethylated (0%) or completely methylated (100%) without true biological variation between subjects, some variability between the measured values of individuals is expected due to technical variation [46]. It is therefore likely that at sites that were on average hypomethylated or hypermethylated in our data, technical variation may account for a large part of the observed variation (although true biological variation may of course also account for part of the variation at these sites). An interesting question that largely remains to be examined is what types of environmental influences can induce changes in DNA methylation and thereby possibly impact on gene expression. Although our study design does not provide insight with regard to which of the observed differences between twins are the result of different environmental exposures and which differences have arisen due to stochastic variation in molecular processes, future studies of MZ twins who are discordant for environmental exposures should allow one to examine the effects of such influences on DNA methylation. Our finding that many CpGs in the genome show dissimilar methylation levels in young and adolescent MZ twins indicates that it is of interest for further studies to specifically search for regions in the genome where differential methylation in MZ twin coincides with differential exposures. As we observed that DNA methylation in MZ twins is overall less similar at CpGs in non-CGI regions, CGI shores and shelves, these regions are of particular interest to studies examining environmental exposures, as these regions may show the strongest effects of environmental influences.

To check whether the lower average MZ twin correlation at hypomethylated sites is not merely related to the distribution of β-values being truncated at zero (and one) by definition, we also ran the analyses on M-values ($M = \log_2\left(\frac{\beta}{1-\beta}\right)$), which have better statistical properties, but reduced biological interpretability, compared to β-values [47]. The MZ twin correlations based on M-values were highly similar to those based on β-values and showed a similar genome-wide average (Table S4) and a similar pattern across regions and mean methylation categories (Figure S3). Irrespective of whether the lower resemblance of MZ twins mainly reflects that these sites harbor more biological variation that is unique to MZ twins or reflects that more variation at these sites is related to measurement error, our findings provide useful information for future heritability and mQTL studies. CpGs that are very weakly correlated between MZ twins are not likely to show high heritability or strong effects of DNA variants on the methylation level.

A limitation of our study is the modest study size, which limited the scope of our analyses to the description of the major patterns (*i.e.*, averages) of twin correlations across the genome. A second limitation is that we did not include DZ twins. The correlation between the phenotypes of MZ twins summarizes the contribution of heritable influences and shared environmental factors to phenotypic variation. It thus remains to be established whether CpGs that were strongly correlated in MZ twin pairs are strongly affected by heritable influences or whether shared environmental influences are also important at these sites. Of interest, a previous twin study of DNA methylation in adipose tissue identified a number of CpGs with evidence for shared environmental effects on DNA methylation [34]. Future studies that include data from both MZ and DZ twin pairs are needed to separate the effects of heritable effects and shared environment on genome-wide DNA methylation profiles in buccal cells. Our results indicate that such studies are worthwhile, as we have shown that methylation at a number of CpGs is strongly correlated between MZ twins in buccal cells.

We studied DNA methylation extracted from buccal samples, which may be easier to collect than blood samples in, e.g., young children, and are therefore well-suited for large-scale studies in humans. A relevant question is how representative DNA methylation extracted from these samples is for DNA methylation variation in other tissues and whether methylation studies of buccal *vs.* blood-derived DNA would lead to similar insights. Although DNA methylation patterns are to a large extent tissue-specific [7] and epigenetic changes arising later in life in one tissue may not be detectable in others, epigenetic variation that is established early in development is more likely to be reflected in multiple tissues [4]. Yet, the methylation patterns of buccal cells are likely to be more informative to the methylation state of other ectoderm-derived tissues, whereas methylation patterns in blood may be more comparable to other mesoderm-derived tissues. Finally, it may be regarded as an advantage that compared to blood, which consists of many different cell types, buccal samples represent a relatively homogenous sample type [48], in the sense that it consists of only two major cell types, which potentially makes correction for cell types more straightforward. On the other hand, an advantage of blood samples is that they may provide more insight into DNA methylation variation related to immune system-mediated processes in the body, which are important in many diseases. To conclude, blood and buccal samples are both valuable for gaining insight into the overall importance of heritable and environmental factors to DNA methylation variation in the genome, and our study showed that the average genome-wide MZ twin correlation for DNA methylation in buccal cells is similar to the average correlation previously reported for peripheral blood [33].

4. Conclusions

To summarize, we computed genome-wide MZ twin correlations for the buccal DNA methylation level at individual CpGs. Methylation levels in MZ twins were moderately to strongly correlated at CpGs with the largest inter-individual variation, which constituted a relatively small proportion of the CpGs that were measured. The average MZ twin correlation across all CpGs was relatively low (mean rho = 0.31), which is similar to findings from previous twin studies [21,33]. Although most CpGs within CGIs were on average hypomethylated, some of them showed large variation in methylation levels. We observed that CpGs with variable methylation levels were more strongly correlated in MZ twins when located in CGIs compared to CpGs in shores and shelves. CpGs in DHS and TFBS were generally hypomethylated, as expected for regulatory active DNA, but CpGs in these regions that were more variably methylated were moderately to strongly correlated in MZ twin pairs, in line with our findings for variably methylated CpGs in general. To conclude, we have shown that in buccal samples from young and adolescent MZ twins, most CpGs show an average methylation level close to zero or 100% and little inter-individual variation, and a subset of CpGs show larger variability with evidence for a familial component (DNA sequence variation or shared environment). These findings are relevant for future heritability studies of DNA methylation and for mQTL studies.

Acknowledgments

This study was funded by Genetics of Mental Illness: A lifespan approach to the genetics of childhood and adult neuropsychiatric disorders and comorbid conditions (European Research Council

(ERC-230374)); and by the Avera McKennan Hospital and University Health Center. We would like to thank the twins who participated in this study.

Author Contributions

Gareth E. Davies, Meike Bartels, and Dorret I. Boomsma conceived of the study. Erik A. Ehli, Zachary M. Weber and Gareth E. Davies performed the laboratory work. Jenny van Dongen and Erik A. Ehli performed the data analysis. Roderick C. Slieker, Bastiaan T. Heijmans, and P. Eline Slagboom provided advice and gave assistance with the data processing and analysis. Jenny van Dongen drafted the manuscript with significant input from all the other authors.

References

1. Hindorff, L.A.; MacArthur, J.; Morales, J.; Junkins, H.A.; Hall, P.N.; Klemm, A.K.; Manolio, T.A. A Catalog of Published Genome-Wide Association Studies. Available online: http://www.genome.gov/gwastudies/ (accessed on 27 January 2014).
2. Hindorff, L.A.; Sethupathy, P.; Junkins, H.A.; Ramos, E.M.; Mehta, J.P.; Collins, F.S.; Manolio, T.A. Potential etiologic and functional implications of genome-wide association loci for human diseases and traits. *Proc. Natl. Acad. Sci. USA* **2009**, *106*, 9362–9367.
3. The ENCODE Project Consortium. An integrated encyclopedia of DNA elements in the human genome. *Nature* **2012**, *489*, 57–74.
4. Mill, J.; Heijmans, B.T. From promises to practical strategies in epigenetic epidemiology. *Nat. Rev. Genet.* **2013**, *14*, 585–594.
5. Goldberg, A.D.; Allis, C.D.; Bernstein, E. Epigenetics: A landscape takes shape. *Cell* **2007**, *128*, 635–638.
6. Jones, P.A. Functions of DNA methylation: islands, start sites, gene bodies and beyond. *Nat. Rev. Genet.* **2012**, *13*, 484–492.
7. Slieker, R.C.; Bos, S.D.; Goeman, J.J.; Bovee, J.V.; Talens, R.P.; van der Breggen, R.; Suchiman, H.E.; Lameijer, E.W.; Putter, H.; van den Akker, E.B.; *et al.* Identification and systematic annotation of tissue-specific differentially methylated regions using the Illumina 450k array. *Epigenetics Chromatin* **2013**, doi:10.1186/1756-8935-6-26.
8. Maunakea, A.K.; Nagarajan, R.P.; Bilenky, M.; Ballinger, T.J.; D'Souza, C.; Fouse, S.D.; Johnson, B.E.; Hong, C.; Nielsen, C.; Zhao, Y.; *et al.* Conserved role of intragenic DNA methylation in regulating alternative promoters. *Nature* **2010**, *466*, 253–257.
9. Bird, A. DNA methylation patterns and epigenetic memory. *Genes Dev.* **2002**, *16*, 6–21.
10. Horvath, S.; Zhang, Y.; Langfelder, P.; Kahn, R.S.; Boks, M.P.; van Eijk, K.; van den Berg, L.H.; Ophoff, R.A. Aging effects on DNA methylation modules in human brain and blood tissue. *Genome Biol.* **2012**, doi:10.1186/gb-2012-13-10-r97.
11. Ziller, M.J.; Gu, H.; Muller, F.; Donaghey, J.; Tsai, L.T.; Kohlbacher, O.; de Jager, P.L.; Rosen, E.D.; Bennett, D.A.; Bernstein, B.E.; *et al.* Charting a dynamic DNA methylation landscape of the human genome. *Nature* **2013**, *500*, 477–481.

12. Kuehnen, P.; Mischke, M.; Wiegand, S.; Sers, C.; Horsthemke, B.; Lau, S.; Keil, T.; Lee, Y.A.; Grueters, A.; Krude, H. An Alu element-associated hypermethylation variant of the POMC gene is associated with childhood obesity. *PLoS Genet.* **2012**, *8*, e1002543.
13. Dempster, E.L.; Pidsley, R.; Schalkwyk, L.C.; Owens, S.; Georgiades, A.; Kane, F.; Kalidindi, S.; Picchioni, M.; Kravariti, E.; Toulopoulou, T.; *et al.* Disease-associated epigenetic changes in monozygotic twins discordant for schizophrenia and bipolar disorder. *Hum. Mol. Genet.* **2011**, *20*, 4786–4796.
14. Talens, R.P.; Boomsma, D.I.; Tobi, E.W.; Kremer, D.; Jukema, J.W.; Willemsen, G.; Putter, H.; Slagboom, P.E.; Heijmans, B.T. Variation, patterns, and temporal stability of DNA methylation: Considerations for epigenetic epidemiology. *FASEB J.* **2010**, *24*, 3135–3144.
15. Bell, J.T.; Pai, A.A.; Pickrell, J.K.; Gaffney, D.J.; Pique-Regi, R.; Degner, J.F.; Gilad, Y.; Pritchard, J.K. DNA methylation patterns associate with genetic and gene expression variation in HapMap cell lines. *Genome Biol.* **2011**, doi:10.1186/gb-2011-12-1-r10.
16. Heijmans, B.T.; Tobi, E.W.; Stein, A.D.; Putter, H.; Blauw, G.J.; Susser, E.S.; Slagboom, P.E.; Lumey, L.H. Persistent epigenetic differences associated with prenatal exposure to famine in humans. *Proc. Natl. Acad. Sci. USA* **2008**, *105*, 17046–17049.
17. Jeffries, A.R.; Perfect, L.W.; Ledderose, J.; Schalkwyk, L.C.; Bray, N.J.; Mill, J.; Price, J. Stochastic choice of allelic expression in human neural stem cells. *Stem Cells* **2012**, *30*, 1938–1947.
18. Waterland, R.A.; Dolinoy, D.C.; Lin, J.R.; Smith, C.A.; Shi, X.; Tahiliani, K.G. Maternal methyl supplements increase offspring DNA methylation at Axin Fused. *Genesis* **2006**, *44*, 401–406.
19. Van Dongen, J.; Slagboom, P.E.; Draisma, H.H.; Martin, N.G.; Boomsma, D.I. The continuing value of twin studies in the omics era. *Nat. Rev. Genet.* **2012**, *13*, 640–653.
20. Ye, K.; Beekman, M.; Lameijer, E.W.; Zhang, Y.; Moed, M.H.; van den Akker, E.B.; Deelen, J.; Houwing-Duistermaat, J.J.; Kremer, D.; *et al.* Aging as accelerated accumulation of somatic variants: Whole-genome sequencing of centenarian and middle-aged monozygotic twin pairs. *Twin Res. Hum. Genet.* **2013**, *16*, 1026–1032.
21. Gordon, L.; Joo, J.E.; Powell, J.E.; Ollikainen, M.; Novakovic, B.; Li, X.; Andronikos, R.; Cruickshank, M.N.; Conneely, K.N.; Smith, A.K.; *et al.* Neonatal DNA methylation profile in human twins is specified by a complex interplay between intrauterine environmental and genetic factors, subject to tissue-specific influence. *Genome Res.* **2012**, *22*, 1395–1406.
22. Martino, D.; Loke, Y.J.; Gordon, L.; Ollikainen, M.; Cruickshank, M.N.; Saffery, R.; Craig, J.M. Longitudinal, genome-scale analysis of DNA methylation in twins from birth to 18 months of age reveals rapid epigenetic change in early life and pair-specific effects of discordance. *Genome Biol.* **2013**, doi:10.1186/gb-2013-14-5-r42.
23. Fraga, M.F.; Ballestar, E.; Paz, M.F.; Ropero, S.; Setien, F.; Ballestar, M.L.; Heine-Suner, D.; Cigudosa, J.C.; Urioste, M.; Benitez, J.; *et al.* Epigenetic differences arise during the lifetime of monozygotic twins. *Proc. Natl. Acad. Sci. USA* **2005**, *102*, 10604–10609.
24. Bocklandt, S.; Lin, W.; Sehl, M.E.; Sanchez, F.J.; Sinsheimer, J.S.; Horvath, S.; Vilain, E. Epigenetic predictor of age. *PLoS One* **2011**, *6*, e14821.
25. Talens, R.P.; Christensen, K.; Putter, H.; Willemsen, G.; Christiansen, L.; Kremer, D.; Suchiman, H.E.; Slagboom, P.E.; Boomsma, D.I.; Heijmans, B.T. Epigenetic variation during the adult lifespan: Cross-sectional and longitudinal data on monozygotic twin pairs. *Aging Cell* **2012**, *11*, 694–703.

26. Boks, M.P.; Derks, E.M.; Weisenberger, D.J.; Strengman, E.; Janson, E.; Sommer, I.E.; Kahn, R.S.; Ophoff, R.A. The relationship of DNA methylation with age, gender and genotype in twins and healthy controls. *PLoS One* **2009**, *4*, e6767.
27. Coolen, M.W.; Statham, A.L.; Qu, W.; Campbell, M.J.; Henders, A.K.; Montgomery, G.W.; Martin, N.G.; Clark, S.J. Impact of the genome on the epigenome is manifested in DNA methylation patterns of imprinted regions in monozygotic and dizygotic twins. *PLoS One* **2011**, *6*, e25590.
28. Gervin, K.; Hammero, M.; Akselsen, H.E.; Moe, R.; Nygard, H.; Brandt, I.; Gjessing, H.K.; Harris, J.R.; Undlien, D.E.; Lyle, R. Extensive variation and low heritability of DNA methylation identified in a twin study. *Genome Res.* **2011**, *21*, 1813–1821.
29. Heijmans, B.T.; Kremer, D.; Tobi, E.W.; Boomsma, D.I.; Slagboom, P.E. Heritable rather than age-related environmental and stochastic factors dominate variation in DNA methylation of the human IGF2/H19 locus. *Hum. Mol. Genet.* **2007**, *16*, 547–554.
30. Ollikainen, M.; Smith, K.R.; Joo, E.J.; Ng, H.K.; Andronikos, R.; Novakovic, B.; Abdul Aziz, N.K.; Carlin, J.B.; Morley, R.; Saffery, R.; *et al.* DNA methylation analysis of multiple tissues from newborn twins reveals both genetic and intrauterine components to variation in the human neonatal epigenome. *Hum. Mol. Genet.* **2010**, *19*, 4176–4188.
31. Wong, C.C.; Caspi, A.; Williams, B.; Craig, I.W.; Houts, R.; Ambler, A.; Moffitt, T.E.; Mill, J. A longitudinal study of epigenetic variation in twins. *Epigenetics* **2010**, *5*, 516–526.
32. Kaminsky, Z.A.; Tang, T.; Wang, S.C.; Ptak, C.; Oh, G.H.; Wong, A.H.; Feldcamp, L.A.; Virtanen, C.; Halfvarson, J.; Tysk, C.; *et al.* DNA methylation profiles in monozygotic and dizygotic twins. *Nat. Genet.* **2009**, *41*, 240–245.
33. Bell, J.T.; Tsai, P.C.; Yang, T.P.; Pidsley, R.; Nisbet, J.; Glass, D.; Mangino, M.; Zhai, G.; Zhang, F.; Valdes, A.; *et al.* Epigenome-wide scans identify differentially methylated regions for age and age-related phenotypes in a healthy ageing population. *PLoS Genet.* **2012**, *8*, e1002629.
34. Grundberg, E.; Meduri, E.; Sandling, J.K.; Hedman, A.K.; Keildson, S.; Buil, A.; Busche, S.; Yuan, W.; Nisbet, J.; Sekowska, M.; *et al.* Global analysis of DNA methylation variation in adipose tissue from twins reveals links to disease-associated variants in distal regulatory elements. *Am. J. Hum. Genet.* **2013**, *93*, 876–890.
35. Bibikova, M.; Barnes, B.; Tsan, C.; Ho, V.; Klotzle, B.; Le, J.M.; Delano, D.; Zhang, L.; Schroth, G.P.; Gunderson, K.L.; *et al.* High density DNA methylation array with single CpG site resolution. *Genomics* **2011**, *98*, 288–295.
36. Van Beijsterveldt, C.E.; Groen-Blokhuis, M.; Hottenga, J.J.; Franic, S.; Hudziak, J.J.; Lamb, D.; Huppertz, C.; de Zeeuw, E.; Nivard, M.; Schutte, N.; *et al.* The Young Netherlands Twin Register (YNTR): Longitudinal twin and family studies in over 70,000 children. *Twin Res. Hum. Genet.* **2013**, *16*, 252–267.
37. Estourgie-van Burk, G.F.; Bartels, M.; Hoekstra, R.A.; Polderman, T.J.; Delemarre-van de Waal, H.A.; Boomsma, D.I. A twin study of cognitive costs of low birth weight and catch-up growth. *J. Pediatr.* **2009**, *154*, 29–32.
38. Willemsen, G.; de Geus, E.J.; Bartels, M.; van Beijsterveldt, C.E.; Brooks, A.I.; Estourgie-van Burk, G.F.; Fugman, D.A.; Hoekstra, C.; Hottenga, J.J.; Kluft, K.; *et al.* The Netherlands Twin Register biobank: A resource for genetic epidemiological studies. *Twin Res. Hum. Genet.* **2010**, *13*, 231–245.

39. Meulenbelt, I.; Droog, S.; Trommelen, G.J.; Boomsma, D.I.; Slagboom, P.E. High-yield noninvasive human genomic DNA isolation method for genetic studies in geographically dispersed families and populations. *Am. J. Hum. Genet.* **1995**, *57*, 1252–1254.
40. R Core Team. R: A language and environment for statistical computing. R Foundation for Statistical Computing, Vienna, Austria. Available online: http://www.R-project.org/ (accessed on 27 January 2014).
41. Chen, Y.A.; Lemire, M.; Choufani, S.; Butcher, D.T.; Grafodatskaya, D.; Zanke, B.W.; Gallinger, S.; Hudson, T.J.; Weksberg, R. Discovery of cross-reactive probes and polymorphic CpGs in the Illumina Infinium HumanMethylation450 microarray. *Epigenetics* **2013**, *8*, 203–209.
42. Boomsma, D.I.; Wijmenga, C.; Slagboom, E.P.; Swertz, M.A.; Karssen, L.C.; Abdellaoui, A.; Ye, K.; Guryev, V.; Vermaat, M.; van Dijk, F.; *et al.* The Genome of the Netherlands: Design, and project goals. *Eur. J. Hum. Genet.* **2014**, *22*, 221–227.
43. Teschendorff, A.E.; Marabita, F.; Lechner, M.; Bartlett, T.; Tegner, J.; Gomez-Cabrero, D.; Beck, S. A beta-mixture quantile normalization method for correcting probe design bias in Illumina Infinium 450k DNA methylation data. *Bioinformatics* **2013**, *29*, 189–196.
44. Kent, W.J.; Sugnet, C.W.; Furey, T.S.; Roskin, K.M.; Pringle, T.H.; Zahler, A.M.; Haussler, D. The human genome browser at UCSC. *Genome Res.* **2002**, *12*, 996–1006.
45. Yuen, R.K.; Jiang, R.; Penaherrera, M.S.; McFadden, D.E.; Robinson, W.P. Genome-wide mapping of imprinted differentially methylated regions by DNA methylation profiling of human placentas from triploidies. *Epigenetics Chromatin* **2011**, doi:10.1186/1756-8935-4-10.
46. Pan, H.; Chen, L.; Dogra, S.; Teh, A.L.; Tan, J.H.; Lim, Y.I.; Lim, Y.C.; Jin, S.; Lee, Y.K.; Ng, P.Y.; *et al.* Measuring the methylome in clinical samples: improved processing of the Infinium Human Methylation450 BeadChip Array. *Epigenetics* **2012**, *7*, 1173–1187.
47. Du, P.; Zhang, X.; Huang, C.C.; Jafari, N.; Kibbe, W.A.; Hou, L.; Lin, S.M. Comparison of Beta-value and M-value methods for quantifying methylation levels by microarray analysis. *BMC Bioinform.* **2010**, doi:10.1186/1471-2105-11-587.
48. Thiede, C.; Prange-Krex, G.; Freiberg-Richter, J.; Bornhauser, M.; Ehninger, G. Buccal swabs but not mouthwash samples can be used to obtain pretransplant DNA fingerprints from recipients of allogeneic bone marrow transplants. *Bone Marrow Transpl.* **2000**, *25*, 575–577.

13

Genetic Profiling for Risk Reduction in Human Cardiovascular Disease

Megan J. Puckelwartz [1] **and Elizabeth M. McNally** [1,2,*]

[1] Department of Medicine, University of Chicago, Chicago, IL 60637, USA; E-Mail: mjroy@uchicago.edu

[2] Department of Human Genetics, University of Chicago, Chicago, IL 60637, USA

* Author to whom correspondence should be addressed; E-Mail: emcnally@uchicago.edu;

Abstract: Cardiovascular disease is a major health concern affecting over 80,000,000 people in the U.S. alone. Heart failure, cardiomyopathy, heart rhythm disorders, atherosclerosis and aneurysm formation have significant heritable contribution. Supported by familial aggregation and twin studies, these cardiovascular diseases are influenced by genetic variation. Family-based linkage studies and population-based genome-wide association studies (GWAS) have each identified genes and variants important for the pathogenesis of cardiovascular disease. The advent of next generation sequencing has ushered in a new era in the genetic diagnosis of cardiovascular disease, and this is especially evident when considering cardiomyopathy, a leading cause of heart failure. Cardiomyopathy is a genetically heterogeneous disorder characterized by morphologically abnormal heart with abnormal function. Genetic testing for cardiomyopathy employs gene panels, and these panels assess more than 50 genes simultaneously. Despite the large size of these panels, the sensitivity for detecting the primary genetic defect is still only approximately 50%. Recently, there has been a shift towards applying broader exome and/or genome sequencing to interrogate more of the genome to provide a genetic diagnosis for cardiomyopathy. Genetic mutations in cardiomyopathy offer the capacity to predict clinical outcome, including arrhythmia risk, and genetic diagnosis often provides an early window in which to institute therapy. This discussion is an overview as to how genomic data is shaping the current understanding and treatment of cardiovascular disease.

Keywords: next generation sequencing; cardiovascular disease; genetic profiling; whole genome sequencing

---

1. Introduction

Next generation sequencing has revolutionized the study of human genome variation and has the capacity to greatly influence health care decision making. The Human Genome Project, as conceived, was to sequence the first human genome in ~15 years at a cost of almost $3 billion using traditional dideoxy chain termination sequencing. In under 10 years, massively parallel next generation sequencing has led to the routine sequencing of exomes and whole genomes. Now achieved in weeks and at a cost that is multiple orders of magnitude less than the first genome, personalized genetic information is now widely available. These rapid advances in sequencing technology require new ways of collecting, analyzing, and disseminating genomic data. Herein, we discuss the ways that genomic information is currently being applied and how that data is shaping the ability to understand and treat cardiovascular disease (CVD).

Genetic variation is considered a contributory component for nearly all disease, whether single gene familial disorders or more common, complex traits with multiple gene involvement. Single gene or "Mendelian" disorders can be attributed to one gene as both necessary and sufficient to cause a large component of the disease phenotype. With complex traits, the gene-gene and gene-environment interactions are multifactorial. CVD consists of both single gene-familial disorders and common, complex disease. CVD is a major health concern affecting over 80,000,000 people in the U.S. alone [1]. CVD extends to heart failure and cardiomyopathy, heart rhythm disorders, atherosclerosis and thromboembolic events, aneurysm and others disorders. Familial aggregation and twin studies demonstrate that most, if not all, of CVD is heavily influenced by a genetic component [2–4].

2. Genetic Variation in CVD

Beginning in the 1980s, family-based linkage analysis was used to identify regions of the genome responsible for monogenic disease. The success of these methods required large families with penetrant phenotypes. Polymorphic genetic markers segregated with the phenotype of interest in large multi-generational families to identify chromosomal regions bearing the causal genes of interest [5]. Such familial linkage studies were highly successful in identifying genes for multiple forms of CVD. In 1989, linkage analysis defined the chromosomal location responsible for hypertrophic cardiomyopathy [6]. The next year, this data was used to identify mutations in the causative gene, *MYH7*, encoding β-myosin heavy chain [7]. Genetic determinants for Long QT syndrome, multiple cardiomyopathies, Marfan's disease, and forms of congenital heart disease were identified highlighting both extensive locus and allelic heterogeneity [8–14]. However, these methods remain limited by the need for large families, a feature often not available since CVD confers survival disadvantage. Furthermore, much of CVD is under the influence of multiple genetic loci, and therefore requires alternative statistical methods and larger phenotypically and genetically characterized cohorts [15]. The HapMap project annotated the location of millions of single-nucleotide polymorphisms (SNPs) and took advantage of the long haplotype structure of the human genome [16]. Concurrently, commercially available platforms such as SNP arrays were developed that allowed simultaneous sampling of hundreds of thousands of SNPs paving the path for genome-wide association studies (GWAS). GWAS, which correlates SNPs with disease phenotypes, does not require a specific mode of inheritance and takes advantage of the extensive linkage disequilibrium (LD) in the human genome. In order to have enough statistical power to detect correlation, these large-scale association studies typically assess thousands to millions of SNPs across the genome in hundreds to thousands of cases and controls. According to the National Human Genome Research Institute (NHGRI) Catalog of Published Genome-Wide Association Studies [17] over 2800 strong SNP associations have been identified ($p < 1 \times 10^{-8}$) to date, and many of these are CVD-associated traits.

2.1. The Overlap between GWAS Hits and Monogenic Disease in CVD

CVD related phenotypes are well suited for GWAS because many CVDs have readily quantifiable discernable traits. Intriguingly, many GWAS "hits" overlap considerably with the same genes already linked to the disease though familial linkage studies. For example, Newton-Cheh and colleagues conducted a meta-analysis of three GWAS from ~14,000 individuals to examine the duration of QT interval from surface electrocardiograms [18]. QT duration reflects electrical depolarization and repolarization of the cardiac ventricles. A long QT interval is a biomarker for arrhythmias and a risk factor for sudden death. Non-familial QT disorders are still highly heritable ($h^2 \approx 0.35$), indicating a genetic component. GWAS identified 10 loci with significant ($p < 5 \times 10^{-8}$) association with QT interval. Five loci were those known to be involved in Mendelian long-QT syndromes, while the other five loci were genes that offer additional insights into variation at the QT interval. In total, the variation at these 10 loci accounted for 5.4%–6.5% of variation in the QT interval, which is quite high by GWAS standards. Genetic testing for the Mendelian form of long QT currently identifies mutations in ~75% of probands, so the additional GWAS loci may represent new candidate genes for mutation screening in familial long QT disorders.

BAG3 (B-cell lymphoma 2-associated athanogene 3) is another example of GWAS results informing rare, Mendelian disease. In 2011, Villard and colleagues performed GWAS to identify loci contributing to sporadic dilated cardiomyopathy. Dilated cardiomyopathy (DCM) is exemplified by left ventricular dilation and systolic dysfunction, and is a major cause of heart failure and the principle indicator for heart transplant [19]. DCM has a high heritable component with 20%–35% of DCM patients having an affected first-degree relative [20]. More than 50 genes have been implicated in familial monogenetic DCM [21–23]. GWAS was performed with DNA from 1179 sporadic (non-familial) DCM patients and 1108 controls using ~500,000 SNPs [19]. The authors identified a DCM-associated non-synonymous SNP (p. C151R) in the coding region of *BAG3*. BAG3 is a co-chaperone that regulates HSP70 [24]. Further analysis of *BAG3* non-synonymous SNPs found another higher frequency SNP also associated with DCM. The authors investigated *BAG3* variation in familial DCM based on both the apparent association of *BAG3* non-synonymous SNPs with sporadic DCM and previously reported linkage with familial DCM between markers on chromosome 10 that include the *BAG3* locus [25]. In a cohort of 168 cases from DCM families, the authors identified additional likely pathogenic mutations in *BAG3*. Features of DCM were identified in 16 of 18 mutation carriers in the cohort [19]. In the same year, Norton and colleagues identified *BAG3* mutations in familial DCM [26]. The authors also created a *BAG3* knockdown zebrafish model that recapitulated the DCM and heart failure found in patients [26]. Together, the GWAS and familial data implicate *BAG3* in DCM and indicate that genes can harbor common variation that influences risk of disease in sporadic cases and rare variation that accounts for familial disease.

2.2. The Missing Heritability of GWAS

Despite the overlap of GWAS findings with monogenic disease, GWAS associations often account for only a small proportion of genetic variation. Also, the vast majority of variants identified by GWAS does not explain the high heritability or reveal the causal mechanism for the cardiovascular phenotype in question [27–29]. Using GWAS, McPherson and colleagues identified an interval on chromosome 9q21 that consistently associated with coronary heart disease in more than 23,000 participants from 6 independent cohorts [30]. Homozygotes for the risk allele have a 30%–40% increased risk for coronary artery disease (CAD). This finding remains perplexing as this region on 9q21 has no annotated genes and is not associated with known CAD risk factors [30]. This same chromosome 9q21 region has also been associated with myocardial infarction in a GWAS with 4587 cases and 12,676 controls [31]. Homozygotes for the allele have 1.64 times increased risk for myocardial infarction compared to noncarriers. Despite the evidence that this region is important for CVD pathophysiology, no disease mechanism has yet been identified.

GWAS is ultimately based on the idea that common diseases are caused by common genetic variants, each with small effect. Through additive and interactive effects, in conjunction with the environment, GWAS variants explain disease [27,32,33]. Recently, the common disease common variant hypothesis has been called into question due to the observation of missing heritability [34,35]. Missing heritability refers to the proportion of genetic variance that is not explained by the effect of common variants identified by GWAS. Several explanations have been suggested to explain missing heritability. It is possible that the number of variants responsible for a trait has been significantly underestimated, and that many more yet identified variants with very small effect sizes must be discovered. Another possibility is the presence of rare variants with larger effect size. Such rare variants are undetectable using present day SNP arrays, which are biased towards SNPs with allele frequencies close to 50%. There is also a possibility that missing heritability arises from variation caused by structural variants in the genome, also difficult to detect with SNP arrays. Lastly, there is also the possibility that gene-gene interactions and gene-environment interactions are of major importance, but are not appropriately modeled by current methods.

GWAS typically interrogate SNPs with a minor allele frequency (MAF) > 5% while largely ignoring variants with lower population-based frequencies (MAF: 0.5%–5%) and those SNPs that are rare (MAF < 0.5%). The rationale for ignoring these variants relates to the strong linkage disequilibrium (LD) in the human genome. The human genome has an estimated mutation rate of approximately 1.4×10^{-8} or approximately 40 new mutations per generation. Projected over the current population of ~7 billion, the world currently has 300–400 million new mutations this generation [36]. Within the coding region alone, there are ~13,000 nonsynonymous variants per genome [37,38]. The National Heart Lung and Blood Institute sponsored exome sequencing project

(NHLBI ESP) sequenced ~15,000 human-protein coding genes in >2000 individuals [39]. This study revealed an abundance of rare variants that were often population specific, potentially offering some support that rare variation explains some component of missing heritability. Alleles that confer high risk of disease are subject to negative selection pressure and would not reach high population frequencies. However, many different rare alleles in the same genes or gene pathways would induce the same phenotype across the population despite the "rare" genetic etiology. Johansen and colleagues recently performed GWAS and resequencing to determine mutational burden of rare variants in individuals with hypertriglyceridemia *versus* control subjects [40]. Hypertriglyceridemia is polygenic in nature and confers risk for cardiovascular disease. In general, GWAS variants explain <10% of variation for lipid traits [41,42]. Loci associated with GWAS signals were assessed for rare variation by focusing on protein-coding sequences of four GWAS-associated genes [40]. The authors identified a significant number of rare variants in individuals with hypertriglyceridemia compared to controls. An additional more restricted study, which analyzed only rare variants unique to cases and removing all reported variants without functional deficits, also confirmed a greater mutational burden in cases compared to controls [40]. These data indicate that GWAS-identified genes may carry rare variation that contributes to the heritability of a complex trait, and rare SNPs with relatively large effects on common disease may not be identified by GWAS studies. The underlying assumption that genotypes can be inferred between common alleles in strong LD may be flawed. Inferring genotypes based on strong LD may misestimate variation across the genome. While haplotype structure may be maintained with several to many common SNPs still inherited together, rare variation may occur between these SNPs, reducing the predictive value of the common SNPs. This rare variation may be the result of a higher than expected mutation rate and/or population structure.

2.3. Rare Variation as a Cause of CVD

Next generation sequencing (NGS) provides a method to identify rare genetic variation. Massively parallel, array-based sequencing dramatically reduced cost and increased efficiency of DNA sequencing. Depending on the platform, sequencing reads range from ~35 base pairs to up to ~1000 base pairs. As the generation of sequence has become far more facile with NGS, large-scale alignment and interpretation has become the rate limiting step. Bioinformatics tools are available for alignment to the referent genome and calling variants (reviewed in [43,44]). A number of efforts are currently underway to catalog human genetic variation. The 1000 Genomes project has sequenced 1092 individuals from 14 populations using a combination of low coverage whole genome sequencing and exome sequencing [45]. It is estimated that this dataset captures 98% of SNPs at a frequency of 1%, finding 38 million SNPs, 59% of which were novel [45]. The NHLBI ESP provided whole exome sequencing on >6500 individuals, including 180,000 exons in 23,000 genes. The ESP is derived from diverse, well-phenotyped populations, including the Atherosclerosis Risk in Communities (ARIC) study, the Coronary Artery Risk Development in Young Adults (CARDIA) study, the Cardiovascular Health Study, the Framingham Heart Study, the Jackson Heart Study, and the Multi-Ethnic Study of Atherosclerosis. The data from this project will examine the genetic contribution to early-onset myocardial infarction, low-density lipoprotein cholesterol, body mass index/type 2 Diabetes mellitus, blood pressure, and ischemic stroke.

These databases are not only providing a rich dataset to identify rare variation, but they are also providing better allele frequencies across diverse populations for alleles once thought to be rare and pathogenic. Jabbari and colleagues examined the NHLBI ESP database (n = 6503) for variants previously associated with Catecholaminergic Polymorphic Ventricular Tachycardia (CPVT) and compared the frequency of those variants with the expected prevalence of CPVT in the population [46]. CPVT is a rare, lethal, hereditary cardiac disease characterized by fatal ventricular arrhythmias in the absence of structural defects of the heart or abnormal electrocardiographic findings [47]. Eleven percent of variants previously associated with CPVT were found in the ESP population, corresponding to a 1:150 prevalence of CPVT in the population, much higher than the known 1:10,000 prevalence. The 1000 Genomes (1KG) database was evaluated for the presence of predicted and previously reported pathogenic variation in three genes associated with cardiomyopathy (*MYH7*, *MYBPC3* and *TTN*) [48]. Nine percent of the population was identified as having a pathogenic variant (9%), which exceeds population prevalence estimates for dilated and hypertrophic cardiomyopathy (0.04%–0.2%, respectively). Similar studies have been performed for other cardiovascular diseases including Brugada Syndrome, cardiac channelopathies and arrhythmogenic right ventricular cardiomyopathy [49–51]. Since in every case the predicted pathogenic variation vastly exceeds the known disease prevalence, it suggests that on their own, these variants are not sufficient to cause disease. Whether these variants represent "at risk" genotypes for developing milder forms of disease is not known and requires prospective studies.

3. The Genetics of Dilated Cardiomyopathy

Cardiomyopathy is marked by a morphologically abnormal ventricle and frequently is associated with heart failure. Cardiomyopathy is divided into four groups: dilated (DCM), hypertrophic (HCM), restrictive (RCM), and arrhythmogenic right ventricular (ARVC), and the most common mode of inheritance is autosomal dominant. Depending on screening methods, nonischemic DCM, defined as DCM not arising from myocardial infarct or ischemia, is familial in 25%–50% of cases [52]. Greater than 50 genes have been implicated in familial DCM, and the majority of mutations are phenocopies as there are no outward clinical signs that predict the specific gene mutation. Identification of the genetic cause of cardiomyopathy is clinically important because of the high incidence of sudden death and clinical progression, which can be medically managed. In 2007, genetic testing relied on commercially available panels that interrogated only 5 genes using traditional sequencing. In 2011, Meder and colleagues developed an array-based panel that enriched for coding regions of 29 known and 18 novel, potential cardiomyopathy genes, followed by next-generation sequencing [53]. Current panels now include >50 genes [54], and the estimated sensitivity in DCM for detecting a pathogenic mutation is under 50% (Partners Healthcare [55]).

3.1. Next Generation Sequencing Identifies TTN as a Major Contributor to DCM

Next-generation sequencing facilitated the screening of *TTN* for DCM-causing variants. The *TTN* gene includes >350 exons and encodes the giant sarcomere protein titin, which ranges in size from ~27,000 to ~33,000 amino acids depending on isoform, making it the largest human protein [56]. Together, two titin molecules span the sarcomere, providing both passive and active contractile forces [57–60].

Previous work has linked *TTN* to dilated cardiomyopathy in families, but extensive screening was limited due to its large size [61–63]. Herman and colleagues developed an array to capture *TTN* exons and sequence *TTN* in patients with DCM (n = 312) [64]. *TTN* truncating mutations accounted for approximately 25% of familial DCM, but had a minimal contribution to hypertrophic cardiomyopathy (~1%). Approximately 30% of *TTN* truncating variants identified were putative splice site disrupting mutations whose effect on function can be difficult to assess *in silico*.

3.2. Beyond Panel Based Sequencing for Cardiomyopathy and Beyond

Despite the inclusion of *TTN*, the sensitivity for detecting a DCM mutation remains at just under 50%. There are several explanations for the missing variation. First, there are likely novel genes not yet associated with cardiomyopathy; second, certain genetic variation may not be readily detectable with NGS and SNP analysis. For example, nucleotide repeat expansions and structural variation is not commonly determined by NGS, as analytic methods are biased toward SNP analysis and small insertions and deletions; Third, pathogenic variation may arise from combinations of pathogenic variation, and analysis is generally biased towards finding a single pathogenic variant. This bias reflects that most families with inherited cardiomyopathy have autosomal dominant inheritance; Fourth, pathogenic variation may be non-coding and at this point, these regions are not captured by gene panels. To combat these problems and provide a more comprehensive variant profile, whole exome sequencing (WES) and whole genome sequencing (WGS) are being applied to identify disease-causing variation for many different diseases (Table 1). WES interrogates the coding portion of the genome; approximately 1%–2% of nuclear DNA, although at higher coverage than-comparably priced WGS [65]. Interrogating only a small portion of the genome, as in WES, is less expensive than WGS, and it is currently the most-readily interpreted, as approximately 85% of Mendelian-disease causing mutations cause changes in the coding sequence of the genome [66]. WES relies on commercially available sequence-capture arrays to enrich for the coding subset of genomic DNA, followed by massively parallel, next-generation sequencing of the enriched fragments. The choice of exome kit is an important consideration as the exome is approximately 30 megabases, and exome capture kits interrogate anywhere from 50 to 100 megabases, depending on the provider. Most of the additional sequences are untranslated regions (UTRs) that may be important for disease pathogenesis.

Table 1. Comparison of Panel, whole exome sequencing (WES) and whole genome sequencing (WGS).

	Panel	WES	WGS
Variation in Known Genes	yes	yes	yes
Novel Gene Identification	no	yes	yes
Structural Variation	no	limited	yes
Non-coding Variation	no	limited	yes
Repeat testing required if first pass negative	yes	yes	no

3.3. Exome Sequencing of Multiple Family Members Improves Identification of Pathogenic Variation

Campbell and colleagues used exome sequencing on three members of a large multi-generational family with classical DCM. After sequencing, variants were filtered by frequency and protein prediction algorithms [67]. Eight potentially causative mutations were shared across the three family members, significantly reducing the potential variants to be considered. Variants were tested for segregation across the other family members and only one variant segregated with disease, *TNNT2* R173W, a known cardiomyopathy gene. These data are particularly convincing as the variant segregates across all affected members of the family, including fourth-degree relatives [67]. The underlying assumption with this analysis is that a single variant accounts for disease in all family members, an assumption that may or may not hold true. In 2013, Wells and others used WES in a large, multi-generational family with DCM of unknown etiology [68]. The proband had undergone extensive unrevealing panel testing for DCM. The authors selected 3 distantly related affected family members for WES. Distantly related family members that are obligate carriers of the same mutation should share fewer variants by descent than closely related members, allowing for easier filtering of potentially causative variants. Variants were then filtered for rarity, functional significance, conservation and autosomal-dominant inheritance [68]. Variants were also prioritized using the VAAST tool, a probabilistic search tool that combines conservation, amino acid substitution chemistry, and frequency data to build a unified likelihood-framework to identify damaged genes and disease-causing variants [69]. Heterozygous, nonsense, nonsynonymous and splice site variants shared between the 3 affected candidates were filtered based on rare frequency in 1 KG and ESP, leaving 26 candidates for analysis. Wells and colleagues then compared these variants to variants identified in ~70 exomes previously sequenced by their laboratory. Variants identified in multiple exomes were removed, leaving 2 putative variants. This comparison allowed for removal of false positives that may be inherent to some sequencing platforms and variant calling pipelines. An *RBM20* variant was identified in an unrelated patient with familial-DCM, consistent with its role in causing disease. The authors confirmed segregation within the larger pedigree of the variant in *RBM20*, a recently identified cardiomyopathy gene [70], providing statistical support for *RBM20* causing DCM. It is interesting to note that this finding is largely based on frequency in both the general population and in a cohort already sequenced by the laboratory. The authors did perform filtering that considered evolutionary conservation (Genomic Evolutionary Rate Profiling) and functional effects (Polyphen2) but these tools did not reduce the list as extensively as population and cohort frequency combined (8 variants *versus* 2 variants, respectively [69,71,72]. These data indicate that exome sequencing followed by extensive filtering, in conjunction with segregation analysis can identify rare, DCM-causing variation in known cardiomyopathy genes.

3.4. Identifying Cardiomyopathy Modifier Loci Using Broad Based Sequencing

With large gene panels or WES/WGS additional, disease-modifying variation can be identified. Intra-familial variability including age of onset, severity and penetrance is a hallmark of DCM, but the loci that modify DCM phenotypes have not been well elucidated [73,74]. Roncarati and colleagues used WES to investigate clinical variability in an extended family with 14 subjects that included four family members with severe DCM that required heart transplant in early adulthood [75]. WES was performed on three severely affected and one unaffected family member. Variants were filtered for rarity, predicted pathogenicity and inheritance. The filtering process left a list of only 28 variants that where further filtered through the Human Phenotype Ontology project, which uses formal ontology to capture phenotypic information to identify relationships between different genes and phenotypes [76]. Through this analysis, the authors identified eight genes with variation associated with Mendelian disease [75], and two of eight *LMNA* and *TTN*, were known to cause DCM. A missense *LMNA* mutation, previously identified in an unrelated DCM patient, was confirmed in all affected family members [77]. A *TTN* variant was identified in five family members, four of whom were severely affected, while the fifth is likely too young to yet be symptomatic. Doubly heterozygous family members had a more severe clinical course than the *LMNA*-only family members, indicating that the *TTN* variant modifies the clinical progression [75]. This mutational stratification is expected to prove useful for assessing clinical risk and guiding treatment. Broad based sequencing through gene panels or WES/WGS is now positioned to outline the contribution of multiple variants to DCM development.

3.5. WES/WGS Can Identify New Genes for Cardiomyopathy

In 2011, Theis and colleagues used genome-wide mapping and exome sequencing in a consanguineous family with autosomal recessive DCM [78]. Genome-wide linkage analysis was performed in nineteen family members and a significant LOD score was identified on chromosome 7q21, in a region containing >250 genes. Exome sequencing was performed on 2 affected siblings, and variants were called and filtered without taking into account the linkage peak on 7q21. Synonymous, intergenic and intronic variants were removed from further consideration. Variants were filtered based on presence in 1 KG, HapMap, and in the authors' collection of exome sequences. Heterozygous SNPs were excluded due to the autosomal recessive inheritance mode. This extensive filtering left only 3 homozygous missense variants and only 1 was not present in unaffected family members, a mutation in *GATAD1* that maps to the already identified linkage region on 7q21. *GATAD1* encodes the GATA zinc finger domain-containing protein 1 which is ubiquitously expressed and is thought to bind to a histone modification site that regulates gene expression [79]. Immunohistochemistry with an antibody to GATAD1 revealed an abnormal staining pattern in the proband heart compared to control heart [78]. These data implicate *GATAD1* in the pathogenesis of DCM indicating that exome sequencing can be used to identify novel DCM genes.

3.6. Limitations of WES

While fruitful, exome sequencing does have limitations. There are inherent technical limitations associated with the method. WES requires a capture step, which is limited by design of capture oligonucleotides. Not all genes or exons are adequately annotated and therefore will not be properly included in the methods to capture exons. Furthermore, there can be inconsistencies in capture resulting in poorly covered exons and off-target sequencing. Capture efficiency only approximates 70%–80% in part due to the high GC content of exonic sequence. Probably the most notable limitation is that only 1%–2% of the entire genome is evaluated. Because approximately 85% of described Mendelian mutations occur in the coding regions of genes, it is assumed that Mendelian disease is more likely to be caused by mutations in protein coding exons than in non-coding sequences.

However, over a third of Mendelian diseases reported in OMIM have no known molecular basis. It is reasonable to conclude that some of these missing mutations are either structural variants or that they occur in non-coding regions of the genome, exome sequencing is not suited to interrogate either of these possibilities.

Copy number variants (CNVs) are regions, >50 bp in length, that differ from the expected diploid status [80]. CNVs are an important component of genomic variation in humans and some contribute to disease including cardiovascular disease [81–83]. A recent study by Norton and colleagues identified a large deletion in *BAG3* in a large multi-generational family with DCM [26]. Whole exome sequencing was performed on 4 of the affected family members and comparative genomic hybridization was performed on the proband to detect copy number variations. The technical limitations of WES prevented identification of the large *BAG3* deletion (>8.7 kb). Algorithms are being created to aid in the use of exome sequencing for the detection of copy number variation [84,85]. However, these methods come with a variety of limitations and cannot detect other types of structural variation such as uniparental disomies or chromosomal rearrangements, both exceedingly important for disease pathogenesis. WGS, as opposed to WES, may offer a better method to detect some structural variants, but may require improved analytic tools for specificity and sensitivity of structural variant detection.

4. WGS as a Tool to Investigate Non-Coding Variation for CVD

Perhaps even more important to understanding disease etiology is the investigation of non-coding variation. Over 98% of the genome is non-coding, and WGS captures nearly 100 fold more of this information compared to WES. However, the interpretation of non-coding variation is currently far more challenging than the interpretation of coding variants. Often referred to as the "dark matter" of the genome, these regions can include microRNAs, long non-coding RNAs, splice variants and regulatory elements that can directly cause or modulate disease phenotypes [37]. miRNAs are important regulators of heart function and recent studies have revealed miRNA misexpression in human cardiac disease and animal models of heart failure [86–88]. For example, miR-208 is encoded by an intron within *MYH6*, which encodes α myosin heavy chain and is in close proximity to *MYH7* [87,88]. mir-208 null mice do not hypertrophy in response to cardiac stress and null mice do not upregulate *Myh7* [89,90]. Silencing of miR-208 reduces cardiac remodeling, deterioration of heart function, and improves survival in a rat model of heart failure, while overexpression of miR-208 in cardiomyocytes leads to cardiomyocyte hypertrophy [90,91]. Another miR, miR-1 is the most highly expressed miRNA in the murine heart [88,92]. miR-1 targets HAND2, a transcription factor important for expansion of ventricular cardiomyocytes. Deletion of miR-1 results in 50% perinatal lethality due to ventricular septal defects [88]. The majority of surviving mice exhibit sudden death due to conduction defects. Overexpression of miR-1 in embryonic cardiomyocytes caused thin-walled ventricles leading to death at embryonic day 13.5 [86]. These data underscore the importance of miRNAs in cardiac phenotypes. Variation in these and other non-coding regions of the genome may play a vital role in the disease process. Annotation of the non-coding genome is currently underway to aid in the interpretation of these variants. The ENCODE project (Encyclopedia of DNA Elements) has assigned biochemical functions to 80% of the genome [93]. Only twelve percent of SNPs identified by GWAS as disease-associated are located in the vicinity of a protein-coding region even though SNPs in coding regions are over-represented on SNP arrays. However, over 60% of disease-associated SNPs identified by GWAS lie within functional, non-coding regions, especially in promoters and enhancers [94].

4.1. WGS Has Greater Sensitivity than WES

WGS is only now emerging as an alternative to WES since higher cost and more complex analysis limited uptake of WGS. Recently, WGS was used to identify a putative causative variant in a family with two children affected by a previously unreported disease defined by cardiomyopathy and progressive muscle weakness [95]. Wang and colleagues performed WES on one sibling and WGS on the other sibling. WES was performed to a depth of 118× with >90% target regions covered by ≥10 reads, while WGS was performed to a depth of 81×. Variants were filtered using ANNOVAR, which relies on frequency and functional variation [96]. After validation with Sanger sequencing and transmission pattern testing, only two genes remained as candidates. One, TAF1L, is homologous to TAF(II)250 and is specifically expressed in testis. The other candidate, *RBCK1*, codes for an E3 ubiquitin-protein ligase. In this family, RBCK1 had two truncating mutations, each inherited from one parent [95]. *RBCK1* was considered a good candidate for both its rarity and the involvement of other ubiquitin-ligase proteins in muscle disease. The WES data set failed to reveal the *RBCK1* variants despite good coverage over the targeted regions including the exons of *RBCK1*. Upon reanalysis, coverage was very low (2 and 4 reads respectively) for the two mutations, with only one read containing a mutation [95]. Further investigation revealed a high GC content surrounding these mutations [97]. Previous studies have also noted that uneven exome coverage has resulted in filtering of disease genes [98]. This study serves as a proof of principle that whole genome sequencing can be used to identify rare Mendelian, cardiomyopathy phenotypes, and, in some instances, may be more sensitive than WES.

4.2. Limitations of WGS

Two of the major limitations of WGS are size and cost. To achieve average coverage ~35–40× with WGS requires approximately 125 Gb of generated sequencing data. Figure 1A compares the amount of data generated from panel (blue), exome (red) and whole genome sequencing (green). WGS produces an order of magnitude more data than WES. All three technologies call variants proportionate to the amount of data generated with WGS calling ~4 million variants per genome, WES calling ~90 K, and gene panels calling far fewer (Figure 1B). At this time, clinical WGS is more expensive than either WES or panel sequencing at approximately \$9000–\$9500 for WGS, ~\$7000 for WES and ~\$4000 for a pan-cardiomyopathy panel. The price for both WES and WGS is considerably more than the cost of a panel. However, this only remains true if a pathogenic mutation is identified in the first panel. The value of each test can be thought of in terms of the cost per variant identified, and with this metric WGS is a better value (Figure 1D). Panel sequencing is ~\$1.70, WES is ~\$0.08 (8¢) and WGS is ~\$0.002 (0.2¢) per variant detected. While only a handful of variants may be germane to identifying the cause of an individual's primary cardiomyopathy, the other sequence data remains available where it may provide useful guidance for life-style and medical decisions.

Figure 1. Size and Cost Considerations of Next-Generation Sequencing. (**A**) The amount of data generated by a typical cardiomyopathy gene panel of ~50 genes (green), whole exome sequencing (red) and whole genome Sequencing (blue) is shown; (**B**) The approximate number of variants produced by each method is indicated; (**C**) The Clinical cost of each method ranges from ~$4000 (cardiomyopathy gene panel, green) to ~$9500 (whole genome sequencing, blue); (**D**) The cost per variant is greatly reduced for WGS ($0.002, blue) *versus* WES ($0.08, red) and gene panel-based sequencing ($1.70, green). Boxes indicate parameters used to calculate values in **A–D** including coverage, base pairs interrogated and total output.

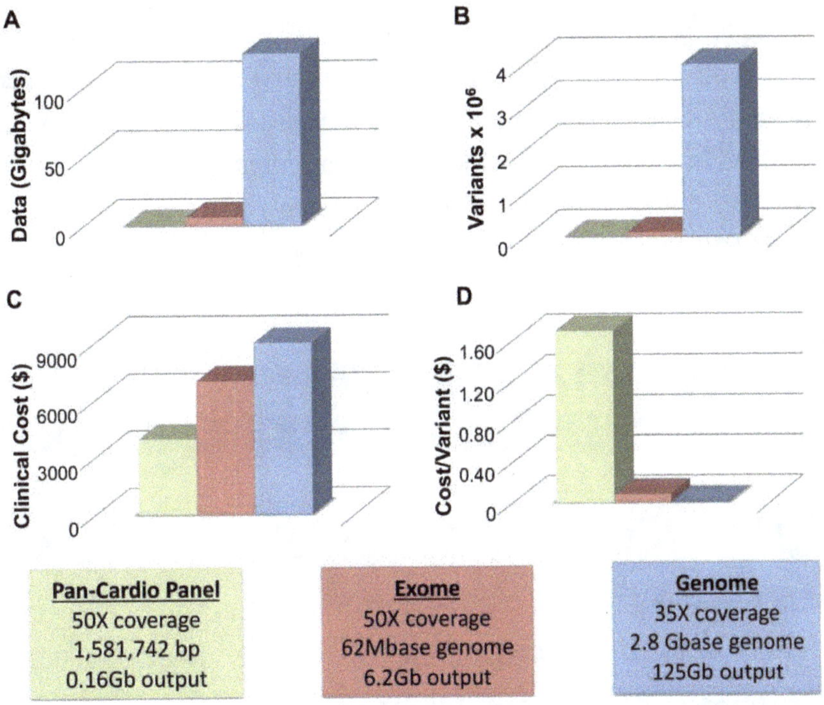

4.3. Multi-Pass Filtering Methods Allow for More Efficient Variant Identification

WGS produces vastly more data than either panels or WES, and this is the double-edged sword of broad based sequence analysis. To cope with this problem we (and others) have adopted stepwise, multi-pass filtering methods (Figure 2). In the first pass, candidate genes are analyzed for the phenotype of interest, in this case cardiomyopathy. Typically, exonic variants in candidate genes are filtered for frequency and for potential protein pathogenicity. A number of tools are freely available that predict the impact of amino acid changes on protein structure and function including PolyPhen-2, GERP, SIFT, PhastCons, Panther and Conseq [72,99–103]. MaxEnt can be used to score the strength of splice site variants [104]. If filtering candidate genes does not produce meaningful variation, the search can be expanded to less attractive candidates or to all rare protein coding variation and finally to non-coding variation. Non-coding variants can be filtered for frequency, however this is more challenging as many frequency databases are biased towards coding regions. Conservation of sequences across multiple species may provide clues about the selection acting on a sequence. The data generated by the ENCODE project will also provide information about variation in functional elements. The complexity of analysis increases with each pass. While this approach approximates panel and WES analysis, it is an improvement in several ways. With this model, WGS only needs to be performed once, while patients that are panel negative will need additional sequencing. In the case of WES, exome capture kits are often updated due to changing gene annotations and the inclusion of newly understood non-coding sequence, requiring retesting with new kits. WGS is less likely to need additional sequencing, and moreover, may provide additional data that over a lifetime can be used to inform not only the primary health concern for which data was collected, but medical choices throughout a patient's life.

Figure 2. Pipeline for WGS Variant Identification. WGS produces ~4 million variants per genome and requires extensive filtering to identify variants of interest. Shown here is a potential pipeline to identify variants. The first pass of the pipeline entails only reviewing variants in the coding regions of genes of interest and filtering by frequency, protein pathogenicity, and mode of inheritance (segregation in available family members). If no variant is identified, a second pass includes the same filtering steps, but on variants in all coding regions. The third pass includes analysis of non-coding variation using frequency, conservation and ENCODE annotation, along with mode of inheritance. The complexity of analysis increases with each pass.

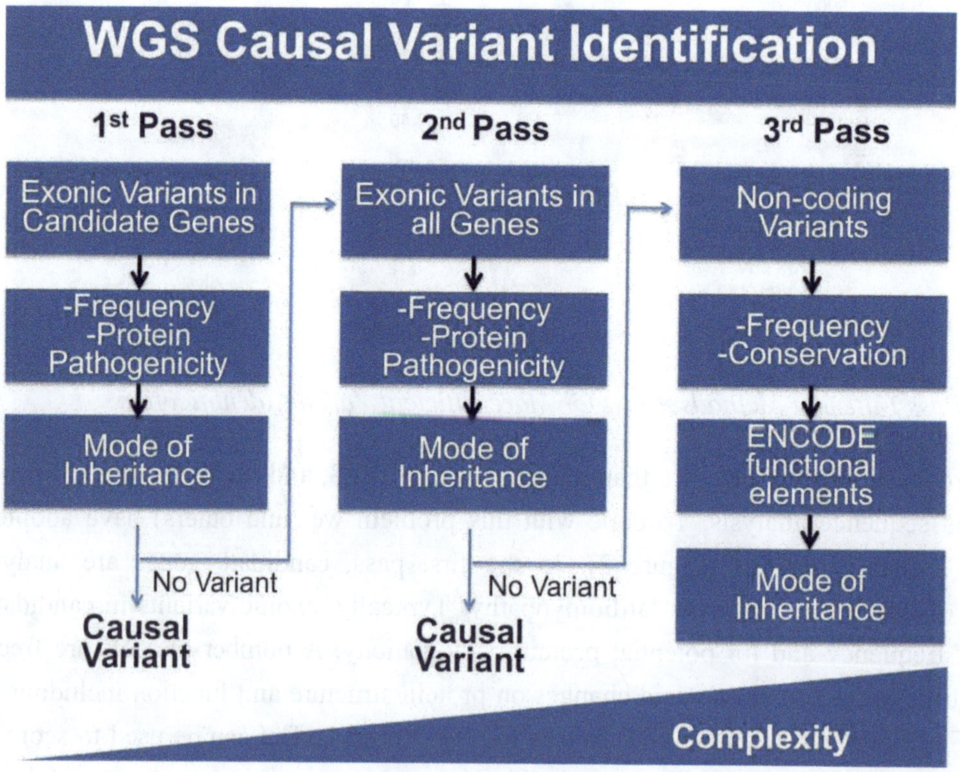

5. Incidental Findings and Their Importance for CVD Related Phenotypes

Incidental findings are a concern with all genetic assessment and especially so for WES and WGS. However, incidental findings are not unique to genetic testing and are part of medical decision making for any mode of testing, including imaging and blood tests. WGS may provide many more incidental findings than any other tests available, and this has led to new recommendations for delivering results of incidental findings from genetic research and testing. In 2006, an NHLBI working recommended reporting research results to study participants when the risk of disease is significant and has important health implications including sudden death or considerable risk of morbidity especially when therapeutic interventions are available [105]. In 2013, the American College of Medical Genetics and Genomics (ACMG) made recommendations for reporting incidental findings in exome and genome sequencing [106]. The ACMG recommended that laboratories performing sequencing should identify and report mutations in genes included on their minimal list. Notably, this list includes 24 phenotypes of which a third are cardiovascular disorders for which penetrance is high and clinical interventions are available [106]. Of the 57 genes for which recommendations were made to report incidental findings, more than half (34) were CVD-associated genes.

6. Conclusions

In the case of CVD genetic profiling, there are often medical management decisions that can reduce risk. This is the case whether the initial genetic profiling was done for assessing CVD risk or whether the genetic profiling was done to assess risk of other inherited diseases. For example, risks for cardiomyopathy and especially arrhythmias can be managed medically, with increased surveillance or even with device insertion. Risks for developing atherosclerosis or aneurysms can be mitigated through drug or even surgical intervention. Importantly, since CVD disorders can be associated with sudden cardiac death, the capacity to intervene based on genetic risk profiles is evident. With the improvement in genetic databases that are accompanied by robust phenotyping, it should be possible to more accurately predict risk for CVD.

Acknowledgments

This work was supported in part by National Institutes of Health NIH AR052646, NIH HL61322, NIH NS072027, and the Doris Duke Charitable Foundation.

Author Contributions

Megan J. Puckelwartz researched the topics and drafted the manuscript. Elizabeth M. McNally revised the manuscript.

References

1. Go, A.S.; Mozaffarian, D.; Roger, V.L.; Benjamin, E.J.; Berry, J.D.; Borden, W.B.; Bravata, D.M.; Dai, S.; Ford, E.S.; Fox, C.S.; *et al.* Executive summary: Heart disease and stroke statistics—2013 update: A report from the American Heart Association. *Circulation* **2013**, *127*, 143–152.
2. Marenberg, M.E.; Risch, N.; Berkman, L.F.; Floderus, B.; de Faire, U. Genetic susceptibility to death from coronary heart disease in a study of twins. *N. Engl. J. Med.* **1994**, *330*, 1041–1046.
3. Post, W.S.; Larson, M.G.; Myers, R.H.; Galderisi, M.; Levy, D. Heritability of left ventricular mass: The Framingham Heart Study. *Hypertension* **1997**, *30*, 1025–1028.

4. Adams, T.D.; Yanowitz, F.G.; Fisher, A.G.; Ridges, J.D.; Nelson, A.G.; Hagan, A.D.; Williams, R.R.; Hunt, S.C. Heritability of cardiac size: An echocardiographic and electrocardiographic study of monozygotic and dizygotic twins. *Circulation* **1985**, *71*, 39–44.
5. Botstein, D.; White, R.L.; Skolnick, M.; Davis, R.W. Construction of a genetic linkage map in man using restriction fragment length polymorphisms. *Am. J. Hum. Genet.* **1980**, *32*, 314–331.
6. Jarcho, J.A.; McKenna, W.; Pare, J.A.; Solomon, S.D.; Holcombe, R.F.; Dickie, S.; Levi, T.; Donis-Keller, H.; Seidman, J.G.; Seidman, C.E. Mapping a gene for familial hypertrophic cardiomyopathy to chromosome 14q1. *N. Engl. J. Med.* **1989**, *321*, 1372–1378.
7. Geisterfer-Lowrance, A.A.; Kass, S.; Tanigawa, G.; Vosberg, H.P.; McKenna, W.; Seidman, C.E.; Seidman, J.G. A molecular basis for familial hypertrophic cardiomyopathy: A beta cardiac myosin heavy chain gene missense mutation. *Cell* **1990**, *62*, 999–1006.
8. Basson, C.T.; Bachinsky, D.R.; Lin, R.C.; Levi, T.; Elkins, J.A.; Soults, J.; Grayzel, D.; Kroumpouzou, E.; Traill, T.A.; Leblanc-Straceski, J.; *et al.* Mutations in human TBX5 [corrected] cause limb and cardiac malformation in Holt-Oram syndrome. *Nat. Genet.* **1997**, *15*, 30–35.
9. Curran, M.E.; Splawski, I.; Timothy, K.W.; Vincent, G.M.; Green, E.D.; Keating, M.T. A molecular basis for cardiac arrhythmia: HERG mutations cause long QT syndrome. *Cell* **1995**, *80*, 795–803.
10. Dietz, H.C.; Cutting, G.R.; Pyeritz, R.E.; Maslen, C.L.; Sakai, L.Y.; Corson, G.M.; Puffenberger, E.G.; Hamosh, A.; Nanthakumar, E.J.; Curristin, S.M.; *et al.* Marfan syndrome caused by a recurrent de novo missense mutation in the fibrillin gene. *Nature* **1991**, *352*, 337–339.
11. Garg, V.; Kathiriya, I.S.; Barnes, R.; Schluterman, M.K.; King, I.N.; Butler, C.A.; Rothrock, C.R.; Eapen, R.S.; Hirayama-Yamada, K.; Joo, K.; *et al.* GATA4 mutations cause human congenital heart defects and reveal an interaction with TBX5. *Nature* **2003**, *424*, 443–447.
12. Garg, V.; Muth, A.N.; Ransom, J.F.; Schluterman, M.K.; Barnes, R.; King, I.N.; Grossfeld, P.D.; Srivastava, D. Mutations in NOTCH1 cause aortic valve disease. *Nature* **2005**, *437*, 270–274.
13. Schott, J.J.; Benson, D.W.; Basson, C.T.; Pease, W.; Silberbach, G.M.; Moak, J.P.; Maron, B.J.; Seidman, C.E.; Seidman, J.G. Congenital heart disease caused by mutations in the transcription factor NKX2-5. *Science* **1998**, *281*, 108–111.
14. Tartaglia, M.; Mehler, E.L.; Goldberg, R.; Zampino, G.; Brunner, H.G.; Kremer, H.; van der Burgt, I.; Crosby, A.H.; Ion, A.; Jeffery, S.; *et al.* Mutations in PTPN11, encoding the protein tyrosine phosphatase SHP-2, cause Noonan syndrome. *Nat. Genet.* **2001**, *29*, 465–468.
15. Lander, E.; Kruglyak, L. Genetic dissection of complex traits: Guidelines for interpreting and reporting linkage results. *Nat. Genet.* **1995**, *11*, 241–247.
16. International HapMap Consortium. The International HapMap Project. *Nature* **2003**, *426*, 789–796.
17. National Human Genome Research Institute (NHGRI) Catalog of Published Genome-Wide Association Studies. Available online: http://www.genome.gov/gwasstudies/ (accessed on 25 February 2014).
18. Newton-Cheh, C.; Eijgelsheim, M.; Rice, K.M.; de Bakker, P.I.; Yin, X.; Estrada, K.; Bis, J.C.; Marciante, K.; Rivadeneira, F.; Noseworthy, P.A.; *et al.* Common variants at ten loci influence QT interval duration in the QTGEN Study. *Nat. Genet.* **2009**, *41*, 399–406.

19. Villard, E.; Perret, C.; Gary, F.; Proust, C.; Dilanian, G.; Hengstenberg, C.; Ruppert, V.; Arbustini, E.; Wichter, T.; Germain, M.; *et al.* A genome-wide association study identifies two loci associated with heart failure due to dilated cardiomyopathy. *Eur. Heart J.* **2011**, *32*, 1065–1076.
20. Jefferies, J.L.; Towbin, J.A. Dilated cardiomyopathy. *Lancet* **2010**, *375*, 752–762.
21. Hershberger, R.E.; Norton, N.; Morales, A.; Li, D.; Siegfried, J.D.; Gonzalez-Quintana, J. Coding sequence rare variants identified in MYBPC3, MYH6, TPM1, TNNC1, and TNNI3 from 312 patients with familial or idiopathic dilated cardiomyopathy. *Circ. Cardiovasc. Genet.* **2010**, *3*, 155–161.
22. Rampersaud, E.; Kinnamon, D.D.; Hamilton, K.; Khuri, S.; Hershberger, R.E.; Martin, E.R. Common susceptibility variants examined for association with dilated cardiomyopathy. *Ann. Hum. Genet.* **2010**, *74*, 110–116.
23. Tiret, L.; Mallet, C.; Poirier, O.; Nicaud, V.; Millaire, A.; Bouhour, J.B.; Roizes, G.; Desnos, M.; Dorent, R.; Schwartz, K.; *et al.* Lack of association between polymorphisms of eight candidate genes and idiopathic dilated cardiomyopathy: The CARDIGENE study. *J. Am. Coll. Cardiol.* **2000**, *35*, 29–35.
24. Takayama, S.; Xie, Z.; Reed, J.C. An evolutionarily conserved family of Hsp70/Hsc70 molecular chaperone regulators. *J. Biol. Chem.* **1999**, *274*, 781–786.
25. Ellinor, P.T.; Sasse-Klaassen, S.; Probst, S.; Gerull, B.; Shin, J.T.; Toeppel, A.; Heuser, A.; Michely, B.; Yoerger, D.M.; Song, B.S.; *et al.* A novel locus for dilated cardiomyopathy, diffuse myocardial fibrosis, and sudden death on chromosome 10q25–26. *J. Am. Coll. Cardiol.* **2006**, *48*, 106–111.
26. Norton, N.; Li, D.; Rieder, M.J.; Siegfried, J.D.; Rampersaud, E.; Zuchner, S.; Mangos, S.; Gonzalez-Quintana, J.; Wang, L.; McGee, S.; *et al.* Genome-wide studies of copy number variation and exome sequencing identify rare variants in BAG3 as a cause of dilated cardiomyopathy. *Am. J. Hum. Genet.* **2011**, *88*, 273–282.
27. Reich, D.E.; Lander, E.S. On the allelic spectrum of human disease. *Trends Genet.* **2001**, *17*, 502–510.
28. Wang, W.Y.; Barratt, B.J.; Clayton, D.G.; Todd, J.A. Genome-wide association studies: Theoretical and practical concerns. *Nat. Rev. Genet.* **2005**, *6*, 109–118.
29. Manolio, T.A.; Collins, F.S.; Cox, N.J.; Goldstein, D.B.; Hindorff, L.A.; Hunter, D.J.; McCarthy, M.I.; Ramos, E.M.; Cardon, L.R.; Chakravarti, A.; *et al.* Finding the missing heritability of complex diseases. *Nature* **2009**, *461*, 747–753.
30. McPherson, R.; Pertsemlidis, A.; Kavaslar, N.; Stewart, A.; Roberts, R.; Cox, D.R.; Hinds, D.A.; Pennacchio, L.A.; Tybjaerg-Hansen, A.; Folsom, A.R.; *et al.* A common allele on chromosome 9 associated with coronary heart disease. *Science* **2007**, *316*, 1488–1491.
31. Helgadottir, A.; Thorleifsson, G.; Manolescu, A.; Gretarsdottir, S.; Blondal, T.; Jonasdottir, A.; Sigurdsson, A.; Baker, A.; Palsson, A.; Masson, G.; *et al.* A common variant on chromosome 9p21 affects the risk of myocardial infarction. *Science* **2007**, *316*, 1491–1493.
32. Lander, E.S. The new genomics: Global views of biology. *Science* **1996**, *274*, 536–539.
33. Pritchard, J.K.; Cox, N.J. The allelic architecture of human disease genes: Common disease-common variant … or not? *Hum. Mol. Genet.* **2002**, *11*, 2417–2423.
34. Gibson, G. Rare and common variants: Twenty arguments. *Nat. Rev. Genet.* **2011**, *13*, 135–145.

35. Manolio, T.A. Cohort studies and the genetics of complex disease. *Nat. Genet.* **2009**, *41*, 5–6.
36. Sun, J.X.; Helgason, A.; Masson, G.; Ebenesersdottir, S.S.; Li, H.; Mallick, S.; Gnerre, S.; Patterson, N.; Kong, A.; Reich, D.; *et al.* A direct characterization of human mutation based on microsatellites. *Nat. Genet.* **2012**, *44*, 1161–1165.
37. Marian, A.J.; Belmont, J. Strategic approaches to unraveling genetic causes of cardiovascular diseases. *Circ. Res.* **2011**, *108*, 1252–1269.
38. Ng, P.C.; Levy, S.; Huang, J.; Stockwell, T.B.; Walenz, B.P.; Li, K.; Axelrod, N.; Busam, D.A.; Strausberg, R.L.; Venter, J.C. Genetic variation in an individual human exome. *PLoS Genet.* **2008**, *4*, e1000160.
39. Tennessen, J.A.; Bigham, A.W.; O'Connor, T.D.; Fu, W.; Kenny, E.E.; Gravel, S.; McGee, S.; Do, R.; Liu, X.; Jun, G.; *et al.* Evolution and functional impact of rare coding variation from deep sequencing of human exomes. *Science* **2012**, *337*, 64–69.
40. Johansen, C.T.; Wang, J.; Lanktree, M.B.; Cao, H.; McIntyre, A.D.; Ban, M.R.; Martins, R.A.; Kennedy, B.A.; Hassell, R.G.; Visser, M.E.; *et al.* Excess of rare variants in genes identified by genome-wide association study of hypertriglyceridemia. *Nat. Genet.* **2010**, *42*, 684–687.
41. Aulchenko, Y.S.; Ripatti, S.; Lindqvist, I.; Boomsma, D.; Heid, I.M.; Pramstaller, P.P.; Penninx, B.W.; Janssens, A.C.; Wilson, J.F.; Spector, T.; *et al.* Loci influencing lipid levels and coronary heart disease risk in 16 European population cohorts. *Nat. Genet.* **2009**, *41*, 47–55.
42. Kathiresan, S.; Willer, C.J.; Peloso, G.M.; Demissie, S.; Musunuru, K.; Schadt, E.E.; Kaplan, L.; Bennett, D.; Li, Y.; Tanaka, T.; *et al.* Common variants at 30 loci contribute to polygenic dyslipidemia. *Nat. Genet.* **2009**, *41*, 56–65.
43. Li, H.; Homer, N. A survey of sequence alignment algorithms for next-generation sequencing. *Brief. Bioinforma.* **2010**, *11*, 473–483.
44. Nielsen, R.; Paul, J.S.; Albrechtsen, A.; Song, Y.S. Genotype and SNP calling from next-generation sequencing data. *Nat. Rev. Genet.* **2011**, *12*, 443–451.
45. Abecasis, G.R.; Auton, A.; Brooks, L.D.; DePristo, M.A.; Durbin, R.M.; Handsaker, R.E.; Kang, H.M.; Marth, G.T.; McVean, G.A. An integrated map of genetic variation from 1092 human genomes. *Nature* **2012**, *491*, 56–65.
46. Jabbari, J.; Jabbari, R.; Nielsen, M.W.; Holst, A.G.; Nielsen, J.B.; Haunso, S.; Tfelt-Hansen, J.; Svendsen, J.H.; Olesen, M.S. New exome data question the pathogenicity of genetic variants previously associated with catecholaminergic polymorphic ventricular tachycardia. *Circ. Cardiovasc. Genet.* **2013**, *6*, 481–489.
47. Priori, S.G.; Napolitano, C.; Memmi, M.; Colombi, B.; Drago, F.; Gasparini, M.; DeSimone, L.; Coltorti, F.; Bloise, R.; Keegan, R.; *et al.* Clinical and molecular characterization of patients with catecholaminergic polymorphic ventricular tachycardia. *Circulation* **2002**, *106*, 69–74.
48. Golbus, J.R.; Puckelwartz, M.J.; Fahrenbach, J.P.; Dellefave-Castillo, L.M.; Wolfgeher, D.; McNally, E.M. Population-based variation in cardiomyopathy genes. *Circ. Cardiovasc. Genet.* **2012**, *5*, 391–399.
49. Risgaard, B.; Jabbari, R.; Refsgaard, L.; Holst, A.G.; Haunso, S.; Sadjadieh, A.; Winkel, B.G.; Olesen, M.S.; Tfelt-Hansen, J. High prevalence of genetic variants previously associated with Brugada syndrome in new exome data. *Clin. Genet.* **2013**, *84*, 489–495.

50. Kapplinger, J.D.; Landstrom, A.P.; Salisbury, B.A.; Callis, T.E.; Pollevick, G.D.; Tester, D.J.; Cox, M.G.; Bhuiyan, Z.; Bikker, H.; Wiesfeld, A.C.; *et al.* Distinguishing arrhythmogenic right ventricular cardiomyopathy/dysplasia-associated mutations from background genetic noise. *J. Am. Coll. Cardiol.* **2011**, *57*, 2317–2327.
51. Andreasen, C.; Refsgaard, L.; Nielsen, J.B.; Sajadieh, A.; Winkel, B.G.; Tfelt-Hansen, J.; Haunso, S.; Holst, A.G.; Svendsen, J.H.; Olesen, M.S. Mutations in genes encoding cardiac ion channels previously associated with sudden infant death syndrome (SIDS) are present with high frequency in new exome data. *Can. J. Cardiol.* **2013**, *29*, 1104–1109.
52. Petretta, M.; Pirozzi, F.; Sasso, L.; Paglia, A.; Bonaduce, D. Review and metaanalysis of the frequency of familial dilated cardiomyopathy. *Am. J. Cardiol.* **2011**, *108*, 1171–1176.
53. Meder, B.; Haas, J.; Keller, A.; Heid, C.; Just, S.; Borries, A.; Boisguerin, V.; Scharfenberger-Schmeer, M.; Stahler, P.; Beier, M.; *et al.* Targeted next-generation sequencing for the molecular genetic diagnostics of cardiomyopathies. *Circ. Cardiovasc. Genet.* **2011**, *4*, 110–122.
54. Zimmerman, R.S.; Cox, S.; Lakdawala, N.K.; Cirino, A.; Mancini-DiNardo, D.; Clark, E.; Leon, A.; Duffy, E.; White, E.; Baxter, S.; *et al.* A novel custom resequencing array for dilated cardiomyopathy. *Genet. Med.* **2010**, *12*, 268–278.
55. Partners Healthcare. Available online: http://pcpgm.partners.org/ (accessed on 25 February 2014).
56. Opitz, C.A.; Kulke, M.; Leake, M.C.; Neagoe, C.; Hinssen, H.; Hajjar, R.J.; Linke, W.A. Damped elastic recoil of the titin spring in myofibrils of human myocardium. *Proc. Natl. Acad. Sci. USA* **2003**, *100*, 12688–12693.
57. Granzier, H.L.; Irving, T.C. Passive tension in cardiac muscle: Contribution of collagen, titin, microtubules, and intermediate filaments. *Biophys. J.* **1995**, *68*, 1027–1044.
58. Horowits, R.; Kempner, E.S.; Bisher, M.E.; Podolsky, R.J. A physiological role for titin and nebulin in skeletal muscle. *Nature* **1986**, *323*, 160–164.
59. Muhle-Goll, C.; Habeck, M.; Cazorla, O.; Nilges, M.; Labeit, S.; Granzier, H. Structural and functional studies of titin's fn3 modules reveal conserved surface patterns and binding to myosin S1—A possible role in the Frank-Starling mechanism of the heart. *J. Mol. Biol.* **2001**, *313*, 431–447.
60. Cazorla, O.; Wu, Y.; Irving, T.C.; Granzier, H. Titin-based modulation of calcium sensitivity of active tension in mouse skinned cardiac myocytes. *Circ. Res.* **2001**, *88*, 1028–1035.
61. Siu, B.L.; Niimura, H.; Osborne, J.A.; Fatkin, D.; MacRae, C.; Solomon, S.; Benson, D.W.; Seidman, J.G.; Seidman, C.E. Familial dilated cardiomyopathy locus maps to chromosome 2q31. *Circulation* **1999**, *99*, 1022–1026.
62. Gerull, B.; Atherton, J.; Geupel, A.; Sasse-Klaassen, S.; Heuser, A.; Frenneaux, M.; McNabb, M.; Granzier, H.; Labeit, S.; Thierfelder, L. Identification of a novel frameshift mutation in the giant muscle filament titin in a large Australian family with dilated cardiomyopathy. *J. Mol. Med.* **2006**, *84*, 478–483.
63. Gerull, B.; Gramlich, M.; Atherton, J.; McNabb, M.; Trombitas, K.; Sasse-Klaassen, S.; Seidman, J.G.; Seidman, C.; Granzier, H.; Labeit, S.; *et al.* Mutations of TTN, encoding the giant muscle filament titin, cause familial dilated cardiomyopathy. *Nat. Genet.* **2002**, *30*, 201–204.
64. Herman, D.S.; Lam, L.; Taylor, M.R.; Wang, L.; Teekakirikul, P.; Christodoulou, D.; Conner, L.; DePalma, S.R.; McDonough, B.; Sparks, E.; *et al.* Truncations of titin causing dilated cardiomyopathy. *N. Engl. J. Med.* **2012**, *366*, 619–628.

65. Ng, S.B.; Turner, E.H.; Robertson, P.D.; Flygare, S.D.; Bigham, A.W.; Lee, C.; Shaffer, T.; Wong, M.; Bhattacharjee, A.; Eichler, E.E.; *et al.* Targeted capture and massively parallel sequencing of 12 human exomes. *Nature* **2009**, *461*, 272–276.

66. Majewski, J.; Schwartzentruber, J.; Lalonde, E.; Montpetit, A.; Jabado, N. What can exome sequencing do for you? *J. Med. Genet.* **2011**, *48*, 580–589.

67. Campbell, N.; Sinagra, G.; Jones, K.L.; Slavov, D.; Gowan, K.; Merlo, M.; Carniel, E.; Fain, P.R.; Aragona, P.; di Lenarda, A.; *et al.* Whole exome sequencing identifies a troponin T mutation hot spot in familial dilated cardiomyopathy. *PLoS One* **2013**, *8*, e78104.

68. Wells, Q.S.; Becker, J.R.; Su, Y.R.; Mosley, J.D.; Weeke, P.; D'Aoust, L.; Ausborn, N.L.; Ramirez, A.H.; Pfotenhauer, J.P.; Naftilan, A.J.; *et al.* Whole exome sequencing identifies a causal RBM20 mutation in a large pedigree with familial dilated cardiomyopathy. *Circ. Cardiovasc. Genet.* **2013**, *6*, 317–326.

69. Yandell, M.; Huff, C.; Hu, H.; Singleton, M.; Moore, B.; Xing, J.; Jorde, L.B.; Reese, M.G. A probabilistic disease-gene finder for personal genomes. *Genome Res.* **2011**, *21*, 1529–1542.

70. Li, D.; Morales, A.; Gonzalez-Quintana, J.; Norton, N.; Siegfried, J.D.; Hofmeyer, M.; Hershberger, R.E. Identification of novel mutations in RBM20 in patients with dilated cardiomyopathy. *Clin. Transl. Sci.* **2010**, *3*, 90–97.

71. Davydov, E.V.; Goode, D.L.; Sirota, M.; Cooper, G.M.; Sidow, A.; Batzoglou, S. Identifying a high fraction of the human genome to be under selective constraint using GERP++. *PLoS Comput. Biol.* **2010**, *6*, e1001025.

72. Adzhubei, I.A.; Schmidt, S.; Peshkin, L.; Ramensky, V.E.; Gerasimova, A.; Bork, P.; Kondrashov, A.S.; Sunyaev, S.R. A method and server for predicting damaging missense mutations. *Nat. Methods* **2010**, *7*, 248–249.

73. Millat, G.; Bouvagnet, P.; Chevalier, P.; Sebbag, L.; Dulac, A.; Dauphin, C.; Jouk, P.S.; Delrue, M.A.; Thambo, J.B.; Le Metayer, P.; *et al.* Clinical and mutational spectrum in a cohort of 105 unrelated patients with dilated cardiomyopathy. *Eur. J. Med. Genet.* **2011**, *54*, e570–e575.

74. Lakdawala, N.K.; Dellefave, L.; Redwood, C.S.; Sparks, E.; Cirino, A.L.; Depalma, S.; Colan, S.D.; Funke, B.; Zimmerman, R.S.; Robinson, P.; *et al.* Familial dilated cardiomyopathy caused by an alpha-tropomyosin mutation: The distinctive natural history of sarcomeric dilated cardiomyopathy. *J. Am. Coll. Cardiol.* **2010**, *55*, 320–329.

75. Roncarati, R.; Viviani Anselmi, C.; Krawitz, P.; Lattanzi, G.; von Kodolitsch, Y.; Perrot, A.; di Pasquale, E.; Papa, L.; Portararo, P.; Columbaro, M.; *et al.* Doubly heterozygous LMNA and TTN mutations revealed by exome sequencing in a severe form of dilated cardiomyopathy. *Eur. J. Hum. Genet.* **2013**, *21*, 1105–1111.

76. Robinson, P.N.; Kohler, S.; Bauer, S.; Seelow, D.; Horn, D.; Mundlos, S. The Human Phenotype Ontology: A tool for annotating and analyzing human hereditary disease. *Am. J. Hum. Genet.* **2008**, *83*, 610–615.

77. Perrot, A.; Hussein, S.; Ruppert, V.; Schmidt, H.H.; Wehnert, M.S.; Duong, N.T.; Posch, M.G.; Panek, A.; Dietz, R.; Kindermann, I.; *et al.* Identification of mutational hot spots in LMNA encoding lamin A/C in patients with familial dilated cardiomyopathy. *Basic Res. Cardiol.* **2009**, *104*, 90–99.

78. Theis, J.L.; Sharpe, K.M.; Matsumoto, M.E.; Chai, H.S.; Nair, A.A.; Theis, J.D.; de Andrade, M.; Wieben, E.D.; Michels, V.V.; Olson, T.M. Homozygosity mapping and exome sequencing reveal GATAD1 mutation in autosomal recessive dilated cardiomyopathy. *Circ. Cardiovasc. Genet.* **2011**, *4*, 585–594.
79. Vermeulen, M.; Eberl, H.C.; Matarese, F.; Marks, H.; Denissov, S.; Butter, F.; Lee, K.K.; Olsen, J.V.; Hyman, A.A.; Stunnenberg, H.G.; *et al.* Quantitative interaction proteomics and genome-wide profiling of epigenetic histone marks and their readers. *Cell* **2010**, *142*, 967–980.
80. Alkan, C.; Coe, B.P.; Eichler, E.E. Genome structural variation discovery and genotyping. *Nat. Rev. Genet.* **2011**, *12*, 363–376.
81. Iafrate, A.J.; Feuk, L.; Rivera, M.N.; Listewnik, M.L.; Donahoe, P.K.; Qi, Y.; Scherer, S.W.; Lee, C. Detection of large-scale variation in the human genome. *Nat. Genet.* **2004**, *36*, 949–951.
82. Tuzun, E.; Sharp, A.J.; Bailey, J.A.; Kaul, R.; Morrison, V.A.; Pertz, L.M.; Haugen, E.; Hayden, H.; Albertson, D.; Pinkel, D.; *et al.* Fine-scale structural variation of the human genome. *Nat. Genet.* **2005**, *37*, 727–732.
83. Kidd, J.M.; Cooper, G.M.; Donahue, W.F.; Hayden, H.S.; Sampas, N.; Graves, T.; Hansen, N.; Teague, B.; Alkan, C.; Antonacci, F.; *et al.* Mapping and sequencing of structural variation from eight human genomes. *Nature* **2008**, *453*, 56–64.
84. Krumm, N.; Sudmant, P.H.; Ko, A.; O'Roak, B.J.; Malig, M.; Coe, B.P.; Quinlan, A.R.; Nickerson, D.A.; Eichler, E.E. Copy number variation detection and genotyping from exome sequence data. *Genome Res.* **2012**, *22*, 1525–1532.
85. Magi, A.; Tattini, L.; Cifola, I.; D'Aurizio, R.; Benelli, M.; Mangano, E.; Battaglia, C.; Bonora, E.; Kurg, A.; Seri, M.; *et al.* EXCAVATOR: Detecting copy number variants from whole-exome sequencing data. *Genome Biol.* **2013**, *14*, R120.
86. Zhao, Y.; Samal, E.; Srivastava, D. Serum response factor regulates a muscle-specific microRNA that targets Hand2 during cardiogenesis. *Nature* **2005**, *436*, 214–220.
87. Thum, T.; Galuppo, P.; Wolf, C.; Fiedler, J.; Kneitz, S.; van Laake, L.W.; Doevendans, P.A.; Mummery, C.L.; Borlak, J.; Haverich, A.; *et al.* MicroRNAs in the human heart: A clue to fetal gene reprogramming in heart failure. *Circulation* **2007**, *116*, 258–267.
88. Zhao, Y.; Ransom, J.F.; Li, A.; Vedantham, V.; von Drehle, M.; Muth, A.N.; Tsuchihashi, T.; McManus, M.T.; Schwartz, R.J.; Srivastava, D. Dysregulation of cardiogenesis, cardiac conduction, and cell cycle in mice lacking miRNA-1-2. *Cell* **2007**, *129*, 303–317.
89. Van Rooij, E.; Sutherland, L.B.; Qi, X.; Richardson, J.A.; Hill, J.; Olson, E.N. Control of stress-dependent cardiac growth and gene expression by a microRNA. *Science* **2007**, *316*, 575–579.
90. Callis, T.E.; Pandya, K.; Seok, H.Y.; Tang, R.H.; Tatsuguchi, M.; Huang, Z.P.; Chen, J.F.; Deng, Z.; Gunn, B.; Shumate, J.; *et al.* MicroRNA-208a is a regulator of cardiac hypertrophy and conduction in mice. *J. Clin. Invest.* **2009**, *119*, 2772–2786.
91. Montgomery, R.L.; Hullinger, T.G.; Semus, H.M.; Dickinson, B.A.; Seto, A.G.; Lynch, J.M.; Stack, C.; Latimer, P.A.; Olson, E.N.; van Rooij, E. Therapeutic inhibition of miR-208a improves cardiac function and survival during heart failure. *Circulation* **2011**, *124*, 1537–1547.
92. Rao, P.K.; Toyama, Y.; Chiang, H.R.; Gupta, S.; Bauer, M.; Medvid, R.; Reinhardt, F.; Liao, R.; Krieger, M.; Jaenisch, R.; *et al.* Loss of cardiac microRNA-mediated regulation leads to dilated cardiomyopathy and heart failure. *Circ. Res.* **2009**, *105*, 585–594.

93. Bernstein, B.E.; Birney, E.; Dunham, I.; Green, E.D.; Gunter, C.; Snyder, M. An integrated encyclopedia of DNA elements in the human genome. *Nature* **2012**, *489*, 57–74.
94. Hindorff, L.A.; Sethupathy, P.; Junkins, H.A.; Ramos, E.M.; Mehta, J.P.; Collins, F.S.; Manolio, T.A. Potential etiologic and functional implications of genome-wide association loci for human diseases and traits. *Proc. Natl. Acad. Sci. USA* **2009**, *106*, 9362–9367.
95. Wang, K.; Kim, C.; Bradfield, J.; Guo, Y.; Toskala, E.; Otieno, F.G.; Hou, C.; Thomas, K.; Cardinale, C.; Lyon, G.J.; *et al.* Whole-genome DNA/RNA sequencing identifies truncating mutations in RBCK1 in a novel Mendelian disease with neuromuscular and cardiac involvement. *Genome Med.* **2013**, *5*, 67.
96. Wang, K.; Li, M.; Hakonarson, H. ANNOVAR: Functional annotation of genetic variants from high-throughput sequencing data. *Nucleic Acids Res.* **2010**, *38*, e164.
97. Benjamini, Y.; Speed, T.P. Summarizing and correcting the GC content bias in high-throughput sequencing. *Nucleic Acids Res.* **2012**, *40*, e72.
98. Sirmaci, A.; Edwards, Y.J.; Akay, H.; Tekin, M. Challenges in whole exome sequencing: An example from hereditary deafness. *PLoS One* **2012**, *7*, e32000.
99. Siepel, A.; Bejerano, G.; Pedersen, J.S.; Hinrichs, A.S.; Hou, M.; Rosenbloom, K.; Clawson, H.; Spieth, J.; Hillier, L.W.; Richards, S.; *et al.* Evolutionarily conserved elements in vertebrate, insect, worm, and yeast genomes. *Genome Res.* **2005**, *15*, 1034–1050.
100. Ng, P.C.; Henikoff, S. SIFT: Predicting amino acid changes that affect protein function. *Nucleic Acids Res.* **2003**, *31*, 3812–3814.
101. Cooper, G.M.; Stone, E.A.; Asimenos, G.; Green, E.D.; Batzoglou, S.; Sidow, A. Distribution and intensity of constraint in mammalian genomic sequence. *Genome Res.* **2005**, *15*, 901–913.
102. Thomas, P.D.; Kejariwal, A.; Campbell, M.J.; Mi, H.; Diemer, K.; Guo, N.; Ladunga, I.; Ulitsky-Lazareva, B.; Muruganujan, A.; Rabkin, S.; *et al.* PANTHER: A browsable database of gene products organized by biological function, using curated protein family and subfamily classification. *Nucleic Acids Res.* **2003**, *31*, 334–341.
103. Ashkenazy, H.; Erez, E.; Martz, E.; Pupko, T.; Ben-Tal, N. ConSurf 2010: Calculating evolutionary conservation in sequence and structure of proteins and nucleic acids. *Nucleic Acids Res.* **2010**, *38*, W529–W533.
104. Yeo, G.; Burge, C.B. Maximum entropy modeling of short sequence motifs with applications to RNA splicing signals. *J. Comput. Biol.* **2004**, *11*, 377–394.
105. Bookman, E.B.; Langehorne, A.A.; Eckfeldt, J.H.; Glass, K.C.; Jarvik, G.P.; Klag, M.; Koski, G.; Motulsky, A.; Wilfond, B.; Manolio, T.A.; *et al.* Reporting genetic results in research studies: Summary and recommendations of an NHLBI working group. *Am. J. Med. Genet. A* **2006**, *140*, 1033–1040.
106. Green, R.C.; Berg, J.S.; Grody, W.W.; Kalia, S.S.; Korf, B.R.; Martin, C.L.; McGuire, A.L.; Nussbaum, R.L.; O'Daniel, J.M.; Ormond, K.E.; *et al.* ACMG recommendations for reporting of incidental findings in clinical exome and genome sequencing. *Genet. Med.* **2013**, *15*, 565–574.

PERMISSIONS

All chapters in this book were first published in MDPI; hereby published with permission under the Creative Commons Attribution License or equivalent. Every chapter published in this book has been scrutinized by our experts. Their significance has been extensively debated. The topics covered herein carry significant findings which will fuel the growth of the discipline. They may even be implemented as practical applications or may be referred to as a beginning point for another development.

The contributors of this book come from diverse backgrounds, making this book a truly international effort. This book will bring forth new frontiers with its revolutionizing research information and detailed analysis of the nascent developments around the world.

We would like to thank all the contributing authors for lending their expertise to make the book truly unique. They have played a crucial role in the development of this book. Without their invaluable contributions this book wouldn't have been possible. They have made vital efforts to compile up to date information on the varied aspects of this subject to make this book a valuable addition to the collection of many professionals and students.

This book was conceptualized with the vision of imparting up-to-date information and advanced data in this field. To ensure the same, a matchless editorial board was set up. Every individual on the board went through rigorous rounds of assessment to prove their worth. After which they invested a large part of their time researching and compiling the most relevant data for our readers.

The editorial board has been involved in producing this book since its inception. They have spent rigorous hours researching and exploring the diverse topics which have resulted in the successful publishing of this book. They have passed on their knowledge of decades through this book. To expedite this challenging task, the publisher supported the team at every step. A small team of assistant editors was also appointed to further simplify the editing procedure and attain best results for the readers.

Apart from the editorial board, the designing team has also invested a significant amount of their time in understanding the subject and creating the most relevant covers. They scrutinized every image to scout for the most suitable representation of the subject and create an appropriate cover for the book.

The publishing team has been an ardent support to the editorial, designing and production team. Their endless efforts to recruit the best for this project, has resulted in the accomplishment of this book. They are a veteran in the field of academics and their pool of knowledge is as vast as their experience in printing. Their expertise and guidance has proved useful at every step. Their uncompromising quality standards have made this book an exceptional effort. Their encouragement from time to time has been an inspiration for everyone.

The publisher and the editorial board hope that this book will prove to be a valuable piece of knowledge for researchers, students, practitioners and scholars across the globe.

LIST OF CONTRIBUTORS

Arianna Moiani, Francesco Gandolfi and Fulvio Mavilio
Genethon, 1bis Rue de l'Internationale, 91020 Evry, France

Julia Debora Suerth and Axel Schambach
Institute of Experimental Hematology, Hannover Medical School, Carl-Neuberg-Str.1, D-30625 Hannover, Germany

Ermanno Rizzi, Marco Severgnini and Gianluca De Bellis
Institute for Biomedical Technologies, Consiglio Nazionale delle Ricerche, Milan 20132, Italy

Nicola Whiffin and Richard S. Houlston
Molecular and Population Genetics Team, Genetics and Epidemiology, The Institute of Cancer Research, Sutton, SM2 5NG, UK

Debra J. H. Mathews
Johns Hopkins Berman Institute of Bioethics, 1809 Ashland Avenue, Baltimore, MD 21205, USA

Leila Jamal
Department of Health Policy and Management, Johns Hopkins Bloomberg School of Health, 615 North Wolfe St., Baltimore, MD 21205, USA
Division of Neurogenetics, Kennedy Krieger Institute, 801 N. Broadway, Rm. 564, Baltimore, MD 21205, USA

Hannelore Ehrenreich and Klaus-Armin Nave
Max Planck Institute of Experimental Medicine, Hermann-Rein-Str.3, 37075 Göttingen, Germany
DFG Center for Nanoscale Microscopy and Molecular Physiology of the Brain (CNMPB), Hermann-Rein-Str.3, 37075 Göttingen, Germany

Sara L. Pulit and Androniki Menelaou
Department of Medical Genetics, Institute for Molecular Medicine, University Medical Center Utrecht, Universiteitsweg 100, 3584 CG, Utrecht, The Netherlands

Maarten Leusink
Department of Medical Genetics, Institute for Molecular Medicine, University Medical Center Utrecht, Universiteitsweg 100, 3584 CG, Utrecht, The Netherlands
Division of Pharmacoepidemiology & Clinical Pharmacology, Utrecht Institute for Pharmaceutical Sciences, Utrecht University, Universiteitsweg 99, 3584 CG, Utrecht, The Netherlands
Julius Center for Health Sciences and Primary Care, University Medical Center Utrecht, Universiteitsweg 100, 3584 CG, Utrecht, The Netherlands

Paul I. W. de Bakker
Department of Medical Genetics, Institute for Molecular Medicine, University Medical Center Utrecht, Universiteitsweg 100, 3584 CG, Utrecht, The Netherlands
Julius Center for Health Sciences and Primary Care, University Medical Center Utrecht, Universiteitsweg 100, 3584 CG, Utrecht, The Netherlands

Michael Kloth and Reinhard Buettner
Institute of Pathology, University Hospital Cologne, Kerpener Str. 62, Cologne D-50937, Germany
Center for Integrated Oncology Cologne-Bonn, Cologne D-50937, Germany

David J. Elliott
Institute of Genetic Medicine, Newcastle University, Newcastle, NE1 3BZ, UK

Jose Russo, Julia Santucci-Pereira and Irma H. Russo
The Irma H. Russo MD Breast Cancer Research Laboratory, Fox Chase Cancer Center, Temple University Health System, 333 Cottman Avenue, Philadelphia, PA 19111, USA

Megan E. Aldrup-MacDonald and Beth A. Sullivan
Department of Molecular Genetics and Microbiology, Duke University Medical Center, Durham, NC 27710, USA
Division of Human Genetics, Duke University, Durham, NC 27710, USA

Albino Bacolla and Karen M. Vasquez
Dell Pediatric Research Institute, Division of Pharmacology and Toxicology, College of Pharmacy, The University of Texas at Austin, 1400 Barbara Jordan Blvd., Austin, TX 78723, USA

David N. Cooper
Institute of Medical Genetics, School of Medicine, Cardiff University, Cardiff CF14 4XN, UK

Dianne F. Newbury
Wellcome Trust Centre for Human Genetics, University of Oxford, Oxford OX3 7BN, UK

Anthony P. Monaco
Tufts University, Ballou Hall, Medford, MA 02155, USA

Silvia Paracchini
School of Medicine, University of St. Andrews, St. Andrews, KY16 9TF, UK

Jenny van Dongen, Meike Bartels and Dorret I. Boomsma
Department of Biological Psychology, VU University Amsterdam, Van der Boechorststraat 1, 1081 BT Amsterdam, The Netherlands

Zachary M. Weber
Avera Institute for Human Genetics, 3720 W. 69th Street, Sioux Falls, SD 57108, USA

Erik A. Ehli and Gareth E. Davies
Avera Institute for Human Genetics, 3720 W. 69th Street, Sioux Falls, SD 57108, USA
Department of Psychiatry, University of South Dakota, 4400 W. 69th Street, Sioux Falls, SD 57108, USA

P. Eline Slagboom, Bastiaan T. Heijmans and Roderick C. Slieker
Department of Molecular Epidemiology, Leiden University Medical Center, 2300 RC Leiden, The Netherlands

Megan J. Puckelwartz
Department of Medicine, University of Chicago, Chicago, IL 60637, USA

Elizabeth M. McNally
Department of Medicine, University of Chicago, Chicago, IL 60637, USA
Department of Human Genetics, University of Chicago, Chicago, IL 60637, USA

Index

Printed in the USA
CPSIA information can be obtained
at www.ICGtesting.com
LVHW080039211023
761596LV00010BA/1824